Mark Grote

mngrote@ucdavis.edu

The IMA Volumes in Mathematics and its Applications

Volume 76

Series Editors
Avner Friedman Willard Miller, Jr.

Springer
New York
Berlin
Heidelberg
Barcelona
Budapest
Hong Kong
London
Milan
Paris
Santa Clara
Singapore
Tokyo

Institute for Mathematics and its Applications
IMA

The **Institute for Mathematics and its Applications** was established by a grant from the National Science Foundation to the University of Minnesota in 1982. The IMA seeks to encourage the development and study of fresh mathematical concepts and questions of concern to the other sciences by bringing together mathematicians and scientists from diverse fields in an atmosphere that will stimulate discussion and collaboration.

The IMA Volumes are intended to involve the broader scientific community in this process.

Avner Friedman, Director
Willard Miller, Jr., Associate Director

* * * * * * * * * *

IMA ANNUAL PROGRAMS

1982–1983	**Statistical and Continuum Approaches to Phase Transition**
1983–1984	**Mathematical Models for the Economics of Decentralized Resource Allocation**
1984–1985	**Continuum Physics and Partial Differential Equations**
1985–1986	**Stochastic Differential Equations and Their Applications**
1986–1987	**Scientific Computation**
1987–1988	**Applied Combinatorics**
1988–1989	**Nonlinear Waves**
1989–1990	**Dynamical Systems and Their Applications**
1990–1991	**Phase Transitions and Free Boundaries**
1991–1992	**Applied Linear Algebra**
1992–1993	**Control Theory and its Applications**
1993–1994	**Emerging Applications of Probability**
1994–1995	**Waves and Scattering**
1995–1996	**Mathematical Methods in Material Science**

IMA SUMMER PROGRAMS

1987	**Robotics**
1988	**Signal Processing**
1989	**Robustness, Diagnostics, Computing and Graphics in Statistics**
1990	**Radar and Sonar (June 18 - June 29)**
	New Directions in Time Series Analysis (July 2 - July 27)
1991	**Semiconductors**
1992	**Environmental Studies: Mathematical, Computational, and Statistical Analysis**
1993	**Modeling, Mesh Generation, and Adaptive Numerical Methods for Partial Differential Equations**
1994	**Molecular Biology**

* * * * * * * * * *

SPRINGER LECTURE NOTES FROM THE IMA:

The Mathematics and Physics of Disordered Media
Editors: Barry Hughes and Barry Ninham
(Lecture Notes in Math., Volume 1035, 1983)

Orienting Polymers
Editor: J.L. Ericksen
(Lecture Notes in Math., Volume 1063, 1984)

New Perspectives in Thermodynamics
Editor: James Serrin
(Springer-Verlag, 1986)

Models of Economic Dynamics
Editor: Hugo Sonnenschein
(Lecture Notes in Econ., Volume 264, 1986)

David Aldous Robin Pemantle
Editors

Random Discrete Structures

With 34 Illustrations

Springer

David Aldous
Department of Statistics
University of California
Berkeley, CA 94720 USA

Robin Pemantle
Department of Mathematics
University of Wisconsin-Madison
Van Vleck Hall
480 Lincoln Drive
Madison, WI 53706 USA

Series Editors:
Avner Friedman
Willard Miller, Jr.
Institute for Mathematics and its
Applications
University of Minnesota
Minneapolis, MN 55455 USA

Mathematics Subject Classifications (1991): 60C05, 60J10, 05C80

Library of Congress Cataloging-in-Publication Data
Random discrete structures/David Aldous, Robin Pemantle.
p. cm. – (IMA volumes in mathematics and its applications: v. 76)
"Based on the proceedings of a workshop that was an integral part of the 1993–94 IMA program on 'Emerging Applications of Probability'"–Fwd.
Includes bibliographical references.
ISBN 0-387-94623-3 (hardcover:alk. paper)
1. Combinatorial probabilities–Congresses. 2. Markov processes–Congresses. 3. Random graphs–Congresses. I. Aldous, D.J. (David J.) II. Pemantle, Robin. III. Series.
QA273.45.R36 1995
519.2′6–dc20 95-44886

Printed on acid-free paper.

Production managed by Hal Henglein; manufacturing supervised by Joe Quatela.
Camera-ready copy prepared by the IMA.
Printed and bound by Braun-Brumfield, Inc., Ann Arbor, MI.
Printed in the United States of America.

9 8 7 6 5 4 3 2 1

ISBN 0-387-94623-3 Springer-Verlag New York Berlin Heidelberg

The IMA Volumes
in Mathematics and its Applications

Current Volumes:

Volume 1: Homogenization and Effective Moduli of Materials and Media
Editors: Jerry Ericksen, David Kinderlehrer, Robert Kohn, and J.-L. Lions

Volume 2: Oscillation Theory, Computation, and Methods of Compensated Compactness
Editors: Constantine Dafermos, Jerry Ericksen, David Kinderlehrer, and Marshall Slemrod

Volume 3: Metastability and Incompletely Posed Problems
Editors: Stuart Antman, Jerry Ericksen, David Kinderlehrer, and Ingo Muller

Volume 4: Dynamical Problems in Continuum Physics
Editors: Jerry Bona, Constantine Dafermos, Jerry Ericksen, and David Kinderlehrer

Volume 5: Theory and Applications of Liquid Crystals
Editors: Jerry Ericksen and David Kinderlehrer

Volume 6: Amorphous Polymers and Non-Newtonian Fluids
Editors: Constantine Dafermos, Jerry Ericksen, and David Kinderlehrer

Volume 7: Random Media
Editor: George Papanicolaou

Volume 8: Percolation Theory and Ergodic Theory of Infinite Particle Systems
Editor: Harry Kesten

Volume 9: Hydrodynamic Behavior and Interacting Particle Systems
Editor: George Papanicolaou

Volume 10: Stochastic Differential Systems, Stochastic Control Theory, and Applications
Editors: Wendell Fleming and Pierre-Louis Lions

Volume 11: Numerical Simulation in Oil Recovery
Editor: Mary Fanett Wheeler

Volume 12: Computational Fluid Dynamics and Reacting Gas Flows
Editors: Bjorn Engquist, M. Luskin, and Andrew Majda

Volume 13: Numerical Algorithms for Parallel Computer Architectures
Editor: Martin H. Schultz

Volume 14: Mathematical Aspects of Scientific Software
Editor: J.R. Rice

Volume 15: Mathematical Frontiers in Computational Chemical Physics
Editor: D. Truhlar

Volume 16: Mathematics in Industrial Problems
by Avner Friedman

Volume 17: Applications of Combinatorics and Graph Theory to the Biological and Social Sciences
Editor: Fred Roberts

Volume 18: q-Series and Partitions
Editor: Dennis Stanton

Volume 19: Invariant Theory and Tableaux
Editor: Dennis Stanton

Volume 20: Coding Theory and Design Theory Part I: Coding Theory
Editor: Dijen Ray-Chaudhuri

Volume 21: Coding Theory and Design Theory Part II: Design Theory
Editor: Dijen Ray-Chaudhuri

Volume 22: Signal Processing: Part I - Signal Processing Theory
Editors: L. Auslander, F.A. Grünbaum, J.W. Helton, T. Kailath, P. Khargonekar, and S. Mitter

Volume 23: Signal Processing: Part II - Control Theory and Applications of Signal Processing
Editors: L. Auslander, F.A. Grünbaum, J.W. Helton, T. Kailath, P. Khargonekar, and S. Mitter

Volume 24: Mathematics in Industrial Problems, Part 2
by Avner Friedman

Volume 25: Solitons in Physics, Mathematics, and Nonlinear Optics
Editors: Peter J. Olver and David H. Sattinger

Volume 26: Two Phase Flows and Waves
Editors: Daniel D. Joseph and David G. Schaeffer

Volume 27: Nonlinear Evolution Equations that Change Type
Editors: Barbara Lee Keyfitz and Michael Shearer

Volume 28: Computer Aided Proofs in Analysis
Editors: Kenneth Meyer and Dieter Schmidt

Volume 29: Multidimensional Hyperbolic Problems and Computations
Editors: Andrew Majda and Jim Glimm

Volume 30: Microlocal Analysis and Nonlinear Waves
Editors: Michael Beals, R. Melrose, and J. Rauch

Volume 31: Mathematics in Industrial Problems, Part 3
by Avner Friedman

Volume 32: Radar and Sonar, Part I
by Richard Blahut, Willard Miller, Jr., and Calvin Wilcox

Volume 33: Directions in Robust Statistics and Diagnostics: Part I
Editors: Werner A. Stahel and Sanford Weisberg

Volume 34: Directions in Robust Statistics and Diagnostics: Part II
Editors: Werner A. Stahel and Sanford Weisberg

Volume 35: Dynamical Issues in Combustion Theory
Editors: P. Fife, A. Liñán, and F.A. Williams

Volume 36: Computing and Graphics in Statistics
Editors: Andreas Buja and Paul Tukey

Volume 37: Patterns and Dynamics in Reactive Media
Editors: Harry Swinney, Gus Aris, and Don Aronson

Volume 38: Mathematics in Industrial Problems, Part 4
by Avner Friedman

Volume 39: Radar and Sonar, Part II
Editors: F. Alberto Grünbaum, Marvin Bernfeld, and
Richard E. Blahut

Volume 40: Nonlinear Phenomena in Atmospheric and Oceanic Sciences
Editors: George F. Carnevale and Raymond T. Pierrehumbert

Volume 41: Chaotic Processes in the Geological Sciences
Editor: David A. Yuen

Volume 42: Partial Differential Equations with Minimal Smoothness and Applications
Editors: B. Dahlberg, E. Fabes, R. Fefferman, D. Jerison, C. Kenig, and J. Pipher

Volume 43: On the Evolution of Phase Boundaries
Editors: Morton E. Gurtin and Geoffrey B. McFadden

Volume 44: Twist Mappings and Their Applications
Editors: Richard McGehee and Kenneth R. Meyer

Volume 45: New Directions in Time Series Analysis, Part I
Editors: David Brillinger, Peter Caines, John Geweke, Emanuel Parzen, Murray Rosenblatt, and Murad S. Taqqu

Volume 46: New Directions in Time Series Analysis, Part II
Editors: David Brillinger, Peter Caines, John Geweke, Emanuel Parzen, Murray Rosenblatt, and Murad S. Taqqu

Volume 47: Degenerate Diffusions
Editors: Wei-Ming Ni, L.A. Peletier, and J.-L. Vazquez

Volume 48: Linear Algebra, Markov Chains, and Queueing Models
Editors: Carl D. Meyer and Robert J. Plemmons

Volume 49: Mathematics in Industrial Problems, Part 5
by Avner Friedman

Volume 50: Combinatorial and Graph-Theoretic Problems in Linear Algebra
Editors: Richard A. Brualdi, Shmuel Friedland, and Victor Klee

Volume 51: Statistical Thermodynamics and Differential Geometry of Microstructured Materials
Editors: H. Ted Davis and Johannes C.C. Nitsche

Volume 52: Shock Induced Transitions and Phase Structures in General Media
Editors: J.E. Dunn, Roger Fosdick, and Marshall Slemrod

Volume 53: Variational and Free Boundary Problems
Editors: Avner Friedman and Joel Spruck

Volume 54: Microstructure and Phase Transitions
Editors: David Kinderlehrer, Richard James, Mitchell Luskin, and Jerry L. Ericksen

Volume 55: Turbulence in Fluid Flows: A Dynamical Systems Approach
Editors: George R. Sell, Ciprian Foias, and Roger Temam

Volume 56: Graph Theory and Sparse Matrix Computation
Editors: Alan George, John R. Gilbert, and Joseph W.H. Liu

Volume 57: Mathematics in Industrial Problems, Part 6
by Avner Friedman

Volume 58: Semiconductors, Part I
Editors: W.M. Coughran, Jr., Julian Cole, Peter Lloyd, and Jacob White

Volume 59: Semiconductors, Part II
Editors: W.M. Coughran, Jr., Julian Cole, Peter Lloyd, and Jacob White

Volume 60: Recent Advances in Iterative Methods
Editors: Gene Golub, Anne Greenbaum, and Mitchell Luskin

Volume 61: Free Boundaries in Viscous Flows
Editors: Robert A. Brown and Stephen H. Davis

Volume 62: Linear Algebra for Control Theory
Editors: Paul Van Dooren and Bostwick Wyman

Volume 63: Hamiltonian Dynamical Systems: History, Theory, and Applications
Editors: H.S. Dumas, K.R. Meyer, and D.S. Schmidt

Volume 64: Systems and Control Theory for Power Systems
Editors: Joe H. Chow, Petar V. Kokotovic, and Robert J. Thomas

Volume 65: Mathematical Finance
Editors: Mark H.A. Davis, Darrell Duffie, Wendell H. Fleming, and Steven E. Shreve

Volume 66: Robust Control Theory
Editors: Bruce A. Francis and Pramod P. Khargonekar

Forthcoming Volumes:

1992 Summer Program: *Environmental Studies: Mathematical, Computational, and Statistical Analysis*

1992–1993: *Control Theory*

Robotics

Nonsmooth Analysis & Geometric Methods in Deterministic Optimal Control

1993-1994: *Emerging Applications of Probability*

Mathematical Population Genetics

Stochastic Problems for Nonlinear Partial Differential Equations

Image Models (and their Speech Model Cousins)

Stochastic Models in Geosystems

Classical and Modern Branching Processes

1994 Summer Program: *Molecular Biology*

1994-1995: *Waves and Scattering*

Computational Wave Propagation

Wavelets, Multigrid and Other Fast Algorithms (Multipole, FFT) and Their Use In Wave Propagation

Waves in Random and Other Complex Media

Inverse Problems in Wave Propagation

FOREWORD

This IMA Volume in Mathematics and its Applications

RANDOM DISCRETE STRUCTURES

is based on the proceedings of a workshop that was an integral part of the 1993–94 IMA program on "Emerging Applications of Probability." We thank David Aldous and Robin Pemantle for organizing this workshop and for editing the proceedings. We also take this opportunity to thank the National Science Foundation, the Air Force Office of Scientific Research, the Army Research Office, and the National Security Agency whose financial support made this workshop possible.

Avner Friedman

Willard Miller, Jr.

PREFACE

The fall quarter of the 1993–1994 IMA Program on Emerging Applications of Probability was focused on discrete probability and its connections with the theory of algorithms. The companion volume "Discrete Probability and Algorithms" (IMA Volume 72) contains proceedings of workshops devoted explicitly to algorithmic applications of probability and to related aspects of the theory of finite Markov chains. This volume, based on papers presented at a Workshop held November 15–19, 1993, has a more mathematical focus. All papers were refereed.

Classical topics in discrete mathematics are represented by the paper of Erdos et al., using probabilistic methods to study triangle-free subgraphs, and by Drew and Johnson, who prove a characterization of completely positive matrices.

Several of the papers in this volume are expositions that condense and make readable some recent work on Markov chains, potential theory, and the second moment method. Salisbury discusses a method of obtaining capacity-theoretic bounds on intersection probabilities for Markov chains by bounding energies of projections of measures. Lyons discusses some results about random walks on Galton-Watson trees obtained in an ergodic-theoretic setting. Northshield, also writing on random walks on trees, exhibits a relation between the Martin boundary of a covering tree of a graph and amenability and recurrence of the original graph. Janson discusses the second moment method as a general method for showing the existence with high probability of combinatorial substructures.

Hildebrand discusses rates of convergence of a certain "autoregressive" random walk on the integers modulo a prime.

Stein's method of obtaining explicit bounds in limit theorems is another topic of much recent interest. Dembo and Rinott give three new examples of normal approximations via Stein's method: nonlinear functionals of correlated normals, exceedances of r-scans, and in a multinomial setting.

Different aspects of random trees are discussed in three papers. Game trees depict two-player games in which the players alternate turns, seeking to maximize or minimize a payoff; Devroye and Kamoun study a model of game trees where the payoff is determined by random edge-weights. Aldous discusses a one-parameter family of random n-leaf trees motivated by evolutionary trees. Dembo and Zeitouni prove a large deviation conjecture relating to continuum trees.

Renewal sequences are a classical topic in discrete applied probability. Diaconis et al. show that a coincidence problem for renewal sequences arises in studying distances between random compositional data. Fristedt discusses intersections of the random sets determined by renewal processes and their continuous-time counterparts.

Finally, Burton and Faris give a rigorous analysis of a Markov chain modeling the self-organizing features of (physiological) neural nets.

It is a pleasure to thank Avner Friedman, Willard Miller, Jr., and the IMA staff for their efficient organization of the workshop and the entire program, and to thank Patricia V. Brick for administering the preparation of this volume.

David Aldous
Robin Pemantle

CONTENTS

PROBABILITY DISTRIBUTIONS ON CLADOGRAMS*

DAVID ALDOUS†

Abstract. By analogy with the theory surrounding the Ewens sampling formula in neutral population genetics, we ask whether there exists a natural one-parameter family of probability distributions on cladograms ("evolutionary trees") which plays a central role in neutral evolutionary theory. Unfortunately the answer seems to be "no" – see Conjecture 2. But we can embed the two most popular models into an interesting family which we call "beta-splitting" models. We briefly describe some mathematical results about this family, which exhibits qualitatively different behavior for different ranges of the parameter β.

1. Probability distributions on partitions and neutral population genetics.

The first few sections give some conceptual background. The reader wishing to "get right to the point" should skim these and proceed to section 3.

For each n there is a finite set of partitions of $\{1, 2, \ldots, n\}$ into unordered families $\{A_1, A_2, \ldots, A_k\}$ of subsets. A one-parameter family $(P_\theta^{(n)})$ of probability distributions on this set of partitions can be defined by

$$P_\theta^{(n)}\{A_1, \ldots, A_k\} = \frac{\Gamma(\theta)}{\Gamma(n+\theta)} \prod_{i=1}^{n} ((i-1)!\ \theta)^{m_i}$$

where m_i is the number of A's with exactly i elements. The right side is a slightly disguised statement of the *Ewens sampling formula* in neutral population genetics, which is central to a mathematically rich and elegant theory which has made an impact in non-mathematical genetics. See e.g. [16,26,11,9]. Our purpose in this paper is to ask whether there is an analogous theory for neutral evolutionary trees. In particular, we study whether the following two results have analogs for evolutionary trees.

(a) *Consistency characterization.* A family $(P^{(n)}; 1 \leq n < \infty)$ satisfies the following three conditions iff it is one of the $(P_\theta^{(n)})$.

(i) Exchangeability. For each n, the distribution is invariant under permutations of the labels $\{1, 2, \ldots, n\}$.

(ii) Sampling invariance. For each n, $P^{(n)}$ induces a distribution on partitions of $\{1, 2, \ldots, n-1\}$ by the action of deleting n: this distribution is $P^{(n-1)}$.

(iii) Subset deletion. For each $j < n$, given that $P^{(n)}$ has $\{j+1, j+2, \ldots, n\}$ as a set in the partition, the remaining partition of $\{1, 2, \ldots, j\}$ has distribution $P^{(j)}$.

(b) *Interpretation via time-evolution.* Suppose there are k neutral alleles (an *allele* is a possible "type" of a gene; *neutral* means to confer no selective advantage or disadvantage). Count the proportions $(X_i(t); 1 \leq i \leq k)$

* Research supported by N.S.F. Grant DMS92-24857.

† Department of Statistics, University of California, Berkeley, CA 94720.

of a population with allele i in generation t. Then under natural models there is a k-dimensional diffusion $(\tilde{X}_i(t))$ representing the limit (as population size tends to infinity and time is rescaled), where randomness comes from the random number of copies of an individual allele which appear in the next generation. If we also allow random mutations to produce new alleles, we get an infinite-dimensional diffusion ("the infinitely-many-alleles model"). A random sample of n individuals from the population can be partitioned into subsets with identical alleles, and this random partition has distribution $P_\theta^{(n)}$, where the parameter θ is related to the mean number of mutations per generation.

2. Phylogenetic trees. A *phylogenetic tree* is a visual representation of an assertion about relationships between species $A, B, C, \ldots$. There are many varieties of such tree, differing in the details of what exactly is being asserted – see Eldredge and Cracraft [10] for an extensive discussion. Figure 1 is a *cladogram*. The basic interpretation is the obvious one: amongst species $\{C, D, E\}$, the most closely related are D and E, and so on. The species are distinguished, i.e. switching A with D gives a different cladogram. But there is no distinction between right and left edges, i.e. switching A with B gives the same cladogram. And there is no explicit time scale.

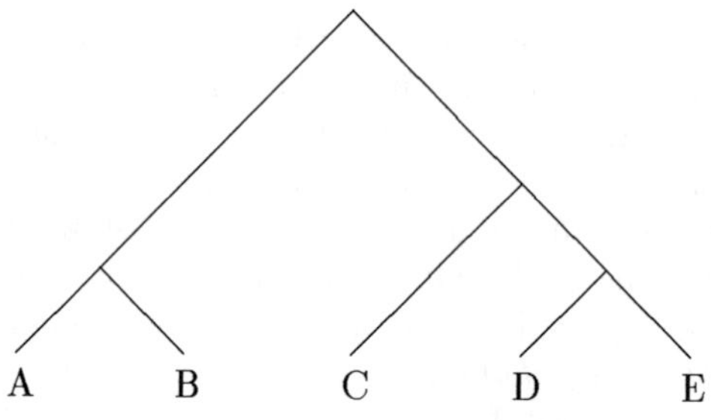

Figure 1

Biologists believe in evolution, and so implicitly believe there is a true "evolutionary tree" linking all living and extinct species, and that classifications of species should be consistent with evolutionary history. The classical Linnaean hierarchy (originally *species, genus, order, class, kingdom* but subsequently extended to many more ranks) remains in practical use, but theoreticians have conducted a vigorous debate about how classification ought to be done.

Figure 2 is one way to picture a true evolutionary tree. Species are represented by vertical lines, from their time of origin to their time of extinction, with dotted horizontal lines indicating the origin of a species from its parent species. Implicit in such a picture are a set of generally-held beliefs about evolution (e.g. that species arise comparatively quickly and then remain largely unchanged until extinction) which I won't go into.

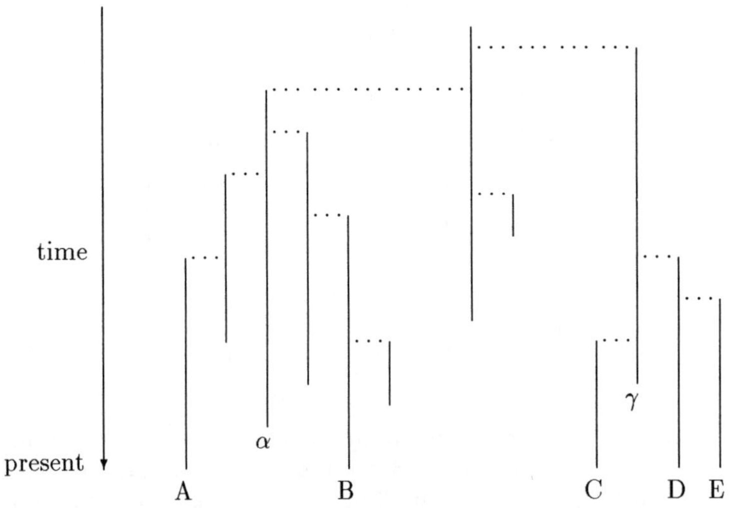

Figure 2

In practice one seldom has enough information about extinct species to be able to draw a tree as detailed as Figure 2, but it is useful to envisage such a tree in order to avoid drawing unwarranted inferences from other representations. To illustrate, consider the cladogram in Figure 1, which is consistent with the evolutionary tree of Figure 2. From the cladogram one might think, loosely speaking, that A and B are more closely related to each other than are C and D. But the evolutionary tree indicates the opposite is possible, if we measure closeness by either time of divergence or number of intermediate species. More dramatically, a cladogram does not indicate ancestor-descendant pairs. In Figure 1, A and B might be "cousins" (as Figure 2 shows), or one might be an offspring of the other (as Figure 2 shows E to be an offspring of D).

A third type of picture, a *phenogram*, is often used, in particular in the context of molecular genetics analysis of living species.

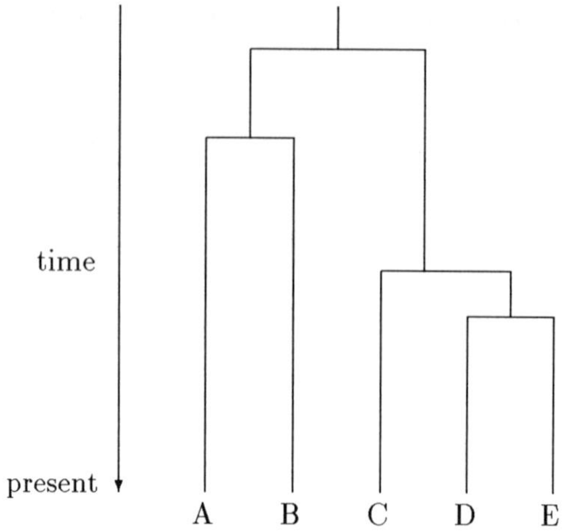

Figure 3

A phenogram contains more information than the cladogram, in that there is an absolute time scale. But the previous warning about casual inferences still holds. From Figure 3 one might assume that the common ancestor α of $\{A, B\}$ lived earlier than the common ancestor γ of $\{C, D, E\}$, whereas Figure 2 shows it is possible that α both originated later than γ and became extinct later than γ.

Minor points. (a) We're going to assume cladograms have only binary splits, although actual cladograms exhibited in the literature usually have some higher-order splits also. One can interpret a higher-order split as a collection of binary splits which cannot be resolved with the data available.

(b) Drawing cladograms with diagonal lines is just a convention, but is useful for distinguishing them from phenograms or other kinds of evolutionary tree.

2.1. Why consider probability models? There are two quite different reasons for considering probability models of phylogenic trees. The first reason concerns technical aspects of reconstructing trees from data. Molecular biologists in general have eschewed probability models in favor of parsimony (deterministic best-fit) methods, which have the advantage of telling you what tree to write down (up to often-serious non-uniqueness and computational tractability issues), but the disadvantage of not indicating quantitative confidence assertions. To implement a more classical statistical methodology involves a complex array of modeling problems, one of which is to specify an *a priori* model of evolutionary history. A related issue is testing the actual algorithms used: given a hypothetical true phylogenetic tree and a model for the mutation process underlying the observed data, how well does the algorithm reconstruct the tree? Since these procedures are computationally intensive anyway, it would seem better to use a "good" *a priori* model (if we could agree on one!) rather than a model chosen purely for mathematical simplicity.

My own motivation comes from the more conceptual question:

> If we had the true evolutionary tree of all species, what could we infer about the relative roles of selectivity and neutrality in the pattern of speciations and extinctions?

Several disparate lines of relevant research appear in the literature. Gould, Raup et al [24,12,23] compared paleontological data with random models (essentially critical branching processes). [1] The recent book by Kauffman [15] contains a wide-ranging study of mathematical models of selectivity effects. See also the conference proceedings [22] and the work cited in the next section. But the bottom line is that (in contrast to neutral population genetics) there is no accepted definite notion, at either the conceptual or mathematical level, of "neutral evolution of species."

One could discuss models of any of the varieties of tree discussed in section 2. For the technical issues mentioned above it is most natural to use phenograms, whereas for our purpose of extracting patterns from published trees we shall use cladograms, which are becoming the most common form of published phylogeny.

2.2. Two particular probability distributions on phylogenetic trees. There is a scattered (and mostly mathematically unsophisticated) biological literature on *a priori* models of random phylogenetic trees. Brief surveys are contained in [14,19]. We describe below the two models which have been most discussed in the biological literature. As noted later these models (under a bewildering variety of different names) have also been extensively studied in other contexts.

The number of different cladograms on n species is

$$c_n = (2n-3)!! = (2n-3)(2n-5)\cdot \quad \cdot 3 \cdot 1.$$

One way to see this is to note that a cladogram on n species has $2n-1$ edges (for this purpose we add an edge upwards from the root) and that each choice of edge at which to add a $n+1$'st species leads to a different cladogram on $n+1$ species, so $c_{n+1} = (2n-1)c_n$. In the uniform model we assume each cladogram is equally likely. The Yule model, considered as a phenogram, is just the elementary continuous-time pure birth process started with one lineage. That is, each lineage persists for a random, exponential(1) time and then splits into two lineages. Continue the process until there are n lineages. The resulting random cladogram has an equivalent description as "random joinings," where we count time backwards from the present. Starting with n lines of descent, we choose uniformly at random one of the $n(n-1)/2$ pairs and join the pair, to make $n-1$ lines of descent, and continue until there is a unique line. In either case, throwing away the time scale leaves a random cladogram.

[1] Their work focused on number of species as a function of time, whereas ours focuses on the combinatorial structure of phylogenetic trees.

2.3. Search trees. Some probability distributions on cladograms can be associated with well-studied random search trees in computer science. Let us explain the connection briefly.

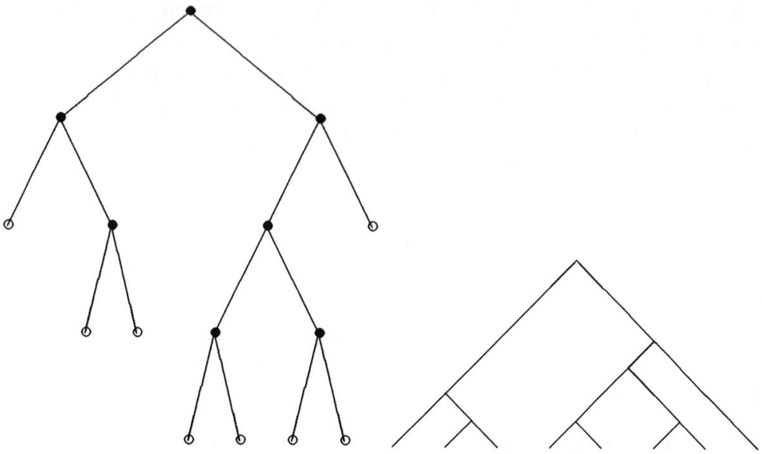

Figure 4

Figure 4 shows a subtree of the infinite binary tree, i.e. a tree in which a node has either zero children, a left child, a right child, or both a left and a right child. Such a "binary tree" can be described using either the $n-1$ *internal* nodes $\bullet$ or, as is customary in computer science, via the n *external* nodes $\circ$, i.e. those nodes outside the subtree whose parent is in the subtree. Such a tree can be mapped to a cladogram on n leaves, as shown in Figure 5.

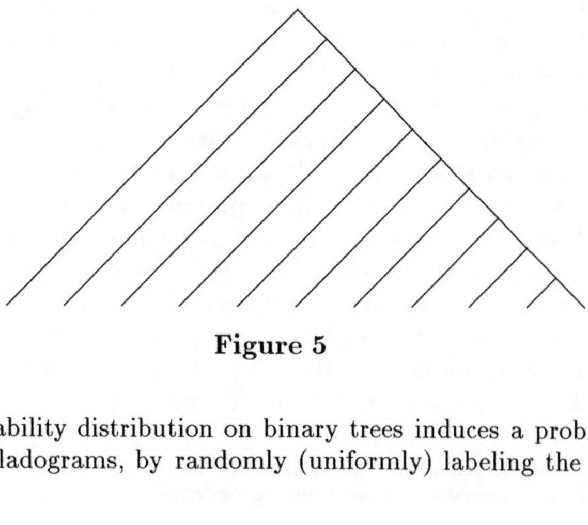

Figure 5

Thus a probability distribution on binary trees induces a probability distribution on cladograms, by randomly (uniformly) labeling the leaves $1, 2, \ldots, n$.

The recent book Mahmoud [20] treats several models of random binary tree from the viewpoint of search trees in computer science. Some of these models – precisely, the induced models on cladograms – occur later.

A wide-ranging abstract study of trees as proximity indicators (but not emphasizing probability models) can be found in Barthelemy and Guenoche [6].

3. Axiomatizing properties of random cladograms. We seek probability models for "neutral evolution of species." By analogy with the results in section 1 for neutral population genetics, it is natural to hope there exists a one-parameter family of probability distributions on cladograms for which

(a) random cladograms $(T_n; n \geq 2)$ are in the family iff they satisfy some specified intrinsic compatibility conditions.

(b) These random cladograms arise from some natural model of species evolving with time.

Let's start with idea (b). Consider the following class of models.

> At each time t there are a finite number of species alive (starting with one species at time 0). From time to time there is an "event" which is either an extinction or a speciation, i.e. either some species B becomes extinct or some species A splits into species A and A'. The time from t until the next event, and the chance the next event is an extinction rather than a speciation, may depend on the past in an arbitrary way. But if the next event is an extinction then each species is equally likely to be the one to become extinct, and if the next event is an speciation then each species is equally likely to be the one to speciate.

At first sight the arbitrariness should allow us to get a family of models, with a parameter representing (say) the ratio of speciation rate to extinction rate. But this is false, because it is easy to show

LEMMA 1. *For any model of the class above which ends with n living species, the cladogram of those species is distributed as the Yule model described in section 2.2.*

Turning to idea (a), the following two compatibility conditions for a family $(T_n; n \geq 2)$ of random cladograms seem to be the natural analogs of those in section 1.

(i) Exchangeability. For each n, the random cladogram is exchangeable in the labels of the n species, i.e. invariant under permutations.

(ii) Group elimination. For each $1 \leq k < n$, conditional on $\{k+1, k+2, \ldots, n\}$ being a group in T_n (i.e. being the set of descendants of some internal vertex), the cladogram restricted to $\{1, 2, \ldots, k\}$ is distributed as T_k.

It is easy to check that the Yule model and the uniform model (described in section 2.2) satisfy these conditions, as does the family of *combs*, i.e. the family with the deterministic "maximally unbalanced" shape below, and with the n species uniformly randomly distributed amongst positions.

Unfortunately we have been unable to construct any more examples, so

CONJECTURE 2. *The three families above are the only families satisfying (i)-(ii).*

Another open question in this "axiomatic" spirit will be mentioned in section 6.3, but the bottom line is that our attempt to define models of "neutral evolution of species" by some close analogy with neutral population genetics seems completely unsuccessful. In the next section we resort to pulling a model out of thin air.

4. The beta-splitting model. Suppose that for each $n \geq 2$ we are given a probability distribution $\mathbf{q}_n = (q_n(i); i = 1, 2, \ldots, n-1)$ which is symmetric $(q_n(i) = q_n(n-i))$. Then we can define probability distributions on cladograms in the obvious way: the root-split has i elements in the left branch and $n-i$ elements in the right branch, where i is chosen at random according to the distribution q_n and where each of the $\binom{n}{i}$ choices of elements for the left branch are equally likely. Repeat recursively in each branch. Interpret the resulting tree as a cladogram by removing the left/right markers. Call these *Markov branching* models.

To specialize this construction, consider a probability density f on $(0,1)$ which is symmetric (that is, $f(x) = f(1-x)$), and define

$$q_n(i) = a_n^{-1} \binom{n}{i} \int_0^1 x^i (1-x)^{n-i} f(x) dx \tag{1}$$

for normalizing constant

$$\begin{aligned} a_n &= \int_0^1 (1 - x^n - (1-x)^n) f(x) dx \\ &= 1 - 2 \int_0^1 x^n f(x) dx . \end{aligned} \tag{2}$$

This specialization has a simple interpretation in terms of splitting intervals (a topic discussed from a different viewpoint by Brennan and Durrett [7,8]). Start with n uniform random "particles" on the unit interval. Split the interval at a random point with density f. Repeat recursively on subintervals, splitting each interval $[a, b]$ at a point $a + X(b-a)$ where the X's are independent with density f, stopping when a subinterval contains only one particle.

Figure 6 illustrates the construction and its interpretation as a cladogram. Note that a subinterval split in which all particles go into one side of the split is suppressed.

Note that for (1, 2) to make sense it is not necessary for f to be a probability density. It is enough to have $f \geq 0$ be symmetric and satisfy $\int_{0+} xf(x)dx < \infty$.

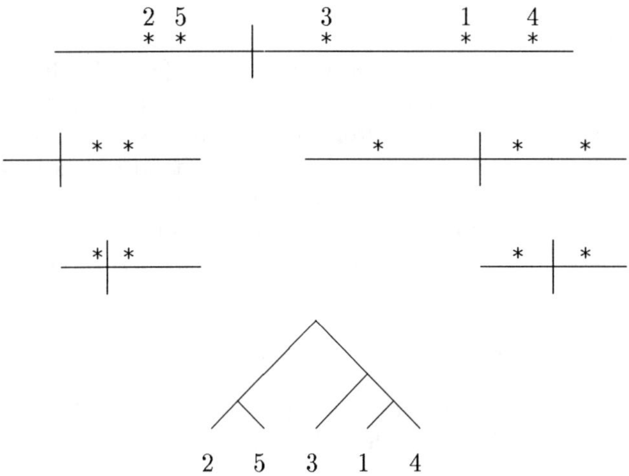

Figure 6

We now specialize further to define a one-parameter family $(T_\beta^{(n)})$ of random cladograms, parametrized by $-2 \leq \beta \leq \infty$. For $-1 < \beta < \infty$ these are obtained by the interval-splitting construction above with the beta density

$$f(x) = \frac{\Gamma(2\beta+2)}{\Gamma^2(\beta+1)} x^\beta (1-x)^\beta, \ 0 < x < 1. \tag{3}$$

Applying (1),

$$q_n(i) = \frac{1}{a_n(\beta)} \frac{\Gamma(\beta+i+1)\Gamma(\beta+n-i+1)}{\Gamma(i+1)\Gamma(n-i+1)}, \ 1 \leq i \leq n-1 \tag{4}$$

where as at (2) $a_n(\beta)$ is the normalizing constant. For $\beta = \infty$ we get the interval-splitting construction where intervals are split deterministically at their center. For $-2 < \beta \leq -1$ the recursive construction with q_n defined at (4) still makes sense, corresponding as mentioned above to the function $f(x) = x^\beta(1-x)^\beta$ for which $\int_{0+} xf(x)dx < \infty$. Finally, for fixed n the $\beta \to -2$ limit is the "comb" described in section 3, so we take the comb as the $\beta = -2$ model.

4.1. Special cases. Three special case have been studied in the literature.

$\beta = \infty$. This is the "symmetric binary trie" studied in computer science, and surveyed in Chapter 5 of [20]. It has been briefly considered in the biological literature [19].

$\beta = 0$. Here we have

$$q_n(i) = \frac{1}{n-1}, \quad 1 \leq i \leq n-1.$$

This is the Yule model of section 2.2. It arises from the "binary search tree" in computer science, surveyed in Chapter 2 of [20]. It also arises from the "coalescent" in mathematical population genetics ([16,26]). And as mentioned in section 2.2 it arises from the Yule process, i.e. the linear pure birth process. All these processes are different as processes indexed by time, but the induced random cladograms are identical.

$\beta = -3/2$. This is the uniform model from section 2.2. To verify, a counting argument shows that the uniform model corresponds to a recursive construction with

$$q_n(i) = \frac{1}{2}\binom{n}{i}\frac{c_i c_{n-i}}{c_n}. \tag{5}$$

where $c_n = (2n-3)!!$ is the number of cladograms on n leaves. But we can write $c_n = 2^{n-1}\Gamma(n-\frac{1}{2})/\Gamma(\frac{1}{2})$, which leads to

$$q_n(i) = \frac{\Gamma(n+1)}{\Gamma(n-\frac{1}{2})\Gamma(\frac{1}{2})}\,\frac{\Gamma(i-\frac{1}{2})\Gamma(n-i-\frac{1}{2})}{\Gamma(i+1)\Gamma(n-i+1)}$$

which is indeed (4) for $\beta = -3/2$.

This is the model of cladograms associated (in the sense of section 2.3) with the uniform random binary tree with n external leaves, which has been studied extensively in combinatorics and computer science (e.g. Knuth [17]). Its asymptotics can be studied via the continuum tree set-up of [2,3].

The beta-splitting model certainly satisfies the exchangeability condition (i) of section 3, but in general does not satisfy the group elimination property (ii). As a specific example, for $\beta = \infty$ the chance of the cladogram

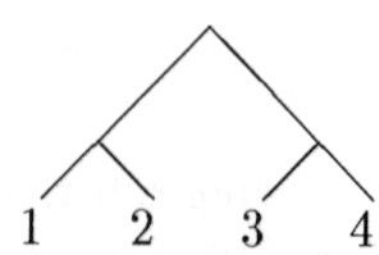

equals 1/7. But if we take the cladogram on 6 leaves and condition on $\{5,6\}$ being a group, then (by an elementary but tedious calculation) the probability that the cladogram restricted to $\{1,2,3,4\}$ is the cladogram above equals $\frac{5\cdot 13}{3\cdot 157}$.

The special case $\beta = -1$ has apparently not been studied before, but turns out to have interesting properties. Note that here we have $q_n(i) = a_n \frac{1}{i(n-i)}$. Because

$$\sum_{i-1}^{n-1} \frac{1}{i(n-i)} = \sum_{i=1}^{n-1} \frac{1}{n}\left(\frac{1}{i} + \frac{1}{n-i}\right) = \frac{2h_{n-1}}{n}$$

where h_{n-1} is the harmonic sum, we can write

$$q_n(i) = \frac{n}{2h_{n-1}} \frac{1}{i(n-i)}, \quad 1 \le i \le n-1. \tag{6}$$

4.2. Is there an underlying process? We are introducing the beta-splitting models as the mathematically most natural way to embed the Yule model and the uniform model into a one-parameter family. To make this convincing for biological applications one would like an underlying continuous-time process of speciation and extinction for which the general beta-splitting model was the associated cladogram. I do not know a natural candidate for such a process.

It has often been asserted in the biological literature [14,19] that the uniform model is unsatisfactory because there is no such underlying process. This is not entirely correct. Consider conditioned critical branching processes, which have been studied in the biological literature ([24,12]) as models for neutral evolution of large groups of species. One of the results from [2,3] is that, for a random sample of n species from a large such group, the cladogram on the sampled species will follow the uniform model.

This observation makes it slightly more plausible that there might be some subtle process underlying the general beta-splitting model.

4.3. A data-set. The most famous datum in the subject (discussed in [14] and by many other writers) is $(9672, 21)$, the split between bird species and crocodilian species. Under the Yule model ($\beta = 0$) the chance of a more unbalanced split is 0.004, whereas under the uniform model ($\beta = -1.5$) the chance is 0.878. Guyer and Slowinski ([14] table 1) consider the sizes of the smaller branch in the root split in 30 large cladograms. It is clear from the raw data (and confirmed in [14] by a test of significance) that the splits are more unbalanced than predicted under the Yule model, but more balanced than predicted under the uniform model. Figure 7 gives a visual display of the data under each model using the quantile transform, (i.e. the bird-crocodile split would be plotted at 0.878 under the uniform model and at 0.004 under the Yule model.) So under a true model we would see 30 independent uniform points.

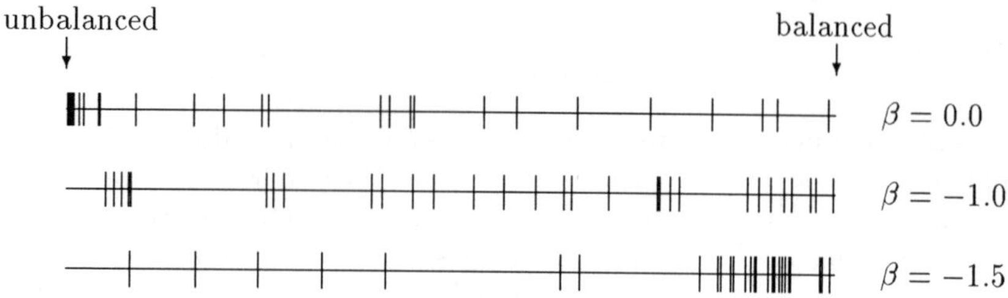

Figure 7

The fit to the $\beta = -1$ model is strikingly better than to the usual models. Is this just a fluke?

Let me also mention that two papers (Savage [25], Guyer and Slowinski [13]) analyze data on small cladograms ($4 \le n \le 7$ and $n = 5$). [25] concludes that the Yule model gives a better fit than the uniform model, whereas [13] concludes the opposite.

5. Some asymptotics for the beta-splitting model. This section outlines some asymptotics for the beta-splitting family. Details of some of the more interesting results may be given elsewhere. In the three special cases ($\beta = -3/2,\ 0,\ \infty$) these results (and much more) are either explicitly known or can be proved by routine methods.

5.1. Asymptotics of the root-split distribution. The top-most split in a cladogram splits it into two branches of sizes i and $n - i$. If we randomly call these "left" and "right" then the size L_n of the left branch has the distribution q_n at (4). And if we write B_n for the size of the branch containing a specified leaf, then by exchangeability

(7) $$P(B_n = i) = \frac{2i}{n}\, q_n(i);\ 1 \le i \le n - 1.$$

It is straightforward to obtain the following asymptotics as $n \to \infty$.

LEMMA 3.

$\boldsymbol{\beta = \infty}$. $n^{-1}L_n \xrightarrow{d} \frac{1}{2}$; $n^{-1}B_n \xrightarrow{d} \frac{1}{2}$.

$\boldsymbol{-1 < \beta < \infty}$. $n^{-1}L_n \xrightarrow{d} X_\beta$ *and* $n^{-1}B_n \xrightarrow{d} Y_\beta$, *where* X_β *has the beta distribution (3) and where* Y_β *has density*

(8) $$f(x) = \frac{\Gamma(2\beta + 3)}{\Gamma(\beta + 1)\Gamma(\beta + 2)}\, x^{\beta+1}(1 - x)^{\beta}.$$

$\boldsymbol{\beta = -1}$.

$$\frac{\log \min(L_n, n - L_n)}{\log n} \xrightarrow{d} U \text{ and } \frac{\log(n - B_n)}{\log n} \xrightarrow{d} U$$

where U has the uniform distribution on $(0,1)$.

$\mathbf{-2 < \beta < -1}$. *For each fixed* $i \geq 1$,

$$P(\min(L_n, n-L_n) = i) \to \gamma_\beta(i) \text{ and } P(B_n = n-i) \to \gamma_\beta(i)$$

where the limit distribution γ_β *is given by*

$$\gamma_\beta(i) = \kappa_\beta^{-1} \int_0^\infty \frac{e^{-t}t^i}{i!} t^\beta \, dt = \kappa_\beta^{-1} \frac{\Gamma(i+1+\beta)}{\Gamma(i+1)}$$

for normalizing constant

(9) $$\kappa_\beta = \int_0^\infty (1-e^{-t})t^\beta dt = \frac{\pi}{\Gamma(b)\sin(\pi(-\beta-1))}.$$

And if $i \sim xn$ *for* $0 < x < 1$,

(10) $$q_n(i) \sim \frac{1}{2}\kappa_\beta^{-1} x^\beta (1-x)^\beta n^\beta.$$

Note that the distribution γ_β can also be expressed as

$$\gamma_\beta(i) = \kappa'_\beta \prod_{j=2}^{i} \left(1 + \frac{\beta}{j}\right), \; i \geq 1$$

for different normalizing constants.

5.2. Depth and height statistics. The *depth* of a leaf in a cladogram is the number of branchpoints on the path from that leaf to the root, where we include the root as a branchpoint. Thus in Figure 1, species A has depth 2 and species E has depth 3. We can define the following three numbers for a cladogram.

- $\bar{d}$, the average depth of the leaves.
- $\tilde{d}$, the depth of the leaf found by starting at the root and recursively choosing the larger branch at each branchpoint (averaging over possibilities if an even split is encountered).
- d^*, the maximal depth of a leaf.

So $\bar{d} \leq \tilde{d} \leq d^*$. In many settings, statistics like these are called *heights*, and in particular one could call d^* the height of the cladogram.

PROPOSITION 4. *Define* $\bar{D}_n, \tilde{D}_n, D_n^*$ *to be the random values of the statistics above for the beta-splitting model. Note that* $E\bar{D}_n = ED_n$, *where* D_n *is the depth of species* 1. *As* $n \to \infty$,

$\mathbf{-1 < \beta \leq \infty}$.

$$(E\bar{D}_n, E\tilde{D}_n, ED_n^*) \sim (\bar{\rho}(\beta), \tilde{\rho}(\beta), \rho^*(\beta)) \; \log n$$

and $D'_n/ED'_n \xrightarrow{d} 1$ *for each of the four statistics, where (for* X_β, Y_β *as in Lemma 3 and* $\beta < \infty$*)*

$$\frac{1}{\bar{\rho}(\beta)} = -E\log Y_\beta = \frac{\Gamma(2\beta+3)}{\Gamma(\beta+1)\Gamma(\beta+2)} \int_0^1 x^{\beta+1}(1-x)^\beta \log(1/x) dx.$$

$$\frac{1}{\tilde{\rho}(\beta)} = -E\log\max(X_\beta, 1-X_\beta) = \frac{2\Gamma(2\beta+2)}{\Gamma^2(\beta+1)}\int_{1/2}^1 x^\beta(1-x)^\beta \log(1/x)dx.$$

$$\rho^*(\beta) = \frac{-\theta}{\log 2EX_\beta^\theta}, \text{ where } \theta = \theta(\beta) \text{ is the solution of}$$

$$EX_\beta^\theta \log X_\beta = \theta^{-1}(EX_\beta^\theta)\log(2EX_\beta^\theta).$$

$\boldsymbol{\beta = -1}$.

$$(E\bar{D}_n, E\tilde{D}_n) \sim (3\pi^{-2}, 6\pi^{-2})\ \log^2 n.$$

$\boldsymbol{-2 < \beta < -1}$.

$$(E\bar{D}_n, E\tilde{D}_n, ED_n^*) \sim (\bar{\rho}(\beta), \tilde{\rho}(\beta), \rho^*(\beta))\ n^{-\beta-1}$$

where

$$\frac{1}{\bar{\rho}(\beta)} = \kappa_\beta^{-1}\int_0^1 x^{\beta+1}(1-x)^\beta(1-x^{-\beta-1})\ dx$$

$$\frac{1}{\tilde{\rho}(\beta)} = \kappa_\beta^{-1}\int_{1/2}^1 x^{\beta}(1-x)^\beta(1-x^{-\beta-1})\ dx$$

for κ defined at (9), and where we do not have a simple expression for $\rho^*(\beta)$. For each of the four statistics we have a non-degenerate limit distribution for D'_n/ED'_n.

Some numerical values are tabulated below. The unexplained decimal numbers were obtained by numerical evaluation of the formulae above, and the others (except for $\rho^*(-1.5)$, discussed later) by exact evaluation.

β	-0.5	0	1	∞
$\bar{\rho}(\beta)$	$\frac{1}{2\log 2-1} = 2.59$	2	$\frac{12}{7} = 1.71$	$\frac{1}{\log 2} = 1.44$
$\tilde{\rho}(\beta)$	4.55	$\frac{1}{1-\log 2} = 3.26$	$\frac{1}{\frac{13}{12}-\log 2} = 2.56$	$\frac{1}{\log 2} = 1.44$
$\rho^*(\beta)$	6.38	4.31	3.19	$\frac{1}{\log 2} = 1.44$

β	-2	-1.75	-1.5	-1.25
$\bar{\rho}(\beta)$	$1/2$	0.952	$\pi^{1/2} = 1.77$	8.13
$\tilde{\rho}(\beta)$	1	1.71	$\frac{2\pi^{1/2}}{2^{3/2}-\log(3+2^{3/2})} = 3.33$	15.5
$\rho^*(\beta)$	1	?	$2\pi^{1/2} = 3.54$	?

Outline of proof.

$\mathbf{-1 < \beta \leq \infty}$. Consider the process of splitting the unit interval into subintervals. For these first-order results the issue is when the "relevant subinterval" has length $\approx 1/n$. For D_n the relevant subinterval is the one containing a prespecified point, and the lengths of such subintervals behave as an i.i.d. product of Y_β's. Similarly for $\tilde{D}_n$, the lengths of the relevant subintervals behave as an i.i.d. product of $\max(X_\beta, 1 - X_\beta)$'s. For D^*_n we are concerned with the maximal-length subinterval after m splits, but here log (length) behaves as (a slight variation of) branching random walk, and the result comes from the usual large deviation analysis of the rightmost walker in branching random walk. (This is an elaboration of the known result ([20] section 2.7) that $1/\rho^*(0)$ is the solution of $2x = \exp(x-1)$).

$\boldsymbol{\beta = -1}$. The recurrence for $t_n = ED_n$ is (for any β)

$$t_n = 1 + \sum_{i=1}^{n-1} \frac{2i}{n} q_n(i) t_i. \tag{11}$$

In the case $\beta = -1$ this becomes

$$0 = 1 + \frac{1}{h_{n-1}} \sum_{i=1}^{n-1} \frac{t_i - t_n}{n-i}.$$

Writing $t_n \sim c \log^2 n$ and approximating the sum by an integral,

$$0 = 1 + 2c \int_0^1 \frac{\log y}{1-y}\, dy,$$

and the integral equals $\pi^2/6$.

The analysis of $E\tilde{D}_n$ is similar, leading to an equation

$$0 = 1 + 2c \int_{1/2}^1 \frac{\log y}{y(1-y)}\, dy,$$

and the integral here equals $\pi^2/12$.

To explain intuitively the order of magnitude in this case, note that D_n is distributed as the number of steps of the following Markov chain, started at n and run until absorption at 1.

$$P(j,i) = \frac{1}{h_{j-1}} \frac{1}{j-i}; \; 1 \leq i \leq j-1.$$

The mean position after one step from j is about $j(1 - \frac{1}{\log j})$, so it takes $O(\log n)$ steps to go from n to around $n/2$, and so it requires $O(\log^2 n)$ steps until absorption.

$-2 < \beta < -1$. In the recurrence (11), writing $t_n \sim cn^{-\beta-1}$ and approximating $q_n(i)$ by the asymptotic values (10),

$$0 \approx 1 + \kappa_\beta^{-1} c n^\beta \sum_{i=1}^{n-1} \left(\frac{i}{n}\right)^{\beta+1} \left(\frac{n-i}{n}\right)^\beta (i^{-\beta-1} - n^{-\beta-1}).$$

Approximating the sum by an integral gives the result for $\bar{\rho}(\beta)$, and the argument for $\tilde{\rho}(\beta)$ is similar. Finally, finiteness of ρ^* and the existence of non-degenerate limit distributions are consequences of the process approximation indicated in section 6.1.

Remarks on Proposition 4 for $\beta = -3/2$. The $\beta = -3/2$ case fits into a long-studied line of work in probabilistic combinatorics, the "simply-generated trees" of Meir and Moon [21]. Many asymptotic results for these models have been proved by generating function methods, in particular the values of $\bar{\rho}(-3/2)$ and $\rho^*(-3/2)$. See [2] for a brief discussion and the interpretation of the limits in terms of Brownian excursion. Curiously, $\tilde{\rho}(-3/2)$ has only recently been investigated in that literature – see the preprints by Vatutin [27] and Luczak [18].

6. Directions for further study.

6.1. More on the beta-splitting model. Two features of the results outlined in section 5 seem sufficiently interesting (from the mathematical, rather than biological, point of view) to warrant more careful study. In all models of random n-vertex trees known to me which deal with "combinatorial" trees (rather than trees with vertices in d-dimensional space), the height statistics (c.f. Proposition 4) grow as either $\Theta(\log n)$ or as $\Theta(n^{1/2})$. So our beta-splitting family exhibits two types of novel behavior. For $\beta = -1$ the mean depth grows as $\Theta(\log^2 n)$, and this immediately raises a host of questions whose answers cannot be immediately guessed by analogy: is the mean height also $\Theta(\log^2 n)$? what are the spreads of the various statistics $D_n, \bar{D}_n, \tilde{D}_n, D_n^*$? are there other models for which these statistics are $\Theta(\log^2 n)$ with different constants? I do not see any elegant probabilistic way of studying this case, but obviously one can try analytic techniques. Secondly, in the case $-2 < \beta < -1$ the mean depth grows as $\Theta(n^{-\beta-1})$. Randomly call edge-splits "left" and "right," thereby inducing a left-to-right ordering on the leaves. Define $H_n(t), 0 \le t \le 1$, by

$$H_n(i/n) = n^{\beta+1}(\text{ depth of leaf } i \text{ in the left-to-right ordering})$$

with linear interpolation. Then one can see heuristically that as $n \to \infty$ there is some limiting stochastic process $H_\infty(t), 0 \le t \le 1$ and that the quantities D_n considered in Proposition 4 have non-degenerate limit distributions expressible in terms of H_∞. For $\beta = -3/2$ the limit process is Brownian excursion – this is part of the circle of ideas discussed in [2,3]. But for general $-2 < \beta < -1$ the limit processes H_∞ seem novel and interesting stochastic processes. Informally, they can be constructed as processes in which intervals split and shrink continuously, as opposed to the discrete-step splitting of section 4 for $\beta > -1$. Details will be given in [4].

6.2. Other one-parameter families. We introduced the beta-splitting models as a mathematically natural way to embed the Yule model and the uniform model into a one-parameter family. One can of course invent other such families. For instance, under the Yule model the chance of a particular cladogram t equals

$$\frac{2^{n-1}}{n!} \prod_{i=3}^{n} (i-1)^{-d_i(t)}$$

where $d_i(t)$ is the number of internal nodes with exactly i descendant species. Thus we can define a family $(T_\gamma^{(n)})$ for which

$$P(T_\gamma^{(n)} = t) = a_n(\gamma) \prod_{i=3}^{n} (i-1)^{\gamma d_i(t)}.$$

So $\gamma = 0$ is the uniform model and $\gamma = -1$ is the Yule model. This is a different family, because (for instance) the $\gamma \to -\infty$ limit is the uniform distribution on maximally-balanced n-cladograms.

The only one-parameter family in the literature which exhibits qualitative change in behavior as the parameter varies are the randomly-growing binary trees discussed in [1] and [5]. These contain the Yule model but not the uniform model.

6.3. Another characterization question. Consider the "Markov branching" models $(P^{(n)}; n \geq 1)$ defined at the start of section 4. These are automatically exchangeable, but do not necessarily have the property (c.f. property (ii) in section 1)

Sampling consistency. For each n, $P^{(n)}$ induces a distribution on cladograms on $\{1, 2, \ldots, n-1\}$ by the action of deleting n: this distribution is $P^{(n-1)}$.

So it is natural to ask

OPEN PROBLEM 1. *Characterize the subclass of Markov branching models which satisfy the sampling consistency property.*

This subclass contains the beta-splitting models, but I suspect it is much larger.

Acknowledgements. I thank John Kingman, Warren Ewens and an anonymous referee for helpful suggestions. The numerical calculations used MATHEMATICA, and I thank Mike Steele for assistance therewith.

REFERENCES

[1] D. Aldous and P. Shields. A diffusion limit for a class of randomly-growing binary trees. *Probab. Th. Rel. Fields*, 79:509–542, 1988.

[2] D.J. Aldous. The continuum random tree II: an overview. In M.T. Barlow and N.H. Bingham, editors, *Stochastic Analysis*, pages 23–70. Cambridge University Press, Cambridge, 1991.

[3] D.J. Aldous. The continuum random tree III. *Ann. Probab.*, 21:248–289, 1993.

[4] D.J. Aldous. A family of self-similar continuous interval-splitting processes. In Preparation, 1995.

[5] M.T. Barlow, R. Pemantle, and E. Perkins. Diffusion-limited aggregation on a tree. Technical report, U. British Columbia, 1994.

[6] J.-P. Barthelemy and A. Guenoche. *Trees and Proximity Representations*. Wiley, New York, 1991.

[7] M.D. Brennan and R. Durrett. Splitting intervals. *Ann. Probab.*, 14:1024–1036, 1986.

[8] M.D. Brennan and R. Durrett. Splitting intervals II: Limit laws for lengths. *Probab. Th. Rel. Fields*, 75:109–127, 1987.

[9] P. Donnelly and P. Joyce. Weak convergence of population genealogical processes to the colescent with ages. *Ann. Probab.*, 20:322–341, 1992.

[10] N. Eldredge and J. Cracraft. *Phylogenic Patterns and the Evolutionary Process*. Columbia University Press, New York, 1980.

[11] W.J. Ewens. Population genetics theory: The past and the future. In S. Lessard, editor, *Mathematical and Statistical Developments of Evolutionary Theory*, pages 177–227, 1990.

[12] S.J. Gould, D.M. Raup, J.J. Sepkoski, T.J.M. Schopf, and D.S. Simberloff. The shape of evolution: a comparison of real and random clades. *Paleobiology*, 3:23–40, 1977.

[13] C. Guyer and J.B. Slowinski. Comparisons between observed phylogenetic topologies with null expectations among three monophyletic lineages. *Evolution*, 45:340–350, 1991.

[14] C. Guyer and J.B. Slowinski. Adaptive radiation and the topology of large phylogenies. *Evolution*, 47:253–263, 1993.

[15] S.A. Kauffman. *The Origins of Order: Self Organization and Selection in Evolution*. Oxford University Press, Oxford, 1993.

[16] J.F.C. Kingman. *Mathematics of Genetic Diversity*. S.I.A.M., Philadelphia, PA, 1980.

[17] D.E. Knuth. *The Art of Computer Programming*, volume 1. Addison-Wesley, 1968.

[18] T. Luczak. A greedy algorithm for estimating the height of random trees. Technical Report 1190, I.M.A., Minneapolis, MN, 1993.

[19] W.P. Maddison and M. Slatkin. Null models for the number of evolutionary steps in a character on a phylogenic tree. *Evolution*, 45:1184–1197, 1991.

[20] H.M. Mahmoud. *Evolution of Random Search Trees*. Wiley, 1992.

[21] A. Meir and J.W. Moon. On the altitude of nodes in random trees. *Canad. J. Math.*, 30:997–1015, 1978.

[22] M.H. Nitecki and A. Hoffman, editors. *Neutral Models in Biology*. Oxford University Press, Oxford, 1987.

[23] D.M. Raup. Mathematical models of cladogenesis. *Paleobiology*, 11:42–52, 1985.

[24] D.M. Raup, S.J. Gould, T.J.M. Schopf, and D.S. Simberloff. Stochastic models of phylogeny and the evolution of diversity. *J. Geology*, 81:525–542, 1973.

[25] H.M. Savage. The shape of evolution: Systematic tree topology. *Biological J. Linnean Soc.*, 20:225–244, 1983.

[26] S. Tavare. Line-of-descent and genealogical processes and their applications in population genetics models. *Theoret. Population Biol.*, 26:119–164, 1984.

[27] V.A. Vatutin. On the height of the primary path of random rooted trees. Technical report, Mathematics, Chalmers Univ., 1993.

STABILITY OF SELF-ORGANIZING PROCESSES

ROBERT M. BURTON* AND WILLIAM G. FARIS†

Abstract. This paper treats the stability of the Markov chain involved in the self-organizing feature maps of Kohonen. These maps have values that give an empirical approximation to a probability distribution while at the same time attempting to preserve a neighborhood topology. Some of the processes have the property that two initial states subjected to the same dynamics approach each other exponentially fast with probability one. This in turn implies a strong form of stability. Thus the initial state or the process is merely a transient effect. The environment alone determines future history.

1. Introduction. The subject is the Markov chain involved in self-organizing feature maps. These maps were first defined by Kohonen [6,4] in the context of neural networks and retinology. A map is an embedding of a finite simple graph in Euclidean space. The evolution of these maps is determined by the effect of a random environment. The maps learn to imitate the environment while also attempting to preserve the graph structure. We give conditions for a strong type of stability in which two initial states approach each other exponentially fast with probability one. Thus the initial state does not matter; the environment determines future history.

First consider a finite connected simple graph $G = (\Lambda, E)$, where Λ is a finite vertex set and the edge set, E is a collection of subsets of Λ of size 2. If $i, j \in \Lambda$ then we set $|i - j|$ equal to the length of the shortest path connecting i and j; in particular $|i - j| = 1$ if and only if $\{i, j\} \in E$. The state space Υ is the set of all one-to-one functions X from Λ to $\mathbf{R}^N$.

To describe the Markov evolution on state space, let X_0 have an initial distribution. Also fix an absolutely continuous probability measure μ on $\mathbf{R}^N$, a *shrinking* parameter α with $0 < \alpha < 1$, and fix an integer *range* parameter $r \geq 0$. Suppose X_n has been defined. Let ω_n be a random variable on $\mathbf{R}^N$ with distribution μ and be independent of all past activity. Then choose the i in Λ that minimizes the Euclidean distance $|X(i) - \omega|$. Consider the *neighborhood* of i consisting of all j in Λ with $|i-j| \leq r$. Then $X_{n+1}(j) = \alpha\omega + (1-\alpha)X(j)$ for all j in this neighborhood, and the other values are unchanged.

The process may be described succinctly by the equations

$$|X_n(i_n) - \omega_n| = \min_k |X_n(k) - \omega_n| \tag{1.1}$$

and

$$X_{n+1}(j) = \alpha\omega_n + (1-\alpha)X_n(j) \qquad \text{for } |j - i_n| \leq r, \tag{1.2}$$

---

* Research supported by AFOSR grant 91–0251 and NSF grant DMS–9103738. Department of Mathematics, Oregon State University, Corvallis, Oregon 97331, USA.

† Research supported by AFOSR grant 88–0189 and NSF grant DMS–900862. Department of Mathematics, University of Arizona, Tucson, Arizona 85721, USA.

with $X_{n+1}(j) = X_n(j)$ for the other j in Λ.

We will call μ the environmental measure and ω_n the environmental sequence. Since the environment measure μ is absolutely continuous, i_n is a.s. uniquely defined as is the Markov chain $\{X_n\}$.

We will always take G with Λ a cubical subset of $\mathbf{Z}^d$ with nearest neighbor edges. For mathematical investigations it is convenient to take μ to be the uniform measure on a rectangular product of intervals. We always do this in this paper.

There are four classes of graphs we consider.

EXAMPLE 1. *Let $N = d = 1$, so we are considering the embedding of a linear graph in the unit interval. We call this the linear process. Notice that if the linear graph is embedded in either an increasing or decreasing way at some time then this order is preserved for all time.*

EXAMPLE 2. *Let $N > 1$ and $d = 1$. In this case we are embedding a linear graph in a higher dimensional cube. The result of the evolution is to approximate a space-filling curve. We call this the space-filling curve process.*

EXAMPLE 3. *Let G be the graph with ℓ points and no edges (or, equivalently take the range parameter $r = 0$). This is a self-organizing clustering process. The idea in this process is to better approximate the underlying measure by avoiding over-fitting, that is, the process will be more dispersed than the empirical measure.*

EXAMPLE 4. *Let $N = d = 2$, so that we have a square grid embedded in the unit square.*

There is a picturesque interpretation of the last example. The set Λ is thought of as composed of cells in the cortex, and the region in $\mathbf{R}^2$ is thought of as the retina. The map X from the cortex to the retina develops in such a way that the points in the retina fall in areas of the retina with extensive stimulation, while at the same time nearby cells in the cortex tend to be connected to nearby points in the retina. This fits with physiological principles in which regions of the retina that were unstimulated during development remain blind later in life. See [6] for detailed explanations.

We consider the *stability* question, formulated as follows. Take two initial states X and Y, and look at their orbits under the same environmental sequence. The question is whether when the initial X and Y are sufficiently close there is a non-zero probability that they stay close for all future time.

If X and Y are close, then the i that makes $X(i)$ closest to ω may also be the i that makes $Y(i)$ closest to ω. In that case, $X(i) - Y(i)$ is replaced by $(1-\alpha)(X(i) - Y(i))$. Thus the two points have been shrunk together. The question is whether this shrinking can persist for all future time.

The ergodic behavior has been studied rigorously in a fundamental paper by Cottrell and Fort [1]. They treat the one-dimensional case $d = 1$ and $N = 1$ with nearest neighbor interaction (range $r = 1$). One interesting feature of this case is that it makes sense to say that a map X is increasing or decreasing. Cottrell and Fort prove that the map is eventually either increasing or decreasing. Once it has reached one or the other status it remains that way. Furthermore for maps of either status they prove convergence to a stationary distribution.

The higher dimensional case has been studied by various authors. The work by Ritter and Schulten [8] presents a diffusion approximation. Other papers by Tolat [9] and by Lo and Bavarian [7] also provide useful insights. A recent paper by Erwin, Obermayer, and Schulten [3] gives an overview of the situation.

2. Markov chains as dynamical systems. The notation in this section is independent of the other sections. Suppose that (Υ, d) is a bounded separable metric space. We assume without loss of generality that the diameter of Υ is 1. One way to define Markov chains on Υ is as a random dynamical system. Suppose that $F_1, F_2, \ldots$ is an iid sequence of random transformations from Υ to itself. A Markov chain $\{X_n\}$ may be defined by setting X_0 to have an initial distribution and setting $X_n = F_n \circ F_{n-1} \circ \cdots F_1(X_0)$ where $\circ$ denotes composition.

Subject to mild regularity conditions all Markov chains may be expressed this way (we do not assume that the transformations are continuous). See Kifer's book [5], page 8, for details. Often this is the natural way to define the chain. This is the case for the self-organizing feature maps; indeed they were defined in exactly this way.

DEFINITION 2.1. *A Markov chain given as a dynamical system with random transformation distribution F is super-stable if (i) there is a constant $\gamma > 0$ so that for every pair of initial random variables X_0 and Y_0 there is a random variable W so that*

$$(2.1) d(F_n \circ F_{n-1} \circ \cdots \circ F_1(X_0), F_n \circ F_{n-1} \circ \cdots \circ F_1(Y_0)) \leq e^{-\gamma(n-W)}$$

and (ii) there is an invariant probability measure for the chain. We say the chain is super-stable with exponential rate if in addition (iii) there are constants $k > 0$ and $0 < \beta < 1$ independent of X_0 and Y_0 so that $\mathbf{Prob}[W = n] \leq k\beta^n$.

Note that this definition is in some sense dual to the property of being a chaotic attractor in which all the randomness is in the initial conditions. Here the initial conditions are irrelevant in the long term; the environment alone gives the behavior.

Let $\mathcal{M}$ be the set of Borel probability measures on Υ and $\mathcal{C}$ be the set of bounded, uniformly continuous real-valued functions on Υ. Define $T : \mathcal{M} \to \mathcal{M}$ by $T(\nu) = \text{dist}(F(X))$ where $\nu = \text{dist}(X)$ and F is distributed as $F_1, F_2, \ldots$. Set $\hat{T} : \mathcal{C} \to \mathcal{C}$ by $\hat{T}f(x) = \mathbf{E}(f(F(x))$. Thus $\int f dT\nu = \int \hat{T} f d\nu$.

We may also put a metric on $\mathcal{M}$, the Wasserstein metric, which we define by $\rho(\nu_1, \nu_2) = \inf\{\mathbf{E}[d(X_1, X_2)|\text{dist}(X_i) = \nu_i, i = 1, 2\}$. The Kantorovich-Rubinstein Theorem together with the assumption that the metric d is bounded implies ρ is a metric which characterizes convergence in distribution, i.e. $\rho(\nu_n, \nu) \to 0$ iff $\nu_n \to \nu$ (dist) [2], p. 329. This is a complete metric if Υ is Polish.

THEOREM 2.1. *Suppose that a Markov chain $\{X_n\}$ defined as a dynamical system on Υ with random transformation distributed as F is super-stable with $\bar{\nu}$ an invariant probability. Then $\bar{\nu}$ is the unique invariant measure; indeed the distribution of X_n converges weakly to $\bar{\nu}$ with respect to the function class $\mathcal{C}$. If the chain is super-stable with exponential rate then this convergence is exponentially fast with respect to the Wasserstein metric.*

Proof: Fix $f \in \mathcal{C}$ and $\nu \in \mathcal{M}$. For each $x, y \in \Upsilon$ we have $\hat{T}^n f(x) - \hat{T}^n f(y) \to 0$ as $n \to \infty$. Integrating with respect to $\nu(dy)$ and using the bounded convergence theorem we have $\hat{T}^n f(x) - \int \hat{T}^n f d\nu \to 0$. Now integrate with respect to $\bar{\nu}(dx)$ and get $\int \hat{T}^n f d\bar{\nu} - \int \hat{T}^n f d\nu \to 0$. Using duality and the invariance of $\bar{\nu}$ with respect to T we see this is equivalent to $\int f d\bar{\nu} - \int \hat{T}^n f d\nu \to 0$. Thus $\int f dT^n\nu \to \int f d\bar{\nu}$, which is to say that $T^n\nu \to \bar{\nu}$ weakly with respect to the class $\mathcal{C}$.

To see convergence in the Wasserstein metric with exponential rate, let Y be a random variable with distribution $\bar{\nu}$ and X have distribution ν so that $T^n\nu$ is the distribution of $F_n \circ \cdots \circ F_1(X)$ and $\bar{\nu}$ is the distribution of $F_n \circ \cdots \circ F_1(Y)$.

$$\rho(T^n(\nu), \bar{\nu}) = \rho(T^n\nu, T^n(\bar{\nu})) \leq \mathbf{E}[d(F_n \circ \cdots \circ F_1(X), F_n \circ \cdots \circ F_1(Y))]$$

$$\leq \mathbf{E}[\min(1, e^{-(n-W)\gamma})]$$

$$= \mathbf{Prob}[W > n] + \mathbf{E}[e^{-(n-W)\gamma}\mathbf{1}_{\{W \leq n\}}]$$

$$= k\beta^{n+1}/(1-\beta) + ke^{-n\gamma}\sum_{w=1}^{n} e^{w\gamma}\beta^w$$

$$\leq k_1[\max(\beta, e^{-\gamma})]^n,$$

where k_1 is another constant.

3. Theorems. We wish to apply Theorem 2.1 to self-organizing Markov chains. Recall that a map $\sigma : \Lambda \to \Lambda$ is a graph automorphism if it is one-to-one, onto and preserves the graph relation, i.e. $\{i, j\} \in E$ if and only if $\{\sigma(i), \sigma(j)\} \in E$. Define a pseudo-metric on Υ by $d(X, Y) = \min\{\sum_i |X(i) - Y(\sigma(i))| : \sigma \text{ is an automorphism of } G\}$. Let $(\bar{\Upsilon}, \bar{d})$ be the metric space obtained by identifying points of distance 0 from each other. $(\bar{\Upsilon}, \bar{d})$ is a separable manifold.

Suppose that $d(X_n, Y_n)$ is small and ω_n is a uniform random variable as in the definition of the self-organizing Markov chain. With high probability $|X_n(i_n) - \omega_n|$ and $|Y_n(i_n) - \omega_n|$ are minimized for the same i_n. In this case we say that X and Y are not split and that i_n was hit. In particular, $d(X_{n+1}, Y_{n+1}) \leq d(X_n, Y_n)$. If after T time units there has been no split and each $i \in V$ has been hit then $d(X_{n+T}, Y_{n+T}) \leq (1-\alpha)d(X_n, Y_n)$ so that X and Y are getting closer together, making the chance of a split even smaller.

In this context several classes of self-organizing maps are known to be super-stable, and we conjecture that many more classes are super-stable. In particular we have,

THEOREM 3.1. *The first three examples from the introduction, namely the linear process, the space-filling process and the clustering process are all super-stable with exponential rate.*

Detailed proofs will appear elsewhere. There are several steps in the argument. First, we find $M, c > 0$ such that for all X_0, Y_0 there is an event A measurable with respect to the σ-field generated by X_0, Y_0 and $\omega_0, \omega_1, \cdots \omega_M$ so that the conditional probability $P(A|X_0, Y_0) > c$, and so that for all outcomes in A we have that $d(X_M, Y_M)$ is small in comparison with $\min\{|X(i) - X(j)| : i \neq j\}$. Now simply observe X_n and Y_n until the condition above is attained. Estimates show that from then on $d(X_n, Y_n)$ converges to zero exponentially quickly with positive probability bounded away from 0. If there is a disaster and convergence is not this quick, then this is observable by a stopping time, and we simply start this procedure again from the beginning until we do get exponential convergence.

REFERENCES

[1] MARIE COTTRELL AND JEAN-CLAUDE FORT, *Étude d'un processus d'auto-organization*, Annales de l'Institut Henri Poincaré, Probabilités et Statistiques, **23** (1987), 1–20.

[2] R.M. DUDLEY, *Real Analysis and Probability*, Wadsworth and Brooks/Cole, Belmont, CA, 1989.

[3] E. ERWIN, K. OBERMAYER, AND K. SCHULTEN, *Self-organizing maps: ordering, convergence properties, and energy functions*, Biological Cybernetics, **67** (1992), 47–55.

[4] J.A. KANGAS, T.K. KOHONEN, AND J.T. LAAKSONEN, *Variants of self-organizing maps*, IEEE Transactions on Neural Networks, **1** (1990), 93–99.

[5] YURI KIFER, *Ergodic Theory of Random Transformations*, Birkhäuser, Boston, 1986.

[6] TEUVO KOHONEN, *Self-Organization and Associative Memory* (Third Edition), Springer-Verlag, Berlin, 1989.

[7] Z.-P. LO AND B. BAVARIAN, *On the rate of convergence in topology preserving neural networks*, Biological Cybernetics, **65** (1991), 55–63.

[8] H. RITTER AND K. SCHULTEN, *Convergence properties of Kohonen's topology conserving maps: fluctuations, stability, and dimension selection*, Biological Cybernetics, **60** (1988), 59–71.

[9] V.V. TOLAT, *An analysis of Kohonen's self-organizing maps using a system of energy functionals*, Biological Cybernetics, **64** (1990), 155–164.

SOME EXAMPLES OF NORMAL APPROXIMATIONS BY STEIN'S METHOD

AMIR DEMBO* AND YOSEF RINOTT†

Abstract. Stein's method is applied to study the rate of convergence in the normal approximation for sums of non-linear functionals of correlated Gaussian random variables, for the exceedances of r-scans of i.i.d. random variables, and in a multinomial setting.

Key words. Stein's method, r-scans, exceedances of Gaussian processes, central limit theorems.

AMS(MOS) subject classifications. 60F05, 60G15

1. Introduction. The purpose of this paper is to demonstrate Stein's Normal approximation method and its potential use, by applying a variant of it to three examples of random variable of the form $W = \sum_{i=1}^{n} \psi(U_i)$. In Section 2 the random variables U_i are correlated and jointly Gaussian, while in Section 3 they have the multinomial distribution, and in both cases ψ is typically a non-linear function satisfying appropriate growth conditions. Because of the global dependence among the summand variables in both examples, we find the normal approximation theorem of [8, Theorem 1.1] most suitable here. The bound on the normal approximation rate of a random variable W appearing in this theorem is expressed in terms of another random variable W^*, whose marginal distribution is the W-size biased distribution. The contribution of this paper is in the explicit construction of the W-size biased variable W^*, which is required for the application of this theorem in each case, and in showing that the method yields a rather tight upper bounds on the rate of convergence (see Theorems 2.3 and 3.1 below). We shall compare our results to the existing literature on these problems. In Section 4 we analyze the number of exceedances of the r-scans generated by n i.i.d. random variables, for the regime in which the Poisson approximation does not apply. In contrast with the previous two examples, here U_i are r-dependent, hence the dependence structure is local, allowing for the direct applicability of Stein's (1986) original normal approximation theorem. The analysis of the r-scan example, follows from a new version of Stein's theorem which provides a useful bound on $|P((W - EW)/\sqrt{VarW} \leq w) - \Phi(w)|$ in situations where $E[\psi(U)] \to 0$ as $n \to \infty$, i.e. on the "boundary" between the Normal and Poisson approximations (see Theorem 4.2 below). We also show how

---

* Department of Statistics and Department of Mathematics, Stanford, CA 94305. This work was supported in part by NSF grants DMS92-09712 and DMS91-06974 and by a US-ISRAEL BSF grant.

† Department of Mathematics, UCSD, La Jolla, CA 92093. This work was supported in part by NSF grant DMS 92-05759.

the rate of convergence in the normal approximation for exceedances of r-spacings on the sphere can then be deduced.

2. Additive functionals of correlated Gaussian random variables. In this section we provide an application of the following normal approximation theorem from [8, Theorem 1.1], which is an extension of an earlier result, [1, Lemma 2.4].

THEOREM 2.1. *Let $W \geq 0$ be a random variable with distribution $dF(w)$ and let W^* be defined on the same probability space as W, and having the marginal distribution $wdF(w)/\lambda$, where $\lambda = EW$, and $\sigma^2 = \text{Var}W$. Then for any piecewise continuously differentiable $h : R \to R$,*

$$|Eh\left(\frac{W-\lambda}{\sigma}\right) - Nh| \leq 2\|h\|\frac{\lambda}{\sigma^2}\sqrt{\text{Var}E(W^* - W|W)} + \|h'\|\frac{\lambda}{\sigma^3}E(W^* - W)^2, \tag{2.1}$$

where $Nh = \int h(x)\phi(x)dx$, $\phi(\cdot)$ denotes the standard normal probability density and $\|\cdot\|$ denotes throughout the supremum-norm.

Note that the distribution of W^* in Theorem 2.1 is well known as the W–size biased (or length biased) distribution in other contexts. See [8] for a detailed discussion and references. For completeness, we provide here a short proof of this theorem.

Proof. Let $Z = \lambda W/\sigma^2$ and $\mu = EZ = \lambda^2/\sigma^2$. Note that $\text{Var}Z = \mu$ and that $Z^* = \lambda W^*/\sigma^2$ has the Z–size biased distribution. Let $Y = Z^* - Z$ and observe that for all $g : \mathbf{R} \to \mathbf{R}$

$$\mu Eg(Z+Y) = EZg(Z)\,. \tag{2.2}$$

In particular, $EY = \text{Var}Z/EZ = 1$. Subtract $\mu Eg(Z)$ from both sides of (2.2) to obtain by a Taylor series expansion of $g(Z+Y) - g(Z)$, that

$$E\left(\frac{Z-\mu}{\mu}g(Z)\right) = E(g(Z+Y)-g(Z)) = E(g'(Z)+(Y-1)g'(Z)+\frac{Y^2}{2}g''(\xi)),$$

for some ξ between Z and $Z+Y$. Consequently,

$$|E(g'(Z) - \frac{Z-\mu}{\mu}g(Z))| \leq \|g'\|E|E[Y-1|Z]| + \frac{1}{2}\|g''\|EY^2\,.$$

For a given continuous and piecewise continuously differentiable $h : \mathbf{R} \to \mathbf{R}$, let f denote the unique solution of the differential equation $f'(x) - xf(x) = h(x) - Nh$ with $\lim_{x\to\infty} f(x)e^{-x^2/2} = 0$, and set $g(z) = \sqrt{\mu}f(\frac{z-\mu}{\sqrt{\mu}})$. Let $\hat{Z} = (Z-\mu)/\sqrt{\mu} = (W-\lambda)/\sigma$ and note that

$$Eh(\hat{Z}) - Nh = E(f'(\hat{Z}) - \hat{Z}f(\hat{Z})) = E(g'(Z) - \frac{Z-\mu}{\mu}g(Z))\,.$$

Since $\|g'\| = \|f'\| \leq 2\|h\|$ ([1, (2.17)]) and $\|g''\| = \|f''\|/\sqrt{\mu} \leq 2\|h'\|/\sqrt{\mu}$ ([20, page 25]), it follows that

$$|Eh(\hat{Z}) - Nh| \leq 2\|h\|E|E[Y-1|Z]| + \frac{\|h'\|}{\sqrt{\mu}}EY^2\,. \tag{2.3}$$

Since $EY = 1$, it follows that $E|E[Y-1|Z]| \le \sqrt{\mathrm{Var}E(Y|Z)}$, by the Cauchy-Schwarz inequality and (2.1) follows from (2.3). □

REMARK 2.1. *Applying (2.2) to $g(z) = z^2$ it follows by simple manipulations and the Cauchy-Schwarz inequality that*

$$\frac{EY^2}{\sqrt{\mu}} \le \mu^{-3/2}|EZ^3 - \mu^3 - 3\mu^2| + 2\sqrt{\mathrm{Var}E(Y|Z)}\,.$$

It is easy to check that the first term on the r.h.s vanishes for $Z \sim N(\mu,\mu)$, and equals $1/\sqrt{\mu}$ for $Z \sim$ Poisson(μ). In the latter case, $Y = 1$ almost surely, with (2.3) implying that $|Eh(\hat{Z}) - Nh| \le \|h'\|/\sqrt{\mu}$. Indeed, insight to the nature of Theorem 2.1 may be gained by comparing (2.3) with the inequality

$$(2.4) \qquad |Eh(Z) - \mathrm{Po}_\mu h| \le 2(1-e^{-\mu})\|h\|E|Y-1|$$

which holds for every $Z \in \mathbb{Z}_+$ with $\mu = EZ$. The inequality (2.4) is the key to the "coupling approach" in the context of the Chen-Stein Poisson approximation method (see [3, Theorem 2.A] or [20, page 93]). With $\|g\|_L = \sup_{z\in\mathbb{N}}|g(z+1) - g(z)|$, a short derivation of (2.4) starts by subtracting $\mu Eg(Z+1)$ from both sides of (2.2), then using the bound $|E(g(Z+Y) - g(Z+1))| \le \|g\|_L E|Y-1|$, and finally considering $g : \mathbb{N} \to \mathbf{R}$ which is the solution of $zg(z) - \mu g(z+1) = h(z) - \mathrm{Po}_\mu h$ (for then $\|g\|_L \le 2(1-e^{-\mu})\|h\|/\mu$, see [3] or [20, page 89]).

In order to apply Theorem 2.1 we have to construct a random variable W^* having the required marginal distribution $wdF(w)/\lambda$ (and with small $\mathrm{Var}E(W^* - W|W)$). For a more general treatment and details of such constructions, see [8]. The following lemma from [8] and the subsequent discussion provide guidelines leading to useful constructions in the setup of this paper.

LEMMA 2.1. *Let $\mathbf{U} = (U_1,\dots,U_n)$ be a random vector, and let ψ_i be nonnegative functions such that $E\psi_i(U_i) < \infty$, $i = 1,\dots,n$. Let $\mathbf{Y}^{(i)} = (Y_1^{(i)},\dots,Y_n^{(i)})$ satisfy $P(\mathbf{Y}^{(i)} \in d\mathbf{y}) = P(\mathbf{U} \in d\mathbf{y})\psi_i(y_i)/E\psi_i(U_i)$. Let I be a random variable taking values in $\{1, \dots, n\}$, distributed independently of all the above variables, with $P(I = i) = E\psi_i(U_i)/\sum_{j=1}^n E\psi_j(U_j)$. Let $W = \sum_{j=1}^n \psi_j(U_j)$ have the distribution F. Then $W^* = \sum_{j=1}^n \psi_j(Y_j^{(I)})$ has the distribution $wdF(w)/\lambda$, where $\lambda = EW$.*

Note that the distribution of $\mathbf{Y}^{(i)}$ can be described as the distribution of $\mathbf{U}$ conditioned on a value of U_i, where U_i has a new marginal distribution, $\psi_i(\cdot)dF_{U_i}(\cdot)/E\psi_i(U_i)$.

Given $W = \sum_{j=1}^n \psi_j(U_j)$, this suggests the following construction of W^*: Let V_i be independent random variables, $V_i \sim \psi_i(u)dF_{U_i}(u)/E\psi_i(U_i)$. If V_i is assigned the value u, let $(Y_1^{(i)},\dots,Y_{i-1}^{(i)},Y_{i+1}^{(i)},\dots,Y_n^{(i)})$ have the conditional distribution of $(U_1,\dots,U_{i-1},U_{i+1},\dots,U_n)$ given $U_i = u$, and set $Y_i^{(i)} = u$. Now choose I and set $W^* = \sum_{j=1}^n \psi_j(Y_j^{(I)})$ as in Lemma 2.1.

Proof of Lemma 2.1. We prove that for any function G, $EG(W^*) = EWG(W)/\lambda$, where $\lambda = EW = \sum_{j=1}^n E\psi_j(U_j)$. Indeed we have,

$$\begin{aligned} EG(W^*) &= E\sum_{i=1}^n G\left(\sum_{j=1}^n \psi_j(Y_j^{(i)})\right) E\psi_i(U_i)/\lambda \\ &= \int \sum_{i=1}^n G\left(\sum_{j=1}^n \psi_j(y_j)\right) P(\mathbf{U}\in d\mathbf{y})\psi_i(y_i)/\lambda \\ &= E\sum_{i=1}^n \psi_i(U_i) G\left(\sum_{j=1}^n \psi_j(U_j)\right) /\lambda = EWG(W)/\lambda. \qquad \square \end{aligned}$$

REMARK 2.2. *(a) For $W = \int_T \psi_x(U_x)dm(x)$ with a finite, non-negative measure $m(\cdot)$ on T, the W-size biased construction $W^* = \int_T \psi_x(Y_x^{(I)})dm(x)$ applies as in Lemma 2.1. Here, I has the law $dP_I(x) = E(\psi_x(U_x))dm(x)/\lambda$ while $P(Y^{(x)} \in dy) = P(U \in dy)\psi_x(y_x)/E\psi_x(U_x)$.*

(b) While Lemma 2.1 is stated and proved for real-valued U_i, it easily extends to $U_i \in \mathbf{R}^d$. In particular, at the cost of somewhat more cumbersome computations, one can extend Theorems 2.3 and 3.1 below, so as to handle variables of the form $W = \sum_{i=1}^n \psi(U_{j(1,i)}, \ldots, U_{j(d,i)})$, with fixed $d > 1$ and $j : \{1,\ldots,d\} \times \{1,\ldots,n\} \to \{1,\ldots,n\}$ of an appropriate structure.

(c) Note that the construction of Lemma 2.1 is restricted to nonnegative functions ψ_i. Functions ψ_i which are bounded below (and not necessarily non-negative), are handled by translation prior to the construction of W^* as in Lemma 2.1. The only effect of such a translation on the conclusion of Theorem 2.1 is in increasing the value of λ. A multivariate version of Theorem 2.1 is given in [8, Theorem 1.3]. From that result one can obtain a bivariate normal approximation for $(\sum_{j=1}^n \psi_j^+(U_j), \sum_{j=1}^n \psi_j^-(U_j))$, where ψ^+, ψ^- denote positive and negative parts. This will lead to a normal approximation for $\sum_{j=1}^n \psi_j(U_j)$ without the assumption that ψ_i are nonnegative. This approach will not be pursued here. However we comment that in some cases it leads to sharp approximation rates, while in other cases, the dependence structure of the positive and negative parts may be different from that of the original functions in a way that leads to unsatisfactory bounds.

The following (somewhat forgotten) fact appears in [15] (for further references see, e.g, [2]).

LEMMA 2.2. *Let $\mathbf{X} = (X_1, \ldots, X_k)$ and $\mathbf{Y} = (Y_1, \ldots, Y_m)$ be random vectors such that the joint distribution of $(\mathbf{X}, \mathbf{Y})$ is multivariate normal. Let $\rho = \rho(\mathbf{X}; \mathbf{Y})$ denote the (first) canonical correlation between $\mathbf{X}$ and $\mathbf{Y}$, that is, $\rho = maxCorr(\mathbf{a}^T\mathbf{X}, \mathbf{b}^T\mathbf{Y})$, where the maximum is taken over all vectors $\mathbf{a} \in R^k$ and $\mathbf{b} \in R^m$. Then $Corr(f(\mathbf{X}), g(\mathbf{Y})) \le \rho$ for any pair of* *real valued functions f and g such that the correlation is defined.*

For completeness we provide the proof of the following easy extension of Lemma 2.2 which is a special case of [4, page 40, Theorem 6.2].

LEMMA 2.3. *Let $(\mathbf{X}, \mathbf{Y})$ be multivariate normal vectors, and $\mathbf{V_1}, \mathbf{V_2}$ independent random vectors which are also independent of $(\mathbf{X}, \mathbf{Y})$. Then, when defined,*

$$\mathrm{Cov}(f(\mathbf{X}, \mathbf{V_1}), g(\mathbf{Y}, \mathbf{V_2})) \leq \rho(\mathbf{X}; \mathbf{Y})\sqrt{\mathrm{Var}\, f(\mathbf{X}, \mathbf{V_1})\mathrm{Var}\, g(\mathbf{Y}, \mathbf{V_2})}. \tag{2.5}$$

Proof. With $f = f(\mathbf{X}, \mathbf{V_1})$, and $g = g(\mathbf{Y}, \mathbf{V_2})$ we have

$$\mathrm{Cov}(f, g) = \mathrm{Cov}(E[f \mid \mathbf{V_1}, \mathbf{V_2}], E[g \mid \mathbf{V_1}, \mathbf{V_2}]) + E[\mathrm{Cov}(f, g \mid \mathbf{V_1}, \mathbf{V_2})]. \tag{2.6}$$

Since f is independent of $\mathbf{V_2}$ and g is independent of $\mathbf{V_1}$, the first term on the r.h.s. of (2.6) vanishes by the independence of $\mathbf{V_1}$ and $\mathbf{V_2}$. Invoking Lemma 2.2 for $f|\mathbf{V_1}, \mathbf{V_2}$ and $g|\mathbf{V_1}, \mathbf{V_2}$, by the independence of the $\mathbf{V_i}$'s we have

$$\begin{aligned} E[\mathrm{Cov}(f, g \mid \mathbf{V_1}, \mathbf{V_2})] &\leq \rho(\mathbf{X}; \mathbf{Y})E[\sqrt{\mathrm{Var}(f|\mathbf{V_1})}]E[\sqrt{\mathrm{Var}(g|\mathbf{V_2})}] \\ &\leq \rho(\mathbf{X}; \mathbf{Y})\sqrt{E[\mathrm{Var}(f|\mathbf{V_1})]E[\mathrm{Var}(g|\mathbf{V_2})]}. \end{aligned}$$

Ignoring the conditioning leads to (2.5). □

We now state our first result, a normal approximation theorem suitable for smooth ψ (see Theorem 2.3 below for the more general case).

THEOREM 2.2. *Let $\mathbf{U} = (U_1, \ldots, U_n)$ have the multivariate normal distribution $N(\mathbf{0}, \Sigma)$, where the diagonal elements of $\Sigma = \{\rho_{ij}\}$ are $\rho_{ii} = 1$. Set $\rho_{ij}^* = \rho_{ij}1_{|\rho_{ij}|<1}$ and*

$$\bar{\rho}_{iji'j'} = \min\left\{2\frac{\max\{|\rho_{ii'}|, |\rho_{ij'}|, |\rho_{ji'}|, |\rho_{jj'}|\}}{\{(1-|\rho_{ij}^*|)(1-|\rho_{i'j'}^*|)\}^{1/2}}, 1\right\}.$$

Let $W = \sum_{i=1}^n \psi(U_i)$ where $\psi \geq 0$ is scaled such that $N\psi = 1$ (hence $EW = n$). Denote $\sigma^2 = \mathrm{Var} W$ and $\phi_\rho(\cdot, \cdot)$ the bivariate normal density with standard marginals and correlation ρ. Define

$$M(\rho) = \int\int\int [\psi(u_1 + \rho(v - u_2)) - \psi(u_1)]^2 \phi_\rho(u_1, u_2)\psi(v)\phi(v)du_1\, du_2 dv.$$

Then for any piecewise continuously differentiable h,

$$\begin{aligned} &\left|Eh\left(\frac{W-n}{\sigma}\right) - Nh\right| \\ &\leq 2\frac{||h||}{\sigma^2}\left\{\sum_{i=1}^n\sum_{j=1}^n\sum_{i'=1}^n\sum_{j'=1}^n \bar{\rho}_{iji'j'}[M(\rho_{ij})M(\rho_{i'j'})]^{1/2}\right\}^{1/2} \\ &\quad + \frac{||h'||}{\sigma^3}\sum_{i=1}^n\left\{\sum_{j=1}^n [M(\rho_{ij})]^{1/2}\right\}^2. \end{aligned} \tag{2.7}$$

REMARK 2.3. *Note that there is no loss of generality in our assumption of $N\psi = 1$. One could easily extend Theorem 2.2 to the case of distinct functions $\psi_i(U_i)$, at the expense of somewhat more cumbersome notation. Similarly, following part (a) of Remark 2.2, Theorem 2.2 extends to $W = \int_T \psi(U_t)dm(t)$ with U_t a centered Gaussian process indexed by a parameter set T and scaled such that $E(U_t^2) = 1$. Here, $N\psi = 1$ results with $n = m(T)$ and the summations in the r.h.s. of (2.7) become integrals with respect to $m(\cdot)$. In particular, one can thus extend Corollary 2.1 below to the current setting (with $r = 0$ in (2.14)), provided that* $\sup_t \int_T |\rho_{ts}|(1-|\rho_{ts}|)^{-1/2}dm(s) = B < \infty$.

Proof. We construct a random variable W^* having the W–size biased distribution as suggested by Lemma 2.1. Specifically, here

$$\begin{aligned}\mathbf{Y}^{(i)} = (&U_1 + \rho_{i1}(V_i - U_i), \dots, U_{i-1} + \rho_{i\,i-1}(V_i - U_i), V_i,\\ &U_{i+1} + \rho_{i\,i+1}(V_i - U_i), \dots, U_n + \rho_{in}(V_i - U_i)),\end{aligned}$$

where $V_i \sim \psi(u)P(U_i \in du)$ independently of all other variables. A direct calculation shows that $\mathbf{Y}^{(i)}$, given $V_i = u$ is distributed like $\mathbf{U}$ given $U_i = u$. (Ignoring the ith component, this conditional distribution is $N(\boldsymbol{\rho}_i u, \Sigma_i - \boldsymbol{\rho}_i\boldsymbol{\rho}_i^T)$ where Σ_i denotes the principal submatrix obtained from Σ by deleting its ith row and column, and $\boldsymbol{\rho}_i^T = (\rho_{i1}, \dots, \rho_{i\,i-1}, \rho_{i\,i+1}, \dots \rho_{in})$.) Consequently, by Lemma 2.1, for I a random variable which is uniform on {1, ..., n} and independent of all other variables, $W^* = \sum_{j=1}^n \psi(Y_j^{(I)})$ and W are defined on a joint space, and W^* has the W–size biased distribution.

In order to apply Theorem 2.1, define $A_{ij} = \psi(U_j + \rho_{ij}(V_i - U_i)) - \psi(U_j)$, and note that

$$\begin{aligned}E[W^* - W|W] &= \frac{1}{n}E\left[\sum_{i=1}^n\sum_{j=1}^n \psi(U_j + \rho_{ij}(V_i - U_i)) - \psi(U_j)|W\right]\\ &= \frac{1}{n}E\left[\sum_{i=1}^n\sum_{j=1}^n A_{ij}|W\right]\end{aligned} \tag{2.8}$$

To compute the variance of this expression note that by ignoring the conditioning on W

$$\mathrm{Var}E\left[\sum_{i=1}^n\sum_{j=1}^n A_{ij}|W\right] \le \sum_{i=1}^n\sum_{i'=1}^n\sum_{j=1}^n\sum_{j'=1}^n \mathrm{Cov}(A_{ij}, A_{i'j'})\,. \tag{2.9}$$

To deal with the typical case of $i \neq i' \neq j \neq j'$ we use Lemma 2.3 to exploit both the possible smallness of the correlations between the terms A_{ij} and $A_{i'j'}$ and the possible smallness of the terms themselves. Indeed, with $\rho(U_i, U_j; U_{i'}, U_{j'})$ denoting the canonical correlation, Lemma 2.3 yields

$$\mathrm{Cov}(A_{ij}, A_{i'j'}) \le \rho(U_i, U_j; U_{i'}, U_{j'})\sqrt{\mathrm{Var}A_{ij}\mathrm{Var}A_{i'j'}}\,. \tag{2.10}$$

Note that $\mathrm{Var}(A_{ij}) \le E(A_{ij}^2) \le M(\rho_{ij})$. A direct calculation shows that $\rho(U_i, U_j; U_{i'}, U_{j'}) \le \bar{\rho}_{iji'j'}$. (See [16] for a similar calculation.) Consequently, by (2.8), (2.9) and (2.10),

$$\text{(2.11)}\quad \mathrm{Var} E[W^* - W|W] \le \frac{1}{n^2} \sum_{i=1}^{n} \sum_{i'=1}^{n} \sum_{j=1}^{n} \sum_{j'=1}^{n} \bar{\rho}_{iji'j'} [M(\rho_{ij}) M(\rho_{i'j'})]^{1/2}.$$

Turning now to bound the term $E(W^* - W)^2$ note that

$$\begin{aligned} E(W^* - W)^2 &= \sum_{i=1}^{n} \frac{1}{n} E\left\{ \sum_{j=1}^{n} [\psi(U_j + \rho_{ij}(V_i - U_i)) - \psi(U_j)] \right\}^2 \\ &= \frac{1}{n} \sum_{i=1}^{n} E\left[\left(\sum_{j=1}^{n} A_{ij} \right)^2 \right]. \end{aligned}$$

By the Cauchy-Schwarz inequality

$$E\left[\left(\sum_{j=1}^{n} A_{ij} \right)^2 \right] = \sum_{j=1}^{n} \sum_{j'=1}^{n} E[A_{ij} A_{ij'}] \le \sum_{j=1}^{n} \sum_{j'=1}^{n} E[A_{ij}^2]^{1/2} E[A_{ij'}^2]^{1/2},$$

implying that

$$E(W^* - W)^2 \le \frac{1}{n} \sum_{i=1}^{n} \left(\sum_{j=1}^{n} E[A_{ij}^2]^{1/2} \right)^2 \le \frac{1}{n} \sum_{i=1}^{n} \left(\sum_{j=1}^{n} M(\rho_{ij})^{1/2} \right)^2.$$
(2.12)

Since $E(W) = n$, combining the bounds of (2.11) and (2.12) with Theorem 2.1 leads to (2.7). □

In Corollary 2.1 below we use the following estimate on $M(\rho)$ to make the bound of Theorem 2.2 more explicit for smooth ψ.

PROPOSITION 2.1. *If $N\psi = 1$ and for some $q < 2$ and $K < \infty$*

$$\text{(2.13)}\qquad |\psi(y) - \psi(x)| \le K|y - x| e^{K(|x|^q + |y|^q)} \quad \forall x, y \in \mathbf{R},$$

then $M(\rho) \le C\rho^2$ for some constant $C(K, q) < \infty$ which depends only on K and q.

REMARK 2.4. *In particular, Proposition 2.1 applies when ψ has a bounded derivative, that is, $|\psi'(u)| \le K$ or more generally is uniformly Lipschitz continuous, i.e. $|\psi(y) - \psi(x)| \le K|y - x|$.*

Proof. Let (U_1, U_2) denote the bivariate normal vector of unit variances and correlation ρ and U an additional standard normal variable independent of (U_1, U_2). Then, by the assumption, and the Cauchy-Schwarz inequality

$$\begin{aligned} M(\rho) &= E\{[\psi(U_1+\rho(U-U_2))-\psi(U_1)]^2\psi(U)\} \\ &\le K^2\rho^2 E[|U-U_2|^2 e^{2K(|U_1|^q+|U-U_2|^q)}|\psi(U)|] \\ &\le K^2\rho^2 E[e^{8K|U_1|^q}]^{1/4} E[|U-U_2|^8 e^{8K|U-U_2|^q}]^{1/4} E[\psi(U)^2]^{1/2}\,. \end{aligned}$$

The r.h.s. is bounded by $\rho^2 C(K,q)$ with $C(K,q)<\infty$ independent of ρ and ψ, since

$$E[\psi(U)^2] = 1+E[(\psi(U)-\psi(U_1))^2]/2 \le 1+K^2E[|U-U_1|^2e^{2K(|U|^q+|U_1|^q)}]/2\,.$$

□

Corollary 2.1. *Assume the conditions of Theorem 2.2, and that* $\max_{i\neq j}|\rho_{ij}|\le r<1$ *and* $\max_i\sum_{j=1}^n|\rho_{ij}|\le B<\infty$. *Let* $\psi\ge 0$ *with* $N\psi=1$ *be such condition (2.13) holds and define* $D=\max\{\|h\|,\|h'\|\}$. *Then, with* $C=C(K,q)<\infty$ *from Proposition 2.1,*

$$\text{(2.14)}\qquad \left|Eh\left(\frac{W-n}{\sigma}\right)-Nh\right| \le \frac{4DB}{\sigma^2}\left\{\frac{2nBC}{1-r}\right\}^{1/2}+\frac{DCB^2n}{\sigma^3}.$$

Proof. This is a direct consequence of Theorem 2.2 and Proposition 2.1. Indeed, we use the simple bound

$$\bar\rho_{iji'j'} \le \frac{2}{1-r}(|\rho_{ii'}|+|\rho_{ij'}|+|\rho_{ji'}|+|\rho_{jj'}|)$$

which applies also when $i=j$ or $i'=j'$ (see the definition of ρ^*_{ij} in Theorem 2.2). Then, since $M(\rho)\le C\rho^2$ the r.h.s. of (2.7) is further bounded by

$$\frac{2D}{\sigma^2}\left\{\frac{2C}{1-r}\sum_{i=1}^n\sum_{j=1}^n\sum_{i'=1}^n\sum_{j'=1}^n|\rho_{ij}||\rho_{i'j'}|(|\rho_{ii'}|+|\rho_{ij'}|+|\rho_{ji'}|+|\rho_{jj'}|)\right\}^{1/2}+$$
$$+\frac{DC}{\sigma^3}\sum_{i=1}^n\left(\sum_{j=1}^n|\rho_{ij}|\right)^2 \le \frac{DB}{\sigma^2}\left\{\frac{8nBC}{1-r}\right\}^{1/2}+\frac{DCnB^2}{\sigma^3}$$

□

For a triangular array or a stationary sequence $\{U_i\}$, we set $B=B_n$ and $\sigma=\sigma_n$ in Corollary 2.1 and assume for simplicity that r and ψ do not depend on n. Then, since $\sigma^2\le nB\mathrm{Var}(\psi(U))$ (by Lemma 2.2), the second term on the r.h.s. of (2.14) dominates, leading to the rate of nB_n^2/σ_n^3. The bound of (2.14) is then non-trivial only when $n^{-1}B_n\to 0$ with asymptotic normality holding for W provided $n^{-1/3}B_n^{-2/3}\sigma_n\to\infty$ (note that unlike B_n, the value of σ_n depends on the function ψ).

Much is known about the asymptotic normality of W for a stationary sequence $\{U_i\}$ and ψ such that $N\psi^2<\infty$. In this context, for $r_i=E(U_0U_i)$

which is both square summable and Fejér summable (i.e., $\sum_{i=1}^{n}(1-i/n)r_i$ has a finite limit), it is known that $n^{-1/2}\sigma_n$ has a finite and typically positive limit and that W is then asymptotically normal (see [21, Theorem 1] and the references therein). In this case Corollary 2.1 supplies the rate of $n^{-1/2}(\sum_{i=0}^{n}|r_i|)^2$ in the Normal approximation for W. In particular, a rate of $n^{-1/2}$ results when $\{r_i\}$ is absolutely summable. When $\sum_i r_i^2 = \infty$, in general the asymptotic normality of W fails if the Hermite rank of $\psi - N\psi$ is not 1 (see [21, Example 1] and the references therein to the earlier works of [22] and [7]). For $\psi - N\psi$ of Hermite rank 1 and $r_i = i^{-\alpha}L(i)$, with $\alpha \in (0,1)$ and $L(i)$ a slowly varying function, it is known (see [22]) that $\sigma_n = O(n^{1-\alpha/2}L(n)^{1/2})$. With $L(n)$ now denoting any slowly varying function we have $B_n = O(n^{1-\alpha}L(n))$, and Corollary 2.1 supplies the rate of $n^{-\alpha/2}L(n)^{1/2}$ for the Normal approximation of W.

The following lemma is key for extending the Normal approximation of Corollary 2.1 to non-smooth ψ (e.g. $\psi(u) = 1_{[a,\infty)}(u)$ corresponding to a count of large exceedances of the Gaussian process).

LEMMA 2.4. *Let* $f(\mathbf{x},z) = \psi(x_1)[\psi(x_3 + z(x_1 - x_2)) - \psi(x_3)]$, *where* $\mathbf{x} = (x_1, x_2, x_3)$, *and* $\psi \geq 0$ *is scaled such that* $N\psi = 1$ *and satisfies* $\psi(u) \leq Ke^{K|u|^q}$ *for some* $0 < K < \infty$ *and* $q < 2$. *Let* $\mathbf{X} = (X_1, X_2, X_3)$ *and* $\mathbf{Y} = (Y_1, Y_2, Y_3)$ *be normal vectors, having zero means and unit variances, such that* X_1 *and* Y_1 *are independent of each other and of* X_2, X_3, Y_2, Y_3. *Define* $r_x = \rho(X_2, X_3)$, $r_y = \rho(Y_2, Y_3)$ *and* $\bar\rho = \max_{i,j}|Corr(X_i, Y_j)|$ *and suppose that* $\max\{|r_x|, |r_y|\} + 2\bar\rho \leq s < 1$. *Then, there exists a constant* $C = C(s, K, q) < \infty$ *such that for any* $a, b \in [-1, 1]$

$$\mathrm{Cov}(f(\mathbf{X},a), f(\mathbf{Y},b)) \leq C\bar\rho|ab|, \qquad \textit{and} \qquad |Ef(\mathbf{X},a)f(\mathbf{Y},b)| \leq C|ab|.$$

Proof. Fixing the (marginal) distributions of the vectors $\mathbf{X}$ and $\mathbf{Y}$, let $\rho_{i,j} = \mathrm{Corr}(X_i, Y_j)$ and let $\phi_\rho(\mathbf{x},\mathbf{y})$ denote the joint normal density of $\mathbf{X}$ and $\mathbf{Y}$, having marginals corresponding to the laws of $\mathbf{X}$ and of $\mathbf{Y}$. Clearly,

$$\mathrm{Cov}(f(\mathbf{X},a), f(\mathbf{Y},b)) = \int f(\mathbf{x},a)f(\mathbf{y},b)[\phi_\rho(\mathbf{x},\mathbf{y}) - \phi_0(\mathbf{x},\mathbf{y})]d\mathbf{x}d\mathbf{y}\,.$$

Let

$$g(\mathbf{x},z) = \int_{x_3}^{x_3+z(x_1-x_2)} \psi(x_1)\psi(t)dt,$$

and observe that $f(\mathbf{x},z) = \frac{\partial}{\partial x_3}g(\mathbf{x},z)$. Note that for any $\xi \in [0,1]$ the minimal eigenvalue of the covariance matrix for $\phi_{\xi\rho}$ is at least $1 - 2\bar\rho - \max\{|r_x|, |r_y|\} \geq (1-s)$. In particular, both ϕ_ρ and ϕ_0 are non-degenerate, and since

$$|g(\mathbf{x},a)| \leq K^2|a||x_1 - x_2|e^{2K(|x_1|+|x_2|+|x_3|)^q}\,, \tag{2.15}$$

boundary terms vanish, and by integration by parts we obtain

$$\operatorname{Cov}(f(\mathbf{X},a),f(\mathbf{Y},b)) = \int g(\mathbf{x},a)g(\mathbf{y},b)\frac{\partial^2}{\partial x_3 \partial y_3}[\phi_\rho(\mathbf{x},\mathbf{y})-\phi_0(\mathbf{x},\mathbf{y})]d\mathbf{x}d\mathbf{y}.$$
(2.16)

Noting that $\frac{\partial^2}{\partial x_3 \partial y_3}\phi_\rho(\mathbf{x},\mathbf{y}) = \frac{\partial}{\partial \rho_{3,3}}\phi_\rho(\mathbf{x},\mathbf{y})$ for any vector ρ (Heat equation) we have that for some $\xi = \xi(\mathbf{x},\mathbf{y})$ such that $0 \le \xi \le 1$

$$\frac{\partial^2}{\partial x_3 \partial y_3}[\phi_\rho(\mathbf{x},\mathbf{y}) - \phi_0(\mathbf{x},\mathbf{y})] = \sum_{i,j} \rho_{ij} \frac{\partial^2}{\partial \rho_{ij} \partial \rho_{3,3}} \phi_{\xi\rho}(\mathbf{x},\mathbf{y}).$$

Fixing ξ and ρ, the ratio $\frac{\partial^2 \phi_{\xi\rho}}{\partial \rho_{ij} \partial \rho_{3,3}} / \phi_{\xi\rho}$ is a polynomial expression of order of at most 4 in $(\mathbf{x},\mathbf{y})$ and of order at most 4 in the entries of the inverse of the covariance matrix for $\phi_{\xi\rho}$. As such it is bounded by $c(1-s)^{-4}(1+|\mathbf{x}|^4+|\mathbf{y}|^4)$ for some universal constant c. It thus follows by (2.15) and (2.16) that for some $c_1 = c_1(K,s) < \infty$,

$$\operatorname{Cov}(f(\mathbf{X},a),f(\mathbf{Y},b)) \le c_1|a||b|\bar{\rho}\int |x_1-x_2||y_1-y_2|(1+|\mathbf{x}|^4+|\mathbf{y}|^4)$$
$$e^{2K((|x_1|+|x_2|+|x_3|)^q+(|y_1|+|y_2|+|y_3|)^q)}\phi_{\xi\rho}(\mathbf{x},\mathbf{y})d\mathbf{x}d\mathbf{y},$$

and since $\phi_{\xi\rho} \le c_2 e^{-(|\mathbf{x}|^2+|\mathbf{y}|^2)/(2(1-s))}$ for some $c_2 = c_2(s) < \infty$, we obtain the stated bound on $\operatorname{Cov}(f(\mathbf{X},a),f(\mathbf{Y},b))$. The bound on $|E(f(\mathbf{X},a)f(\mathbf{Y},b)|$ results by similar (and somewhat simpler) considerations. □

THEOREM 2.3. *Assume the conditions of Theorem 2.2, and moreover assume that* $\max_{i\neq j}|\rho_{ij}| \le r < 1/3$ *and* $\max_i \sum_{j=1}^n |\rho_{ij}| \le B < \infty$. *Let* $\psi \ge 0$ *with* $N\psi = 1$ *be such that* $\psi(u) \le Ke^{K|u|^q}$ *for some* $0 < K < \infty$ *and* $q < 2$, *and define* $D = \max\{\|h\|,\|h'\|\}$. *Then, for some* $C(r,K,q) < \infty$,

$$(2.17) \qquad \left|Eh\left(\frac{W-n}{\sigma}\right) - Nh\right| \le \frac{4DB(nBC)^{1/2}}{\sigma^2} + \frac{DCB^2 n}{\sigma^3}.$$

REMARK 2.5. *Comparing (2.17) with (2.14) we see that the conclusions in the discussion below Corollary 2.1 apply also to non-smooth* ψ.

Proof. From the proofs of Theorem 2.2 and Corollary 2.1 it is clear that to obtain (2.17) it suffices to show that

$$(2.18) \quad \operatorname{Cov}(A_{ij},A_{i'j'}) \le C(|\rho_{ii'}|+|\rho_{ij'}|+|\rho_{ji'}|+|\rho_{jj'}|)|\rho_{ij}||\rho_{i'j'}|$$

and

$$(2.19) \qquad |E(A_{ij}A_{i'j'})| \le C|\rho_{ij}||\rho_{i'j'}|\,.$$

For $i \neq i' \neq j \neq j'$ we obtain these bounds with $C = C(3r,K,q) < \infty$ from Lemma 2.4, by applying this lemma for $a = \rho_{ij}$ and $b = \rho_{i'j'}$. While the cases in which not all four indices (i, j, i', j') are distinct require special considerations, the bounds of (2.18) and (2.19), perhaps with a different value for C, can be proved by properly adapting the proof of Lemma 2.4. For example, one may assume that $i \neq j$ and $i' \neq j'$ in (2.18) by following the same argument as in (3.3) and (3.4). We omit the details here. □

Remark. The assumption that $r < 1/3$ has been made in order to simplify the proof of Theorem 2.3 and may be somewhat relaxed. For example, in case of a stationary sequence $\{U_i\}$ with $B = B_n$ such that $n^{-1}B_n \to 0$, clearly $|r_i| \to 0$. Hence, for each i there are no more than $L < \infty$ values of j with $|\rho_{ij}| = |r_{i-j}| \geq 1/3$. By considering the corresponding A_{ij} separately, one can deduce a bound similar to (2.17) even when $L > 0$.

3. An example concerning the multinomial distribution. Our next example deals with the multinomial distribution where the numbers of balls n and the number of cells N are of the same rate. For simplicity, we shall state a result for $n = N$ and equally likely cells, though the method readily lends itself to generalizations. In this case the number of balls in each cell is asymptotically Poisson(1), and there is negative dependence (e.g., negative association), between the cell counts. However, if we denote the cell counts by $U_1, \ldots, U_N$ then negative association of $\psi(U_1), \ldots, \psi(U_N)$ is guaranteed only for a monotone ψ. Even in the monotone case, Newman's CLT ([18]) does not apply in general for $W = \sum_{i=1}^{N} \psi(U_i)$. The case where W counts the number of empty cells was extensively studied in [14]. The present approach can easily be extended to the case of non-equal but commensurate cell probabilities, and for $W = \sum_{i=1}^{n} \psi_i(U_i)$, with different $\psi_i(\cdot)$. Limit theorems for such a W, without rates, were studied for example in [17], and in a more general setting in [9] where the first step is to represent the multinomial distribution by independent Poisson variables conditioned on their sum. More recently, this approach and characteristic function methods, were taken in [10] and [11] to assess the convergence rate. See also [12] for statistical applications and further references. We comment briefly that for $W = \sum_{i=1}^{n} \psi_i(U_i)$, where $(U_1, \ldots, U_n)$ are jointly distributed as independent variables conditioned on their sum, a W-size biased coupling variable can be constructed naturally, extending the ideas of the discussion below.

THEOREM 3.1. *Let* $\mathbf{U} = (U_1, \ldots, U_n)$ *have the multinomial distribution* $P(U_1 = k_1, \ldots, U_n = k_n) = \binom{n}{k_1, \ldots k_n}(\frac{1}{n})^n$. *Let* $W = \sum_{i=1}^{n} \psi(U_i)$, *where* $\psi \geq 0$ *is scaled such that* $E\psi(U) = 1$ *(hence* $\lambda = EW = n$*), and is such that* $\psi(u) \leq Ke^{Ku}$ *for some* $K < \infty$ *and all* $u \geq 0$. *Let* $\sigma^2 = \mathrm{Var}W$ *and* $D = \max\{||h||, ||h'||\}$. *Then there exists a constant* $C = C(K) < \infty$ *such that*

$$|Eh\left(\frac{W-\lambda}{\sigma}\right) - Nh| \leq \frac{CD}{\sigma}\max\left\{1, \frac{n}{\sigma^2}\right\} \tag{3.1}$$

REMARK 3.1. *Since U_1 given $U_2 = k_2$ is Binomial $(n-k_2, 1/(n-1))$ we have the expansion $P(U_1 = k_1, U_2 = k_2)/P(U_1 = k_1)P(U_2 = k_2) = 1-(k_1k_2 - k_1 - k_2 + 1)/n + o(1/n)$. Applying this to $\mathrm{Cov}(\psi(U_1), \psi(U_2)) = \sum\sum \psi(k_1)\psi(k_2)\,[P(U_1 = k_1, U_2 = k_2)/P(U_1 = k_1)P(U_2 = k_2) - 1]P(U_1 = k_1)P(U_2 = k_2)$ and $n^{-1}\sigma^2 = \mathrm{Var}\psi(U) + (n-1)\mathrm{Cov}(\psi(U_1), \psi(U_2))$, we conclude that for fixed ψ and large n*

$$n^{-1}\sigma^2 = \mathrm{Var}\psi(U) - \mathrm{Cov}(\psi(U), U)^2/\mathrm{Var}U + o(1)\,.$$

Hence, $n^{-1}\sigma^2$ is bounded away from zero for any non-linear ψ, and the rate obtained in (3.1) is of order $n^{-1/2}$ as expected.

By translation, Theorem 3.1 applies also to ψ which is possibly negative but bounded below. Since we know that $P(\max_i U_i > \log n) \le c/n$ (see the discussion below (3.2)), we may accommodate ψ such that $\psi(u) \ge -u^k$ for some k by the n-dependent translation $\psi + (\log n)^k$, at the cost of a somewhat slower rate of convergence.

Proof. Let $V_i \sim \psi(u)P(U_i \in du)$, be i.i.d. random variables independent of all other variables. Let $\mathbf{Z^{(i)}} = (Z_1^{(i)}, \ldots, Z_n^{(i)})$ be independent conditional on $\mathbf{U}$, with $Z_j^{(i)}$ denoting the net change in the number of balls in the jth cell when removing with equal probabilities m balls from the balls in all but the ith cell, and adding them to the ith cell if $V_i - U_i = m \ge 0$, while for $V_i - U_i = -m < 0$ removing m balls from cell i, and distributing them with equal probabilities into the remaining cells. To construct the random variable W^* needed for applying Theorem 2.1, choose a random index I uniformly in $\{1, \ldots, n\}$ and let $Y_j^{(I)} = U_j + Z_j^{(I)}$. Invoking Lemma 2.1, the random variable $W^* = \sum_{j=1}^n \psi(Y_j^{(I)})$ has the W–size biased distribution.

We turn now to bound the r.h.s. of (2.1). Our assumptions on $\psi(\cdot)$ imply that for all k,

$$E(e^{kV}) \le KE(e^{(K+k)U}) \le c(k, K) \tag{3.2}$$

where $c = c(k, K) < \infty$ is independent of n. By Markov's inequality $P(V_i > \log n) \le c/n^k$ and also $P(U_j > \log n) \le c/n^k$. Since h is bounded, we may and shall assume that $V_i \le \log n$ for all i, and $U_j \le \log n$ for all j, thereby neglecting events whose probability is of smaller order than our approximation rate. Let now $A_{ij} = \psi(U_j + Z_j^{(i)}) - \psi(U_j)$ and note that

$$E[W^* - W|W] = \frac{1}{n}E\left[\sum_{i=1}^n A_{ii}|W\right] + \frac{1}{n}E\left[\sum_{i=1}^n \sum_{j\ne i} A_{ij}|W\right],$$

hence, using $\mathrm{Var}(A+B) \le 2\mathrm{Var}A + 2\mathrm{Var}B$ and ignoring the conditioning on W in the second term,

$$\mathrm{Var}E[W^* - W|W] \le \frac{2}{n^2}\mathrm{Var}E\left[\sum_{i=1}^n A_{ii}|W\right] + \frac{2}{n^2}\mathrm{Var}\sum_{i=1}^n\sum_{j\ne i} A_{ij}\,. \tag{3.3}$$

Since $A_{ii} = \psi(V_i) - \psi(U_i)$ and $\sum_{i=1}^n \psi(V_i)$ is independent of W we see that

$$\text{Var}E\left[\sum_{i=1}^n A_{ii}|W\right] = \text{Var}W = \sigma^2 . \tag{3.4}$$

By the symmetry in the problem,

$$\begin{aligned}
\frac{1}{n^2}\text{Var}\sum_{i=1}^n\sum_{j\neq i} A_{ij} &= \frac{1}{n^2}\sum_{i=1}^n\sum_{i'=1}^n\sum_{j\neq i}\sum_{j'\neq i'} \text{Cov}(A_{ij}, A_{i'j'}) \\
&= \frac{n-1}{n}\sum_{i'=1}^n\sum_{j'\neq i'} \text{Cov}(A_{12}, A_{i'j'}) \\
&\leq 2E(A_{12}^2) + nE|A_{12}A_{13}| + nE|A_{12}A_{32}| + 2nE|A_{12}A_{23}| \\
&\quad + n^2|\text{Cov}(A_{12}, A_{34})|. \qquad (3.5)
\end{aligned}$$

Define $L_{ij} = 2(U_j + 1)|V_i - U_i|$, and $T_{ij} = Ke^{K(U_j + |V_i - U_i|)}$. Our assumptions on $\psi(\cdot)$ imply that $|A_{ij}| \leq T_{ij}$. Moreover, for all $j \neq i$ we have

$$P(A_{ij} \neq 0|\mathbf{U}, \mathbf{V}) \leq P(|Z_j^{(i)}| \geq 1|\mathbf{U}, \mathbf{V}) \leq E[|Z_j^{(i)}||\mathbf{U}, \mathbf{V}] \leq \frac{L_{ij}}{n} ,$$

where for the last inequality to hold we assumed that $U_i < n/2$. Consequently,

$$E(A_{12}^2) = E[E(A_{12}^2|\mathbf{U}, \mathbf{V})] \leq E[T_{12}^2 P(A_{12} \neq 0|\mathbf{U}, \mathbf{V})] \leq \frac{1}{n}E[T_{12}^2 L_{12}] \leq \frac{c_1}{n},$$

where $c_1 = c_1(K) < \infty$ is independent of n (the last inequality is obtained by a calculation similar to that of (3.2)). Conditionally on $\mathbf{U}, \mathbf{V}$, the random variable A_{12} is independent of $A_{i'j'}$, for all $i' \neq 1$. Therefore,

$$E|A_{12}A_{32}| = E[E[|A_{12}|\ |\mathbf{U}, \mathbf{V}]E[|A_{32}|\ |\mathbf{U}, \mathbf{V}]] \leq \frac{1}{n^2}E[T_{12}L_{12}T_{32}L_{32}] \leq \frac{c_2}{n^2} ,$$

and similarly also $E|A_{12}A_{23}| \leq c_3/n^2$. Assuming with no loss of generality that $V_2 < n/2$ and $V_3 < n/2$, it is easy to check that

$$P(A_{12} \neq 0 \text{ and } A_{13} \neq 0|\mathbf{U}, \mathbf{V}) \leq E[|Z_2^{(1)}Z_3^{(1)}|\ |\mathbf{U}, \mathbf{V}] \leq \frac{1}{n^2}L_{12}L_{13} ,$$

and thus to obtain the bound $E|A_{12}A_{13}| \leq c_4(K)/n^2$. Let P^* be the product of the laws of the triplets (V_1, U_1, U_2) and (V_3, U_3, U_4). Denote by f_n the probability ratio dP/dP^*, and by $E^*[\cdot]$ expectations according to the distribution P^*, so that $E[\cdot] = E^*[\cdot f_n]$. Define $B_{ij} = E[A_{ij}|V_i, U_i, U_j]$, observing that $E[A_{12}]E[A_{34}] = E^*[B_{12}B_{34}]$ and by the conditional independence alluded to above, also $E[A_{12}A_{34}] = E[B_{12}B_{34}]$. Thus,

$$\text{Cov}(A_{12}, A_{34}) = E[A_{12}A_{34}] - E[A_{12}]E[A_{34}] = E^*[(f_n - 1)B_{12}B_{34}].$$

As we have seen, $n|B_{ij}| \le T_{ij}L_{ij}$ implying that

$$(3.6) \qquad n^2|\mathrm{Cov}(A_{12}, A_{34})| \le n^{-1}E^*[n|f_n - 1|T_{12}L_{12}T_{34}L_{34}].$$

Clearly, f_n depends only on the value of $U_1 + U_2$ and that of $U_3 + U_4$, with

$$f_n(x,y) = \frac{P(U_1 = k_1, U_2 = x - k_1, U_3 = k_3, U_4 = y - k_3)}{P(U_1 = k_1, U_2 = x - k_1)P(U_3 = k_3, U_4 = y - k_3)} \text{ for all } k_1,\ k_3\,.$$

By straightforward calculations, for all $n \ge \max(6, x+y)$,

$$(3.7) \qquad -2(xy+6) \le n(f_n(x,y) - 1) \le 3(x+y)2^{(x+y)}\,.$$

Thus, with $U_j < n/4$, we deduce from (3.6) and (3.7) that $|\mathrm{Cov}(A_{12}, A_{34})| \le c_5/n^3$. Combining all these bounds with (3.3)-(3.5), we see that for some $C_1 = C_1(K) < \infty$

$$(3.8) \qquad \mathrm{Var}E[W^* - W|W] \le \frac{2\sigma^2}{n^2} + \frac{C_1}{n}.$$

Turning now to bound the term $E(W - W^*)^2$, by the symmetry of the problem,

$$\begin{aligned} E[(W^* - W)^2] &= \frac{1}{n}\sum_{i=1}^{n} E\left[\left(\sum_{j=1}^{n} A_{ij}\right)^2\right] = E\left[\left(\sum_{j=1}^{n} A_{1j}\right)^2\right] \\ &= \sum_{j=1}^{n}\sum_{j'=1}^{n} E[A_{1j}A_{1j'}] \\ &\le E[A_{11}^2] + 2nE[|A_{11}A_{12}|] + nE[A_{12}^2] + n^2E[|A_{12}A_{13}|]. \end{aligned}$$

Since $E[A_{11}^2] \le E[T_{11}^2] \le c_7(K)$ and $nE[|A_{11}A_{12}|] \le E[T_{11}T_{12}L_{12}] \le c_8(K)$, it now follows that $E[(W^* - W)^2] \le C_2$ for some $C_2 = C_2(K) < \infty$. Combining the latter with (2.1) and (3.8), one can easily verify that (3.1) holds for $C = 2\sqrt{2 + C_1} + C_2$. □

4. Exceedances of the r-scans process and the r-spacings on the sphere. Let $X_1, X_2, \ldots, X_{n+r-1}$ be i.i.d. random variables with $R_i = \sum_{k=0}^{r-1} X_{i+k}$, $i = 1, 2, \ldots, n$ denoting their r-scans process. We are interested in counting the exceedances of $\{R_i\}$, that is, $W = \sum_{i=1}^{n} J_i$ where $J_i = 1_{R_i \le a}$. Let $p = P(R_1 \le a)$. Clearly, $\lambda = EW = np$ and (for $n \ge r$),

$$(4.1) \qquad \sigma^2 = \mathrm{Var}W = \lambda[1 - P(R_1 \le a) + 2\sum_{\Delta=1}^{r-1}(1 - \frac{\Delta}{n})\psi(\Delta)]$$

where $\psi(\Delta) = P(R_{\Delta+1} \le a | R_1 \le a) - P(R_1 \le a) \ge 0$. In particular, for non-negative X's whose distribution function is strictly positive on $(0, \infty)$

and continuous at zero we have $\sigma^2/\lambda \to 1$ when $a \to 0$ suggesting that in this regime W is approximately Poisson(λ). Applying the Chen-Stein Poisson approximation method, this result is proved in [6, Theorem 1] which also supplies the total-variation error bound

$$||\mathcal{L}(W) - \text{Poisson}(\lambda)||_{var} \leq 4 \sum_{\Delta=1}^{\infty} P\left(\sum_{k=1}^{\Delta} X_i \leq a\right).$$

The study in [6] of the asymptotics of W when $n \to \infty$ is motivated by its applicability to the evaluation of the significance of observed inhomogeneities in the distribution of markers (e.g. restriction sites) along the length of long DNA sequences (see also [13] for more detailed biomolecular applications). Note that for non-negative X's the Poisson approximation is not valid if $\lambda/n \to 0$ and a/r is bounded away from 0, for then $\sigma^2/\lambda - 1$ is also bounded away from 0.

In the following theorem, we remove the restriction of X's being non-negative and approximate the law of W by $N(\lambda, \sigma^2)$ in the Kolmogorov-Smirnov distance. While the Poisson approximation for the count $\hat{W} = \sum_{i=1}^{n}(1 - J_i)$ of large exceedances $1_{R_i > a}$ might fail depending on the tails of X (see [6, Theorem 2]), the $N(\lambda, \sigma^2)$ approximation to W presented below is equivalent to the $N(n - \lambda, \sigma^2)$ approximation to $\hat{W}$.

THEOREM 4.1. *For all* w,

$$(4.2) \quad \left|P\left(\frac{\sum_{i=1}^{n} J_i - \lambda}{\sigma} \leq w\right) - \Phi(w)\right| \leq 14\frac{(2r-1)^2}{\sigma} \leq 14\frac{(2r-1)^2}{\sqrt{np(1-p)}},$$

where Φ *denotes the standard normal distribution function.*

For expectations of smooth functions of W, applying [20, page 110] directly with calculations similar to those appearing around (4.6), leads to

$$(4.3) \quad |E[h((W - \lambda)/\sigma)] - Nh| \leq 2\frac{(2r-1)^2}{\sigma}(||h'|| + \frac{4}{\sqrt{2r-1}}||h||).$$

The Normal approximations in (4.2) and (4.3) have an error rate of order $r^2/\sqrt{n \min(p, 1-p)}$. Consequently, if $p = P(\sum_{i=1}^{r} X_i \leq a)$ is held bounded away from 0 and 1 (e.g., by taking $a = a(r)$ such that $a(r)/r \to E(X)$ at the appropriate rate), then asymptotic normality follows provided $r(n)n^{-1/4} \to 0$. Also, if $c = \limsup_r a(r)/r < E(X)$ is such that $P(X < c) > 0$ then $p \to 0$ but $pe^{rI(c)}$ is bounded below for some $I(c) < \infty$ and we have asymptotic normality when $\limsup_n r(n)/\log n < 1/I(c)$. For X's non-negative of positive density at 0 and a fixed, $pr!e^{rK(a)}$ is bounded below for some $K(a) < \infty$ and hence $\limsup_n r(n) \log\log n/\log n < 1$ implies asymptotic normality.

Theorem 4.1 is a corollary of the next theorem.

THEOREM 4.2. *Let* $Y_1, \ldots, Y_n$ *be random variables satisfying* $|Y_i - E(Y_i)| \leq B$ *a.s.,* $i = 1, \ldots, n$, $E\sum_{i=1}^{n} Y_i = \lambda$, $\text{Var}\sum_{i=1}^{n} Y_i = \sigma^2 > 0$ *and*

$\frac{1}{n}E\sum_{i=1}^{n}|Y_i - E(Y_i)| = \mu$. *Let* $S_i \subset \{1,\ldots,n\}$ *be such that* $j \in S_i$ *if and only if* $i \in S_j$ *and* (Y_i, Y_j) *is independent of* $\{Y_k\}_{k \notin S_i \cup S_j}$ *for* $i, j = 1, \ldots, n$. *Then, for* $D = \max_{1\le i\le n}|S_i|$,

$$\left|P\left(\frac{\sum_{i=1}^{n} Y_i - \lambda}{\sigma} \le w\right) - \Phi(w)\right| \le 7\frac{n\mu}{\sigma^3}(DB)^2 \tag{4.4}$$

Remark. In triangular arrays for which $\mu \ll B$, the above is an improvement of [19, Theorem 2.2] (which in turn is based on [20, page 110]). By combining Stein's method with the study of characteristic functions, the Normal approximation with error term comparable to that of [19, Theorem 2.2] is provided in [23, Theorem 5] for Y_i any strictly stationary D-dependent sequence of random variables with finite third moment.

Proof of Theorem 4.1. The result follows from (4.4) by setting $S_i = \{i - r + 1, \ldots, i + r - 1\}$ with $D = 2r - 1$ and $B = 1$ and noting that $\sigma^2 \ge \lambda(1 - p) = n\mu/2$ (see (4.1)). □

Proof of Theorem 4.2. Without loss of generality assume that $E(Y_i) = 0$, hence $|Y_i| \le B$ a.s., $i = 1, \ldots, n$ and $\mu = E(|Y_I|)$, where the random index I is uniformly distributed over $\{1, \ldots, n\}$ independent of the Y's. Set $A = DB/\sigma$, $X_i = Y_i/\sigma$, $i = 1, \ldots, n$, and $W = \sum_{i=1}^{n} X_i$. The proof of (4.4) parallels that of [19, Theorem 2.1], departing from the latter by using the conditional (upon $I = i$) independence of X_I and $W^* = W - \sum_{j\in S_I} X_j$ which is not available in the more general framework considered there.

For completeness we review the basic steps of the proof, leaving out some of the technical calculations.

For a given function $h : R \to [0, 1]$, continuous and piecewise continuously differentiable, let f denote the unique solution of the differential equation $f'(w) - wf(w) = h(w) - Nh$ with $\lim_{w\to\infty} f(w)e^{-w^2/2} = 0$. Since $E[nX_I f(W^*)] = 0$, simple manipulations and a Taylor series expansion of $f(W) - f(W^*)$ (with integral remainder) yield

$$Eh(W) - Nh = E\{f'(W)[1 - nX_I(W - W^*)]\} + E\int_{W^*}^{W} nX_I(t - W^*)df'(t).$$

With the bound $\|f'\| \le 1$ (see [1, (2.17)]), we obtain

$$|E\{f'(W)[1 - nX_I(W - W^*)]\}| \le \sqrt{E\left\{\sum_{i=1}^{n}\sum_{j\in S_i}(X_iX_j - EX_iX_j)\right\}^2}$$

We now apply the above to the function

$$h(x) = \begin{cases} 1 & \text{if } x \le w \\ 1 - \frac{1}{\epsilon}(x - w) & \text{if } w \le x \le w + \epsilon \\ 0 & \text{otherwise} \end{cases}.$$

Since $Nh \leq \Phi(w) + \frac{\epsilon}{2\sqrt{2\pi}}$, and $P(W \leq w) \leq Eh(W)$, we obtain

$$P(W \leq w) - \Phi(w) \leq \frac{\epsilon}{2\sqrt{2\pi}} + \sqrt{E\left\{\sum_{i=1}^{n}\sum_{j\in S_i}(X_iX_j - EX_iX_j)\right\}^2} + E\int_{W^*}^{W} nX_I(t - W^*)df'(t)\,. \tag{4.5}$$

Let $U_{i,j} = X_iX_j - EX_iX_j$, so that

$$E\left\{\sum_{i=1}^{n}\sum_{j\in S_i}(X_iX_j - EX_iX_j)\right\}^2 = \sum_{i=1}^{n}\sum_{j\in S_i}\sum_{k=1}^{n}\sum_{\ell\in S_k} EU_{i,j}U_{k,\ell}\,. \tag{4.6}$$

Observe that per fixed i there are at most $4D^3$ non-zero terms in the r.h.s. of (4.6), each of which is bounded by $2(B/\sigma)^3E|X_i|$. Consequently, the r.h.s. of (4.6) is bounded by $8n\mu A^3/\sigma$.

Let ψ denote the indicator function of the interval $[w, w+\epsilon]$. Then

$$E\int_{W^*}^{W} nX_I(t-W^*)df'(t) = E\int_{W^*}^{W} nX_I(t-W^*)\{f(t)+tf'(t)-\frac{1}{\epsilon}\psi(t)\}dt. \tag{4.7}$$

The first term in (4.7) is readily bounded by $\frac{1}{2}||f||En|X_I|(W - W^*)^2 \leq \frac{1}{2}\sqrt{\pi/2}\,n\mu A^2/\sigma$, using the estimate $||f|| \leq \sqrt{\pi/2}||h - Nh|| \leq \sqrt{\pi/2}$ ([20, page 25]), and noting that $|W - W^*| \leq A$, and $E|X_I| = \frac{\mu}{\sigma}$. For the second term in (4.7), simply replacing t by $t - W^* + W^*$ and integrating, lead to the bound $\frac{1}{3}||f'||nE|X_I(W - W^*)^3| + \frac{1}{2}||f'||nE|X_I(W - W^*)^2W^*|$. From $||f'|| \leq 1$ (see [1, (2.17)]), we obtain $\frac{1}{3}||f'||nE|X_I(W - W^*)^3| \leq \frac{1}{3}n\mu A^3/\sigma$. Next we also use $E|W| \leq \sqrt{EW^2} = 1$, and the conditional independence of X_I and W^* given I, to conclude that

$$\begin{aligned}
&\frac{1}{2}||f'||nE|X_I(W - W^*)^2W^*| \leq \frac{n}{2}A^2E|X_IW^*| \\
&= \frac{A^2}{2}\sum_{i=1}^{n}E|X_i|E[|W^*|\,|I = i] \leq \frac{A^2}{2}\sum_{i=1}^{n}E[|X_i|]E[|W| + A\,|I = i] \\
&= \frac{nA^2}{2}E|X_I|(E|W| + A) \leq n\mu\frac{A^2}{2}(1 + A)/\sigma.
\end{aligned}$$

Returning to (4.7) we now study the term $\frac{1}{\epsilon}E\int_{W^*}^{W} nX_I(t - W^*)\psi(t)dt$. It can be written as $\frac{1}{\epsilon}E\int_{W^*}^{W} nX_I(t - W^*)\psi(t)dt\mathcal{I}_Q$, where $Q = \{(W \wedge W^*,\ W \vee W^*) \cap (w, w+\epsilon) \neq \emptyset\}$ and $\mathcal{I}_Q$ denotes the indicator of the event Q. Since

$$Q = \{(W \wedge W^*, W \vee W^*) \cap (w, w + \epsilon) \neq \emptyset\} \subseteq \{|W^* - w| \leq A + \epsilon\} = R,$$

we have

$$|\frac{1}{\epsilon}E\int_{W^*}^{W} nX_I(t-W^*)\psi(t)dt| \le \frac{n}{2\epsilon}E[|X_I|(W-W^*)^2\mathcal{I}_R] \le$$

$$\frac{nA^2}{2\epsilon}E[|X_I|\mathcal{I}_R] = \frac{A^2}{2\epsilon}\sum_{i=1}^{n} E[|X_i|\mathcal{I}_R \mid I=i] = \frac{A^2}{2\epsilon}\sum_{i=1}^{n} E[|X_i|]P(\mathcal{I}_R|I=i)$$

$$\le \frac{nA^2}{2\epsilon}E[|X_I|]P(|W-w|\le 2A+\epsilon).$$

As in [19], setting $\Delta = \sup_w |P(W\le w)-\Phi(w)|$ it follows that

$$\begin{aligned} P(|W-w|\le 2A+\epsilon) &\le \Phi(w+2A+\epsilon)-\Phi(w-2A-\epsilon)+2\Delta \\ &\le 2(2A+\epsilon)/\sqrt{2\pi}+2\Delta\ . \end{aligned}$$

From (4.5) and the above bounds we obtain:

$$P(W\le w)-\Phi(w) \le \frac{\epsilon}{2\sqrt{2\pi}} + \sqrt{8n\mu A^3/\sigma} + i\frac{1}{2}\sqrt{\pi/2}\,n\mu A^2/\sigma +$$

$$+\frac{1}{3}n\mu A^3/\sigma + \frac{n}{2}\mu A^2(1+A)/\sigma + \frac{n\mu A^2}{\sigma\epsilon}\{(2A+\epsilon)/\sqrt{2\pi}+\Delta\}\ .$$

A corresponding lower bound is computed similarly. Note that

$$\sigma^2 = \sum_{i=1}^{n}\sum_{j\in S_i} E(Y_iY_j) \le \sum_{i=1}^{n}\sum_{j\in S_i} BE|Y_i| \le BDn\mu$$

and therefore $An\mu/\sigma \ge 1$. The latter fact is useful in the final calculations. In particular it implies that the bound in (4.4) is trivial unless $A = (DB/\sigma) \le 1/7$. Choosing $\epsilon = 7n\mu A^2/\sigma$, the bound (4.4) follows by straightforward calculations. □

Next, we briefly study the problem of normal approximation for counts of exceedances of r-spacings, in which the dependence structure is more involved, and is not restricted to "local" dependence.

Let $U_0,\ldots,U_{n-1}$ be the adjacent arc lengths on the unit sphere, corresponding to its partition by independent random points, distributed uniformly. Let $\tilde{W} = \sum_{i=0}^{n-1} 1_{T_i\le a/n}$ where $T_i = \sum_{j=0}^{r-1} U_{(i+j)\text{mod}n}$.

Taking $R_i = \sum_{j=0}^{r-1} X_{(i+j)\text{mod}n}$ to be the r-scans of i.i.d. Exponential(1) random variables $\{X_j\}$ and $S_n = \sum_{i=0}^{n-1} X_i$, it is well known that one can generate the sequence $\{T_i\}$ by the formula $T_i = R_i/S_n$. Hence, one approach to the approximation of $\tilde{W}$ is to use the $N(\lambda,\sigma^2)$ approximation already obtained for $W = \sum_{i=0}^{n-1} 1_{R_i\underset{\sim}{<}a}$ (when $r^2/\sigma \to 0$) and use the above mentioned coupling to show that $\tilde{W}$ and W are not far apart. Indeed, for every $h: R\to R$ bounded, continuous and piecewise continuously differentiable function

$$
\begin{aligned}
&|E[h((\tilde{W}-\lambda)/\sigma)] - Nh| \\
&\leq |E[h(\frac{\tilde{W}-\lambda}{\sigma})] - E[h(\frac{W-\lambda}{\sigma})]| + |E[h(\frac{W-\lambda}{\sigma})] - Nh| \\
&\leq \frac{1}{\sigma}||h'||E|W-\tilde{W}| + 2\frac{(2r-1)^2}{\sigma}(||h'|| + \frac{4}{\sqrt{2r-1}}||h||).
\end{aligned}
$$

The first term on the r.h.s. is further bounded by

$$
\frac{||h'||}{\sigma}E|\sum_{i=1}^{n-r+1}(1_{R_i \leq a\frac{S_n}{n}} - 1_{R_i \leq a})| \leq \frac{n||h'||}{\sigma}E|1_{R_1 \leq a\frac{S_n}{n}} - 1_{R_1 \leq a}| = \Delta_n \;.
$$

Some calculations based on the fact that S_n is a Gamma$(n,1)$ random variable, suggest that Δ_n is of order at most $\max(r,a)\sqrt{\min(p,1-p)\log n} + \frac{1}{\sigma}$.

An alternative approach to the Normal approximation of $\tilde{W}$ is by applying Theorem 2.1. This approach is expected to have a better approximation rate for choices of a, r, n for which the difference between the first two moments of W and of $\tilde{W}$ is non-negligible. To construct the $\tilde{W}$-size biased random variable $\tilde{W}^*$, choose independently $I \in \{0, \ldots, n-1\}$ uniformly and $V \sim P(T \in dy | T \leq a/n)$ both independent of all other variables. For $I = i$, if $T_i > a/n$ multiply each of the r arc lengths contributing to T_i by V/T_i and each of the remaining $(n-r)$ arc lengths by $(1-V)/(1-T_i)$. Otherwise, leave the arc lengths unchanged. In both cases let Z_j be the r-spacings corresponding to the newly obtained arc lengths and $\tilde{W}^* = \sum_{j=0}^{n-1} 1_{Z_j \leq a/n}$. We shall not pursue this approach further.

REFERENCES

[1] Baldi, P., Rinott, Y. and Stein, C. (1989). *A normal approximations for the number of local maxima of a random function on a graph*, In Probability, Statistics and Mathematics, Papers in Honor of Samuel Karlin. T. W. Anderson, K.B. Athreya and D. L. Iglehart eds., Academic Press, New York, 59-81.

[2] Breiman, L. and Friedman, J. H. (1985). *Estimating optimal transformations for multiple regression and correlation*, J. Amer. Stat. Assoc 80, 580-598.

[3] Barbour, A. D., Holst, L. and Janson, S. *Poisson Approximation.* Oxford University Press, Oxford, 1992.

[4] Csaki, P., and Fischer, J. (1963). *On the general notion of maximal correlation.* Magyar Tud. Akad. Mat. Kutato Int. Kozl. 8, 27-51.

[5] Chen, L. H. Y. (1991). *The rate of convergence in the Central Limit Theorem for dependent random variables with arbitrary index set.* Preprint.

[6] Dembo, A. and Karlin, S. (1992). *Poisson Approximations for r-scan processes.* Annals of Applied Probability 2, 329-357.

[7] Dobrushin, R. L. and Major, P. (1979). *Non-central limit theorems for non-linear functionals of Gaussian fields.* Z.W. verb. Geb. 50, 27-52.

[8] Goldstein, L. and Rinott, Y. (1994). *Multivariate normal approximations by Stein's method and size bias couplings.* To appear in Advances in Applied Probability.

[9] Holst, L. (1981) *Some conditional limit theorems in exponential families*, Annals of Probability 9, 818-830.

[10] Ivchenko, G. I., and Medvedev, Yu. I. (1987). *Estimation of the rate of convergence in limit theorems for divisible statistics*, Theory of Probability and its Applications 31, 81-87.

[11] Ivchenko, G. I., and Mirakhmedov, Sh. A. (1988). *On the asymptotic normality of randomized separable statistics in a generalized allocation scheme*, Theory of Probability and its Applications 33, 749-755.

[12] Ivchenko, G. I., and Tsukanov, S. V. (1988). *New statistical tests in a multinomial scheme and their asymptotic properties.* Theory Probability Appl. 33, 187-191.

[13] Karlin, S., and Brendel V. (1992). *Chance and statistical significance in protein and DNA sequence analysis.* Science 257, 39-49.

[14] Kolchin, V. F., Sevast'yanov, B.A., and Chistyakov, V. P. (1978). *Random Allocations.* Winston & Sons Washington D.C.

[15] Kolmogorov, A.N., and Rosanov, Yu.A. (1960). *On strong mixing conditions for stationary Gaussian Processes*, Theory of Probability and its Applications 5, 204-208.

[16] Mardia, K.V. Kent, J.T., and Bibby, J.M. *Multivariate Analysis.* Academic Press, 1979.

[17] Morris, C. (1975) *Central limit theorems for multinomial sums*, Annals of Statistics 3, 165-188.

[18] Newman, C. M. (1984) *Asymptotic independence and limit theorems for positively and negatively dependent random variables.* In Inequalities in Statistics and Probability. Y. L. Tong, ed., IMS, 127-140.

[19] Rinott, Y. (1992). *On normal approximation rates for certain sums of dependent random variables.* To appear in Journal of applied and Comp. Math.

[20] Stein, C. *Approximate Computation of Expectations.* IMS, Hayward, Calif., 1986.

[21] Sun, T. and Ho, H. *On central and non-central limit theorems for non-linear functions of a stationary Gaussian process.* In E. Eberlein and M. Taqqu, editors, *Dependence in Probability and Statistics*, 3-19. Birkhäuser, Basel, Switzerland, 1986.

[22] Taqqu, M. (1975) *Weak convergence to fractional Brownian motion and to the Rosenblatt process.* Z.W. verb. Geb. 31, 287-302.

[23] Tikhomirov, A. N. (1980) *On the convergence rate in the central limit theorem for weakly dependent random variables.* Theory Probab. and Appl. XXV, 790-809.

LARGE DEVIATIONS FOR RANDOM DISTRIBUTION OF MASS

AMIR DEMBO* AND OFER ZEITOUNI†

Abstract. We study the large deviations for the model of random mass distribution proposed by Aldous. Based on a suitable approximation argument, we prove Aldous' conjecture concerning the large deviations behavior.

Key words. Large deviations, random trees.

AMS(MOS) subject classifications. Primary: 60F10; Secondary: 60F15.

1. Introduction. In [2], Aldous considered an $\mathbb{R}^d$ embedding of the Brownian Continuum Random Tree (BCRT) as a model for random distributions of mass. He conjectured certain large deviations behavior for his model. In this note, we provide a proof for Aldous' conjecture.

Aldous' BCRT $\mathbb{R}^d$ embedding may be constructed as follows from first principles (we describe later his random tree construction, which will also be useful for us. An almost identical construction may be found in [5]). Let $g(\cdot) : [0,1] \to [0,\infty)$ be a stochastic process on $(\Omega, \mathcal{F}, P)$ with continuous sample path such that $g(0) = g(1) = 0$ and $g(t) > 0$ $\forall t \in (0,1)$ (a.s.). For any $t_1 \le t_2 \in [0,1]$, let $m^g(t_1,t_2) = \inf_{t_1 \le s \le t_2} g(s)$. Conditioned on $\mathcal{F}_g$ now define the $[0,1]$-indexed, $\mathbb{R}^d$-valued Gaussian process $z^g(\cdot)$ by $E_g(z^g(\cdot)) = 0$ and, denoting by w_d a d-dimensional Brownian Motion, independent of g,

$$\begin{aligned} E_g(z^g(t_1)(z^g(t_2))^T) &= E_g(w_d(m^g(t_1,t_2))(w_d(m^g(t_1,t_2)))^T) \\ &= m^g(t_1,t_2)I \end{aligned}$$

(Hereafter, v^T denotes the transpose of a vector v and $\langle v,u\rangle \stackrel{\Delta}{=} v^T u$). That is, z^g is the stochastic process with the characteristic function of its finite dimensional distributions being

$$E\left(\exp\left(i\sum_{\ell=1}^{k}\langle z^g(t_\ell), v_\ell\rangle\right)\right) = E\left(\exp\left(-\frac{1}{2}\sum_{i,j=1}^{\ell}\langle v_i, v_j\rangle m^g(t_i,t_j)\right)\right)$$

It is easy to check, by Kolmogorov's criterion (see [4, Theorem 2.2.8]) and conditioning on $g(\cdot)$, that if $g(\cdot)$ is a.s. globally Hölder continuous then $z^g(\cdot)$ possesses a separable, continuous version. Aldous' $\mathbb{R}^d$-embedded BCRT is merely the process $z^{\mathrm{BE}}(\cdot)$, where $BE(\cdot)$ is the standard Brownian excursion

---

* Department of Mathematics and Department of Statistics, Stanford University, Stanford, CA 94305, U.S.A.
Partially supported by NSF DMS92-09712 grant and by a US-ISRAEL BSF grant.
† Department of Electrical Engineering, Technion, Haifa 32000, Israel.
Partially supported by US-ISRAEL BSF grant and Fund for promotion of research at the Technion.

on $[0,1]$, and $\mathcal{F}_{\text{BE}}$ denotes the σ-field generated by the Brownian excursion. Let S denote the closure of $\{z^{\text{BE}}(s), 0 \le s \le 1\}$, and, more generally, for any set $A \subset [0,1]$, let S_A denote the closure of $\{z^{\text{BE}}(s), s \in A\}$. We refer to S as to the support of the (random) mass distribution. Let now $\underline{x} = \{x_1, \cdots, x_n\}$ be n disjoint deterministic points in $\mathbb{R}^d$, and let $B_\delta(x_i)$ denote the ball of radius δ centered at x_i. An embedded binary tree in $\mathbb{R}^d$ is a binary tree whose vertices are points in $\mathbb{R}^d$. Let $ST(\underline{x})$ be the length of the Steiner tree containing 0 and $\underline{x}$ (i.e., the infimum of the sum of edge length over all embedded binary trees containing 0 and $x_1, \cdots, x_n$). For a random set $A = A(\omega)$, let $A \bigcap_{i=1}^n B_\delta(x_i) \neq \emptyset$ denote the set $\{\omega : \cap_{i=1}^n \{A \cap B_\delta(x_i) \neq \emptyset\}\}$, that is, the event that $A(\omega)$ intersects each of the balls $B_\delta(x_i)$. Our main interest here is to prove the following theorem.

THEOREM 1.1.

$$\lim_{\delta \to 0} \limsup_{\epsilon \to 0} \epsilon^{2/3} \log P\left(\sqrt{\epsilon} S \bigcap_{i=1}^n B_\delta(x_i) \neq \emptyset\right) =$$

$$\lim_{\delta \to 0} \liminf_{\epsilon \to 0} \epsilon^{2/3} \log P\left(\sqrt{\epsilon} S \bigcap_{i=1}^n B_\delta(x_i) \neq \emptyset\right) \tag{1.1}$$

$$= -3 \cdot 2^{-5/3} \text{ST}(\underline{x})^{4/3}$$

2. Proof of the Theorem. In order to see where the Steiner tree arises from in (1.1), it is useful to recall parts of the formulation, due to Aldous, of the BCRT as a random tree. We deal here only with finite binary trees, referring the reader to [3] for a complete exposition to Aldous' limiting results for such trees. Let $J = \{0\} \cup \{\text{j} : \exists n \in \mathbb{N} \text{ such that j} \in 0 \times \{0,1\}^n\}$. Let $j_T \in J$ be obtained from $j \in J$ by truncating the rightmost digit, (with $0_T = \phi \notin J$) and let j_0 (j_1) be obtained from j by concatenating 0 (1 respectively) to it. A *finite binary tree* T is a finite, ordered list of pairs (j_i, ℓ_i), $i = 1, \cdots, |T|$, where $\ell_i > 0$ is the length of the i-th edge connecting the vertex $j_i \in J$ with $(j_i)_T$ and ϕ is a fictitious reference point at which the tree is rooted. The collection j_i, $i = 1, 2, \cdots, |T|$, is such that $j_i \in T, j_i \neq 0 \Rightarrow (j_i)_T \in T$ and either $(j_i)_0 \notin T$ and $(j_i)_1 \notin T$ or $(j_i)_0 \in T$ and $(j_i)_1 \in T$. If $j_i \in T$ but $(j_i)_0 \notin T$, we say that j_i is a "leaf". Note that the number of edges is $|T|$ and the number of leaves is $(|T|+1)/2$ (in particular, $|T|$ is thus an odd number). The leaves of T correspond to a given ordered list of "abstract points" $\underline{x} = (x_1, \cdots, x_m)$ by associating the vertex j_{2i-1} with x_i. We shall think of T as generated from the null tree by sequentially adding the vertices $j_1, j_3, \cdots, j_{2m-1}$ with the vertex j_{2i+1} added by creating the branch point j_{2i} at an existing edge. The shape $\underline{t}$ of T thus records (only) which existing edge each new leaf and its edge are attached to, i.e. $\underline{t} = (t_1, \cdots, t_{m-1})$ where $t_i \le (2i-1)$ is the index of the edge in the reduced sub-tree corresponding to the first i leaves at which the branch point j_{2i} is located. In particular, the number of different shapes of trees with exactly m (distinguished) leaves equals $\prod_{i=1}^{m-1}(2i-1)$, and two trees corresponding to the same abstract points $\underline{x}$ are regarded identical if they have the same shape and the same edge length vector $\underline{\ell} = (\ell_1, \cdots, \ell_{2m-1})$.

Now let $\underline{X} = (X_1, \cdots, X_m)$ be i.i.d. $U[0,1]$ random variables, independent of $BE(\cdot)$. Starting from $BE(\cdot)$ and $\underline{X}$, one may construct a random finite binary tree $T_{\rm BE}^{\underline{X}}$ with $2m-1$ vertices, of shape $\underline{t}$ and associated edge length $\underline{\ell}$ (For a precise description of such a construction, see [2, section 3, and 3, proof of Theorem 13]; for a pictorial description, see [2, Figure 1].)

The following lemma is crucial to our analysis.

LEMMA 2.1. *[3, Lemma 21 and Corollary 22] For any* $A \in \mathcal{B}_{\mathbb{R}^{2m-1}}$,

$$(2.1)\ P(T_{\rm BE}^{\underline{X}} = \underline{t}, \underline{\ell} \in A) = \int_{\underline{\ell} \in A} \left(\sum_{i=1}^{2m-1} \ell_i \right) e^{-\left(\sum_{i=1}^{2m-1} \ell_i\right)^2/2} d\ell_1 \cdots d\ell_{2m-1}$$

($\underline{t}$ corresponds to a particular shape of the binary tree).

Note that (2.1) implies that the distribution of the shapes is uniform over the $\prod_{i=1}^{m-1}(2i-1)$ different possibilities, and that the law of $\underline{\ell}$, conditioned on a particular tree, is independent of the shape of the tree (and of the order of the leaves) with density $(\sum_{i=1}^{(2m-1)} \ell_i) e^{-(\sum_{i=1}^{2m-1} \ell_i)^2/2} \prod_{i=1}^{m-1}(2i-1)$.

The vertices of the finite binary tree $T_{\rm BE}^{\underline{X}}$ may be embedded into $\mathbb{R}^d$ in the following way. Let W^j, $j \in T$ be independent, d-dimensional Brownian motions. Let $Y_\phi = 0$ and $Y_{j_i} = Y_{(j_i)_T} + W^{j_i}(\ell_i) \in \mathbb{R}^d$ for each $j_i \in T$. Obviously, the joint law of $\ell_1, \cdots, \ell_{2m-1}$ and $Y_{j_1}, \cdots, Y_{j_{2m-1}}$, for a particular tree T depends only on the shape of the tree and possesses the density

$$f(\ell_1, \cdots, \ell_{2m-1}, Y_{j_1}, \cdots, Y_{j_{2m-1}}) = \prod_{i=1}^{m-1}(2i-1)$$

$$\prod_{i=1}^{2m-1} (2\pi\ell_i)^{-d/2} \left(\sum_{i=1}^{2m-1} \ell_i \right) \exp\left(-\frac{1}{2}\left(\sum_{i=1}^{2m-1} \ell_i \right)^2 - \frac{1}{2} \sum_{i=1}^{2m-1} \frac{|Y_{j_i} - Y_{(j_i)_T}|^2}{\ell_i} \right). \tag{2.2}$$

The reason for the usefulness of the tree representation lies in the following fact. Let $X_1, \cdots, X_m$ be as above. Let π be the rearranging permutation for $X_1, \cdots, X_m$. That is, $X_{\pi(1)} < X_{\pi(2)} < \cdots < X_{\pi(m)}$; Let $\bar{X}_i \in (X_{\pi(i)}, X_{\pi(i+1)})$ be such that $BE(\bar{X}_i) < BE(s)$ for all $s \in (X_{\pi(i)}, X_{\pi(i+1)})$, $s \neq \bar{X}_i$ (such a point $\bar{X}_i$ exists a.s.). Let $L_m^Y = \frac{1}{2m-1} \sum_{i=1}^{2m-1} \delta_{Y_i}$, and let

$$L_m = \frac{1}{2m-1} \left(\sum_{i=1}^{m} \delta_{z^{BE(X_i)}} + \sum_{i=1}^{m-1} \delta_{z^{BE(\bar{X}_i)}} \right).$$

Then, [2 and 3], $L_m^Y \stackrel{d}{=} L_m$. (Actually, much more is known, but we will not use it here.) In particular, for any $m > 0$

$$P\left(\operatorname{supp} L_m^{\epsilon^{1/2}Y} \bigcap_{i=1}^{n} B_\delta(x_i) \neq \emptyset\right) =$$

$$P\left(\epsilon^{1/2} S_{\{X_1,\cdots,X_m,\bar{X}_1,\cdots,\bar{X}_{m-1}\}} \bigcap_{i=1}^{n} B_\delta(x_i) \neq \emptyset\right) \tag{2.3}$$

$$\leq P\left(\epsilon^{1/2} S \bigcap_{i=1}^{n} B_\delta(x_i) \neq \emptyset\right)$$

The next step is a finite dimensional computation. For a sketch of it, see [2].

LEMMA 2.2. *For all $m \geq n$ and every $\delta > 0$*

$$\begin{aligned} -3\cdot 2^{-5/3} ST(\underline{x})^{4/3} &\leq \liminf_{\epsilon\to 0} \epsilon^{2/3} \log P\left(\operatorname{supp} L_m^{\epsilon^{1/2}Y} \bigcap_{i=1}^{n} B_\delta(x_i) \neq \emptyset\right) \\ &\leq \limsup_{\epsilon\to 0} \epsilon^{2/3} \log P\left(\operatorname{supp} L_m^{\epsilon^{1/2}Y} \bigcap_{i=1}^{n} B_\delta(x_i) \neq \emptyset\right) \\ &\leq -3\cdot 2^{-5/3}(ST(\underline{x}) - n\delta)^{4/3} \end{aligned}$$

Proof. We begin by proving the lower bound, the idea being to fix a particular shape which is an embedding of a binary tree which contains vertices corresponding to $x_1, \ldots, x_n$, and consider the probability of an ϵ-neighborhood of it using the density (2.2). To this end, fix $\eta > 0$. For all $m \geq n$ there exists an embedded binary tree with root $y_\phi = 0$ and $2m-1$ disjoint vertices $y_{j_i} \neq 0$, such that $\{x_i\} \subset \{y_{j_i}\}$ and $\sum_i |y_{j_i} - y_{(j_i)_T}| \leq ST(\underline{x}) + \eta$. Taking hereafter $y_\phi = 0$, let

$$\begin{aligned} h_\epsilon(\underline{\ell}, \underline{y}) = &\prod_{i=1}^{2m-1} \ell_i^{-d/2} \left(\sum_{i=1}^{2m-1} \ell_i\right) \\ &\exp\left(-\frac{1}{2\epsilon^{2/3}}\left(\left(\sum_{i=1}^{2m-1} \ell_i\right)^2 + \sum_{i=1}^{2m-1} \frac{|y_{j_i} - y_{(j_i)_T}|^2}{\ell_i}\right)\right). \end{aligned} \tag{2.4}$$

Note that, rescaling (2.2), it follows that for every $\underline{\ell}$ and all $\epsilon \leq \delta$

$$\begin{aligned} &P(\operatorname{supp} L_m^{\epsilon^{1/2}Y} \cap B_\delta(x_i) \neq \emptyset) \geq \\ &k_m \epsilon^{\beta_m} \int_{\substack{|\bar{y}_{j_i} - y_{j_i}| < \epsilon \\ |\bar{\ell}_i - \ell_i| < \epsilon}} h_\epsilon(\underline{\bar{\ell}}, \underline{\bar{y}}) d\bar{y}_{j_1} \cdots d\bar{y}_{j_{2m-1}} d\bar{\ell}_1 \cdots d\bar{\ell}_{2m-1} \\ &\geq c_m \epsilon^{\alpha_m} \prod_i (\ell_i + \epsilon)^{-d/2} \left(\sum_i \ell_i - 2m\epsilon\right) \\ &\quad \exp\left[-\frac{1}{2\epsilon^{2/3}}\left(\left(\sum_i \ell_i + 2m\epsilon\right)^2 + \sum_i \frac{(|y_{j_i} - y_{(j_i)_T}| + 2\epsilon)^2}{\ell_i - \epsilon}\right)\right] \end{aligned}$$

where k_m, c_m, β_m and α_m are some finite positive constants. Choosing $\ell_i = \frac{|y_{j_i} - y_{(j_i)_T}|}{(2(ST(\underline{x})+\eta))^{1/3}}$ it follows that for every $\delta > 0$

$$\liminf_{\epsilon\to 0} \epsilon^{2/3} \log P\left(\operatorname{supp} L_m^{\epsilon^{1/2}Y} \bigcap B_\delta(x_i) \neq \emptyset\right) \geq -3 \cdot 2^{-5/3}(ST(\underline{x}) + \eta)^{4/3}.$$

Since η can be made arbitrarily small, the lower bound follows.

In contrast with the lower bound, when proving the upper bound one has to consider all shapes with the desired intersection properties. Let

$$ST^\delta(\underline{x}) = \inf\{ST(\underline{z}) : z_i \in B_\delta(x_i), i = 1, \ldots, n\}.$$

Fix m and note that for all δ and every shape $\underline{t}$ of the finite binary tree

$$\inf_{\{y_{j_i}\}\cap_i B_\delta(x_i)\neq\emptyset} \sum_{i=1}^{2m-1} |y_{j_i} - y_{(j_i)_T}| \geq ST^\delta(\underline{x}) \geq ST(\underline{x}) - n\delta \,.$$

Hence,

$$P\left(\operatorname{supp} L_m^{\epsilon^{1/2}Y} \bigcap B_\delta(x_i) \neq \emptyset\right) \leq P\left(\sum_{i=1}^{2m-1} |W^{j_i}(\ell_i)| \geq \epsilon^{-1/2}(ST(\underline{x}) - n\delta)\right)$$

(2.5)

$$= P\left(\sum_{i=1}^{2m-1} \sqrt{\ell_i}|V_i| \geq \epsilon^{-1/2}(ST(\underline{x}) - n\delta)\right),$$

where V_i are i.i.d. standard d-dimensional Gaussian variables, which are independent of $\underline{\ell}$. Since for all $\theta > 0$

$$E[e^{\theta|V|}] \leq 2^d e^{\theta^2/2}(1 + c_d\theta^{d-1})$$

where $c_d < \infty$ is some dimensional constant, it follows by applying Chebycheff's inequality that for any $z > 0$ and all $\theta = \theta(\underline{\ell}, z) > 0$

$$P(\sum_{i=1}^{2m-1} \sqrt{\ell_i}|V_i| \geq z) \leq k_d E[e^{-\theta z + \theta^2(\sum_i \ell_i)/2} \prod_{i=1}^{2m-1} (1 + c_d(\theta\sqrt{\ell_i})^{d-1})] \,,$$

where $k_d = k_d(m)$ is again some dimensional constant. In particular, taking $\theta = z/(\sum_i \ell_i)$ we have

$$P\left(\sum_{i=1}^{2m-1} \sqrt{\ell_i}|V_i| \geq z\right) \leq$$

$$k_d E\left[\left(1 + c_d\left(z^2/\left(\sum_i \ell_i\right)\right)^{(d-1)/2}\right)^{2m-1} e^{-z^2/(2\sum_i \ell_i)}\right]$$

$$= \tilde{k}_d \int_0^\infty s^{2m-1}(1 + c_d(z/\sqrt{s})^{d-1})^{2m-1} e^{-\frac{1}{2}(s^2+z^2/s)} ds \,,$$

with $\tilde{k}_d = \tilde{k}_d(m)$ again denoting a dimensional constant taking into account the number of shapes, and the equality due to (2.1). For $z_\epsilon = \epsilon^{-1/2}(ST(\underline{x}) - n\delta)$ we get by first rescaling ($s = \epsilon^{-1/3}u$), then applying Laplace's method, that for all m, and all δ

$$\limsup_{\epsilon\to 0} \epsilon^{2/3} \log P\left(\sum_{i=1}^{2m-1} \sqrt{\ell_i}|V_i| \geq z_\epsilon\right) \leq -3 \cdot 2^{-5/3}(ST(\underline{x}) - n\delta)^{4/3}$$

and the upper bound of the lemma follows by (2.5). □

Due to (2.3) and Lemma 2.2, it only remains to prove the upper bound in the theorem. To this end, let $X_1, \cdots, X_m$ be i.i.d. $U[0,1]$ independent of $z^{BE}(\cdot)$. For every $\Delta > 0$ let $(\tilde{X}_1^{m,\Delta}, \cdots, \tilde{X}_m^{m,\Delta})$ be random variables with the law of $(X_1, \cdots, X_m)$ conditioned on the event

$$E_{m,\Delta} = \left\{\omega : \exists i(j) : \left\{0, 1, \cdots, \left[\frac{1}{\Delta}\right]\right\} \to \{1, \cdots, m\} \text{ s.t. } X_{i(j)} \in [j\Delta, j\Delta + \Delta)\right\}.$$

Note that

$$P\left(\epsilon^{1/2} S \bigcap_{i=1}^{n} B_\delta(x_i) \neq \emptyset\right) \leq P\left(\epsilon^{1/2} S_{\{\tilde{X}_1^{m,\Delta}, \cdots, \tilde{X}_m^{m,\Delta}\}} \bigcap_{i=1}^{n} B_{2\delta}(x_i) \neq \emptyset\right)$$

$$+ P\left(\sup_{\substack{s=j\Delta \\ 0<t-s<\Delta}} \epsilon^{1/2}|z^{BE}(t) - z^{BE}(s)| > \delta\right). \tag{2.6}$$

Since $P(E_{m,\Delta}) \geq 1/2$ for all $m \geq m_0(\Delta)$ large enough, we have then

$$P\left(\epsilon^{1/2} S_{\{X_1, \cdots, X_m, \bar{X}_1, \cdots, \bar{X}_{m-1}\}} \bigcap B_{2\delta}(x_i) \neq \emptyset\right) \geq \frac{1}{2} P\left(\epsilon^{1/2} S_{\{\tilde{X}_1^{m,\Delta}, \cdots, \tilde{X}_m^{m,\Delta}\}} \bigcap B_{2\delta}(x_i) \neq \emptyset\right).$$

Thus, by Lemma 2.2, for $m(\Delta) \geq \max\{m_0(\Delta), n\}$

$$\lim_{\delta\to 0} \limsup_{\epsilon\to 0} \epsilon^{2/3} \log P(\epsilon^{1/2} S_{\{\tilde{X}_1^{m,\Delta}, \cdots, \tilde{X}_m^{m,\Delta}\}} \cap B_{2\delta}(x_i) \neq \emptyset) \leq -3 \cdot 2^{-5/3} ST(\underline{x})^{4/3}.$$

Therefore, using (2.6), the theorem will be proved once we show that for fixed $\delta > 0$

$$\lim_{\Delta\to 0} \limsup_{\epsilon\to 0} \epsilon^{2/3} \log P\left(\sup_{\substack{s=j\Delta \\ 0<t-s<\Delta}} \epsilon^{1/2}|z^{BE}(t) - z^{BE}(s)| > \delta\right) = -\infty. \tag{2.7}$$

Note that due to the continuity of $z^{BE}(\cdot)$,

$$(2.8)\qquad P\left(\sup_{\substack{s=j\Delta\\ 0<t-s<\Delta}} \epsilon^{1/2}|z^{BE}(t)-z^{BE}(s)|>\delta\right) \leq \sum_{j=0}^{[1/\Delta]} P\left(\sup_{i,k}\epsilon^{1/2}|z^{BE}(j\Delta+a_{i,k})-z^{BE}(j\Delta)|>\delta\right),$$

where $a_{i,k}=i\Delta/2^k$ for $i=1,\ldots,2^k-1$, $k\geq 1$.

Let $\|\cdot\|_\alpha$ denote the α-Hölder norm, that is $\|\phi\|_\alpha=\sup_{s,t\in[0,1]}|\phi(t)-\phi(s)|/|t-s|^\alpha$. In order to proceed, we need the following lemma, whose proof we postpone.

LEMMA 2.3. *Let BE be the Brownian excursion on* $[0,1]$. *Then, for* $\alpha<1/2$,

$$(2.9)\qquad \limsup_{\epsilon\to 0}\epsilon^{2/3}\log P\left(\|BE\|_\alpha>\frac{K}{\epsilon^{1/3}}\right)\leq -K^2/8.$$

Proceeding with the proof of the theorem, fix $\alpha<1/2$ and j, and for $i=0,\cdots,2^{k-1}-1$, let

$$Z_i^k=z^{BE}(j\Delta+a_{2i+1,k})-z^{BE}(j\Delta+a_{2i,k}),$$

noting that

$$\sup_{i,k}|z^{BE}(j\Delta+a_{i,k})-z^{BE}(j\Delta)|\leq\sum_{k=1}^{\infty}\sup_i|Z_i^k|.$$

Observe that $Z_i^k\overset{d}{=}w_d(\Theta_{i,k})$ where $w_d(\cdot)$ is a d-dimensional Brownian motion independent of $BE(\cdot)$, and $\Theta_{i,k}\in\mathcal{F}_{BE}$ is non-negative and bounded above by $2\|BE\|_\alpha(\Delta/2^k)^\alpha$. Thus, for $c_k=\delta/(2k^2)$ (so that $\sum_{k=1}^\infty c_k\leq\delta$)

$$P\left(|Z_i^k|>\frac{c_k}{\epsilon^{1/2}}\,\Big|\,\|BE\|_\alpha\leq\frac{K}{\epsilon^{1/3}}\right)\leq P\left(\sup_{t\leq 2(\frac{\Delta}{2^k})^\alpha\frac{K}{\epsilon^{1/3}}}|w_d(t)|>\frac{c_k}{\epsilon^{1/2}}\right)\leq 4de^{-\frac{c_k^2 2^{\alpha k}}{4dK\epsilon^{2/3}\Delta^\alpha}}$$

Thus, for some $\alpha'<\alpha$, $C<\infty$ and all k

$$P\left(\sup_i|Z_i^k|>\frac{c_k}{\epsilon^{1/2}}\,\Big|\,\|BE\|_\alpha\leq\frac{K}{\epsilon^{1/3}}\right)\leq d2^{k+1}e^{-\frac{c_k^2 2^{\alpha k}}{4dK\epsilon^{2/3}\Delta^\alpha}}\leq Ce^{-\frac{\delta 2^{\alpha' k}}{K\epsilon^{2/3}\Delta^\alpha}}$$

Hence, for some $0<C_1,\,C_2<\infty$

$$P(\sup_{i,k}\epsilon^{1/2}|z^{BE}(j\Delta+a_{i,k})-z^{BE}(j\Delta)|>\delta\,|\,\|BE\|_\alpha\leq\frac{K}{\epsilon^{1/3}})$$
$$\leq\sum_{k=1}^{\infty}P(\sup_i|Z_i^k|>\frac{c_k}{\epsilon^{1/2}}\,|\,\|BE\|_\alpha\leq\frac{K}{\epsilon^{1/3}})\leq C_2e^{-\frac{C_1\delta}{K\epsilon^{2/3}\Delta^\alpha}}.$$

With this inequality holding for all j, it follows from (2.8) that

$$(2.10)\qquad P\left(\sup_{\substack{s=j\Delta\\0<t-s<\Delta}} \epsilon^{1/2}|z^{BE}(t)-z^{BE}(s)|>\delta\right)\le$$
$$C_2\left(\frac{1}{\Delta}+1\right)e^{-\frac{C_1\delta}{K\epsilon^{2/3}\Delta^\alpha}}+P\left(\|BE\|_\alpha>\frac{K}{\epsilon^{1/3}}\right)$$

Now let $K=K(\Delta)\to\infty$ be such that $K\Delta^\alpha\to 0$ and use (2.9), (2.10) to get (2.7) and conclude the proof. □

Proof of Lemma 2.3. We use the fact that if BR is a Brownian bridge on $[0,1]$ and U is the instant when it attains its minimum value on $[0,1]$, then (cf. [7])

$$BE(\cdot)\overset{d}{=}BR((\cdot+U)\text{mod}1)-BR(U).$$

Consequently,

$$\begin{aligned}\sup_{0\le t\ne s\le 1}&\frac{|BE(t)-BE(s)|}{|t-s|^\alpha}\\ &\overset{d}{=}\sup_{0\le t\ne s\le 1}\frac{|BR((t+U)\text{mod}1)-BR((s+U)\text{mod}1)|}{|t-s|^\alpha}\\ &\le 2\sup_{0\le t\ne s\le 1}\frac{|BR(t)-BR(s)|}{|t-s|^\alpha}.\end{aligned}$$

Hence, suffices to prove the claim for BR which is a centered Gaussian process, with covariance $\rho(s,t)=(s\wedge t)-st$ (cf. [4, page 359]). To this end, define for $t\ne s$

$$Z_{s,t}=\frac{BR(t)-BR(s)}{|t-s|^\alpha},\quad \alpha<1/2,$$

with $Z_{t,t}=0$. Then $E(Z_{s,t})=0$, and

$$E(Z_{s,t}^2)=\frac{t+s-s^2-t^2-2(s\wedge t-st)}{|t-s|^{2\alpha}}=\frac{|t-s|-|t-s|^2}{|t-s|^{2\alpha}}.$$

Hence, $\sup_{s,t}E(Z_{s,t}^2)=\sigma^2\le 1$. Moreover, since the sample paths of BR are a.s. $(\frac{1}{2}-\epsilon)$-Hölder continuous, the sample paths of $Z_{s,t}$ are a.s. bounded. Therefore, by [6, Theorem 2.5], (see also [1, Theorem 6.9.2]),

$$\lim_{\epsilon\to 0}\epsilon^{2/3}\log P\left(\sup_{s,t}|Z_{s,t}|>\frac{K}{\epsilon^{1/3}}\right)=-\frac{K^2}{2\sigma^2},$$

and the conclusion (2.9) follows. □

REFERENCES

[1] Adler, R. *The Geometry of Random Fields*, Wiley, New York, 1981.

[2] Aldous, D.J., Tree-Based Models for Random Distribution of Mass, *J. Statist. Phys.* 1993 **73** 625-641.

[3] Aldous, D.J., The continuum random tree III, *Ann. Probab.* 1993 **21** 248-289.

[4] Karatzas I. and Shreve S.E., *Brownian Motion and Stochastic Calculus*, Springer-Verlag, New York, 1988.

[5] Le Gall, J.F., Brownian excursions, trees and measure valued branching processes, *Ann. Probab.* 1991 **19** 1399-1439.

[6] Marcus M.B. and Shepp L.A., Sample behaviour of Gaussian processes, *Proc. Sixth Berkeley Symp. Math. Statist. Prob.* 1972 **2** 423-442. Univ. of California Press, Berkeley.

[7] Vervaat, W., A relation between Brownian bridge and Brownian excursion. *Ann. Probab.* 1979 **7** 141-149.

RANDOM MINIMAX GAME TREES

LUC DEVROYE* AND OLIVIER KAMOUN*

Abstract. In this paper, we study random minimax trees of the incremental type. These are complete b-ary trees with n levels of edges, in which we associate independent identically distributed random variables with the edges. The value of a leaf is the sum of the edge values on the path to the root. The value of each internal node is obtained at alternating levels by taking the minimum or maximum value of the values of the children. We are interested in the behavior of the value of the root, V_n. For bounded generic edge random variable X, we show that V_n/n tends to a limit almost surely as $n \to \infty$. The limit is a highly nonlinear function of the distribution of X. For the case of a Bernoulli random variable with parameter p, the limit is a continuous function that is zero for p near zero, one for p near one, $1/2$ for p in an interval around $1/2$, and nonlinear inbetween. A comparison is made with a random minimax tree model studied by Pearl, in which the leaf values are independent.

Keywords and phrases. Game tree. Minimax tree. Probabilistic analysis. Asymptotic behavior. Random trees.

1. Introduction. Random MIN-MAX trees play a major role in the understanding of search strategies in game trees. They can be used to explain why certain search algorithms are preferable in some situations. For example, Pearl [Pearl80] considered a complete b-ary tree T with n levels of edges and associated with the leaves independent random variables all distributed as a given random variable X. To each internal node u with A_u its set of children, he associated a value $V(u)$ according to the standard MIN-MAX rule:

$$V(u) = \begin{cases} \max_{v \in A_u} \{V(v)\} & \text{if } u \text{ at even level in } T \\ \min_{v \in A_u} \{V(v)\} & \text{if } u \text{ at odd level in } T \end{cases} . \tag{1}$$

The parity of a level is with respect to distance from the root, not the leaves. The root is thus a maximizing node in all cases. The values of the nodes thus obtained should be thought of as values obtained by an evaluation function in a game. The trees obtained in this manner will be called Pearl trees. Pearl continued the study of this model in his book [Pearl84], in which several search alorithms are compared with respect to their average case behavior. The simplicity of Pearl's model leads to a beautiful and uncluttered analysis. Some results are quite striking: for example, if V_n is the value of a root node n levels removed from the leaf level, and if the generic random variable X is uniformly distributed on $[0, 1]$, then

$$\lim_{n\to\infty} V_{2n} = 1 - \xi_b \text{ almost surely ;}$$
$$\lim_{n\to\infty} V_{2n+1} = \xi_b \quad \text{almost surely ,}$$

* School of Computer Science, McGill University, Montreal, Canada H3A 2A7.

where ξ_b is the positive solution of $x^b + x - 1 = 0$. For $b = 2$, $\xi_b = (\sqrt{5}-1)/2 = 0.618\ldots$ is related the golden ratio. If X has distribution function F, one can obtain the correct limit result from this by applying the probability integral transform.

The limiting behavior of V_n depends upon whether the bottommost operation is a MIN or MAX. Hence the disturbing bi-asymptotic result mentioned above. If we had always started with a MIN operation from the leaf level up, then V_n would not have oscillated. Nau [Nau82a,Nau82b] pointed out this pathology in Pearl's model. It can all be traced back to the independence assumption for the leaf values. Incremental models are more responsive to this criticism. They effectively incorporate the notion that siblings in the tree should have highly correlated values: for an early survey, we refer to Newborn [Newb77]. Knuth and Moore [KM75] and Fuller [Newb77] have studied particular models of this kind. In this paper, we follow the model developed by Nau [Nau82a,Nau82b,Nau83]: with every edge in a b-ary complete tree T with n levels of edges, we associate an independently drawn random variable distributed as X. A leaf is given as value the sum of the edge values found on the path from the leaf to the root. Values of internal nodes are obtained by the rules laid out in (1). We take the liberty of calling our trees incremental trees, even though authors such as Nau [Nau82a,Nau82b] reserve this terminolgy for broader classes of trees.

The main question here is: how does the root's value (V_n) vary with n? The answer is both interesting and to some extent surprising. In discovering the answer, we will learn a lot about the rich structure of the tree. In any case, even for the simplest kind of edge random variable X, the analysis is much more involved than for the Pearl trees. Besides the general incremental tree with edge random variable X, we will also study the informative Bernoulli tree, which is an incremental tree in which X is a Bernoulli random variable with

$$\boldsymbol{P}\{X = 1\} = 1 - \boldsymbol{P}\{X = 0\} = p \ .$$

The first reassuring observation is that the oscillatory behavior of V_n seen for Pearl trees has disappeared. Our main result states that for all nonnegative edge variables X, there exists a constant $\mathcal{V}(X)$ such that

$$\lim_{n\to\infty} \frac{V_n}{n} = \mathcal{V}(X) \text{ almost surely } .$$

This result confirms that the root's value increases linearly with n, the number of levels. This was to be expected, since with $X \equiv 1$, we would have $V_n \equiv n$. We would of course like to know how $\mathcal{V}(X)$ is related to X. The relationship with the distribution of X is spectacularly nonlinear. For example, for Bernoulli trees with X Bernoulli (p), it is not true that $\mathcal{V}(X) \equiv p$, even though nearly all leaf values are very close to np. Consider now Pearl trees with n levels and in which leaf values are i.i.d. binomial (n, p) with $p \in (0,1)$ fixed. We have seen above that the root's value is either the ξ_b or $1-\xi_b$ quantile of the leaf distribution. For the binomial (n, p) distribution, both converge rapidly to np, and therefore, V_n/n is very close to p. This difference in behavior shows that incremental models are much more colorful and structurally interesting.

For the Bernoulli model in particular, we will give more details about the nonlinear behavior of $\mathcal{V}(X)$ as a function of p. For example, when p is near zero, $\mathcal{V}(X) = 0$. In this case, it will be shown that V_{2n} has a limit distribution function F_∞:

$$\lim_{n\to\infty} \boldsymbol{P}\{V_{2n} \leq i\} = F_\infty(i) \ , \ i \geq 0 \ .$$

For $p > \alpha$, which is a fixed threshold depending upon b only, the behavior of V_n changes abruptly, as V_n starts to grow linearly with n: $\mathcal{V}(X) > 0$. The limit value $\mathcal{V}(X)$ varies nonlinearly with p. While the limit is continuous in p, it is constant and equal to $1/2$ on $[\beta, 1-\beta]$, an interval centered at $1/2$. As $b \to \infty$, this central interval extends to $(0,1)$, making the root's value basically independent of p.

We will also establish how close V_n is to $\boldsymbol{E}V_n$. The results in this paper are based upon the thesis of the second author [Kam92]. They are a first limited step towards the understanding of incremental models. Hopefully, the methods used below will be useful in the study of the performance of search algorithms on incremental trees. For more details and more elaborate simulations than those reported here, we refer to the thesis.

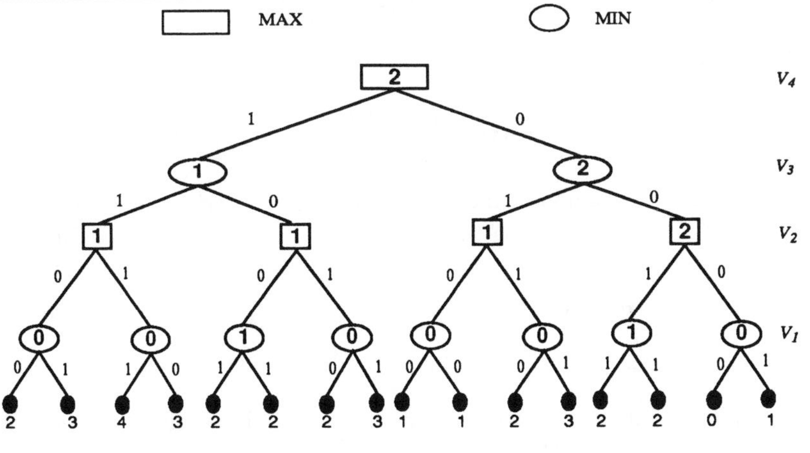

A SUM model.

2. Another construction of the incremental model. We can look at the incremental model with edge random variable X in an equivalent manner that is better suited for analysis. Again, we consider an n-level complete b-ary tree T. Let u be an internal node, and let A_u be the set of its children. For all $v \in A_u$, we associate with the edge (u,v) an independent drawing $E(u,v)$ of a given random variable X. Let F be the distribution function of X:

$$F(x) = \mathbf{P}\{X \leq x\} .$$

With each node u we associate a value according to the following recurrence: if u is a leaf, then $V(u) = 0$. The level of a node is determined by its distance from the leaf level. For an internal node u we define

$$(2) \quad V(u) = \begin{cases} \max_{v \in A_u} \{V(v) + E(u,v)\} & \text{if } u \text{ at even level} \\ \min_{v \in A_u} \{V(v) + E(u,v)\} & \text{if } u \text{ at odd level} \end{cases} .$$

All nodes at path distance n from the leaf level are independent and identically distributed. A generic random variable of this kind is denoted by V_n. It is easy to see that this is the value of the root of a tree of height n which follows an incremental model with edge distribution F. Thus, $V_0 \equiv 0$. Clearly, we have the following distributional identities:

$$V_n \stackrel{\mathcal{L}}{=} \begin{cases} \max_{1 \leq j \leq b} \{V_{n-1,j} + X_j\} & \text{if } n \text{ is even} \\ \min_{1 \leq j \leq b} \{V_{n-1,j} + X_j\} & \text{if } n \text{ is odd} \end{cases} ,$$

where X_j denotes an independent copy of the random variable X, and $V_{n-1,j}$ denotes an independent copy of V_{n-1}. Let F_n be the distribution function of V_n:

$$F_n(x) = \mathbf{P}\{V_n \leq x\} .$$

Clearly, we see that

$$F_0(i) = \begin{cases} 0 & \text{if } i < 0 \\ 1 & \text{if } i \geq 0 \end{cases} .$$

When X is a continuous random variable, the distribution function of $V_{2n-1,j}+X_j$ is the convolution of F and F_{2n-1}. Thus, we have the following relations:

$$(3) \quad \begin{aligned} F_{2n}(x) &= \left(\int F_{2n-1}(x-t)dF(t)\right)^b , \\ F_{2n+1}(x) &= 1 - \left(1 - \int F_{2n}(x-t)dF(t)\right)^b . \end{aligned}$$

When X is an integer-valued random variable, and f is the discrete probability density ($f(j) = \mathbf{P}\{X = j\}$),

$$(4) \quad \begin{aligned} F_{2n}(i) &= \left(\sum_j f(j)\ F_{2n-1}(i-j)\right)^b , \\ F_{2n+1}(i) &= 1 - \left(1 - \sum_j f(j)\ F_{2n}(i-j)\right)^b . \end{aligned}$$

3. Limit of $F_{2n}(0)$ as n tends to infinity. The behavior of incremental trees when the edge random variables are mostly zero is peculiar and forms the basis of further analysis in future sections. We assume that X is an integer-valued random variable. Let F be the distribution function of X and k be the smallest non-zero value taken by X. We define $p = \mathbf{P}\{X > 0\} = F(k)$. Then V_n is clearly stochastically bigger than kV'_n where V'_n is the root's value of the Bernoulli tree with parameter p. We prove the following theorem.

THEOREM 1. *For all b there exists $\alpha \in (0,1)$ such that, for $p \in [0,\alpha]$,*

$$\lim_{n\to\infty} F_{2n}(0) > 0,$$

and

$$\lim_{n\to\infty} F_{2n+1}(0) > 0.$$

And for $p > \alpha$

$$\lim_{n\to\infty} F_{2n}(0) = \lim_{n\to\infty} F_{2n+1}(0) = 0 .$$

Finally,

$$\alpha \le 1 - b^{-1/(b+1)} \overset{b\to\infty}{\longrightarrow} 0.$$

Proof. We first prove Theorem 1 for X Bernoulli(p). Then using the remark given at the beginning of the chapter the theorem follows immediately for general discrete positive random variable. Let $p \in [0,1)$. Use recurrences (4) for $i = 0$:

$$\begin{aligned} F_0(0) &= 1, \\ F_{2n}(0) &= \Big((1-p)F_{2n-1}(0)\Big)^b, \\ F_{2n+1}(0) &= 1 - \Big(1-(1-p)F_{2n}(0)\Big)^b . \end{aligned}$$

Combining all this, we note that for $n \ge 1$,

$$F_{2n}(0) = (1-p)^b \left(1 - \Big(1-(1-p)F_{2n-2}(0)\Big)^b\right)^b \stackrel{\text{def}}{=} G_0(F_{2n-2}(0)) ,$$

where $G_0(x) = (1-p)^b\Big(1-\big(1-(1-p)x\big)^b\Big)^b$. This is a simple functional iteration, the solution of which depends upon the behavior of the mapping G_0. $G_0(x)$ is an order b^2 polynomial that is a strictly increasing mapping: $[0,1] \to [0,1]$, since $G_0(0) = 0$ and $G_0(1) \le 1$. Then $F_{2n}(0)$ is decreasing and converges to L_0, the greatest fixed point on $[0,1]$. Define the set of p such that $G_0(x)$ has a non-zero fixed point on $[0,1]$:

$$\Gamma = \{p \in [0,1] | L_0 > 0\}.$$

Define also

$$h(x,p) = \frac{G_0(x) - x}{x}.$$

As $G_0(0) = 0$, h is a (b^2-1)-th order polynomial function of p and x. As $G_0'(0) = 0$, the derivative of $G_0(x) - x$ is equal to -1 at $x = 0$, and thus zero is a simple root of $G_0(x) - x$ and it is not a root of h. Thus we have,

$$\Gamma = \{p \in [0,1] | h(x,p) \text{ has a root in } [0,1]\}.$$

Since h is continuous, the inverse image of $\{0\}$ is a closed set of $\Re^2$, and Γ too is a closed set. Since G_0 is decreasing in p, h is also decreasing in p. We also have that $0 \in \Gamma$ since $h(1,0) = 0$. We will prove that there exists $\alpha \in \Re$ such that $\Gamma = [0,\alpha]$. We already know that Γ is a closed set containing zero. Thus, we just have to prove that Γ is convex. Assume that $p \in \Gamma$. Thus there exists $L_0 > 0$ such that $h(L_0,p) = 0$. Then for all $p' \in (0,p]$ we have,

$$h(L_0,p') \geq 0, \; h(1,p') \leq 0.$$

Thus $h(x,p')$ has a non-zero root in $[L_0,1]$ and $p' \in \Gamma$. This implies that Γ is convex. Thus Theorem alphatheo is proved for $p \leq \alpha$.

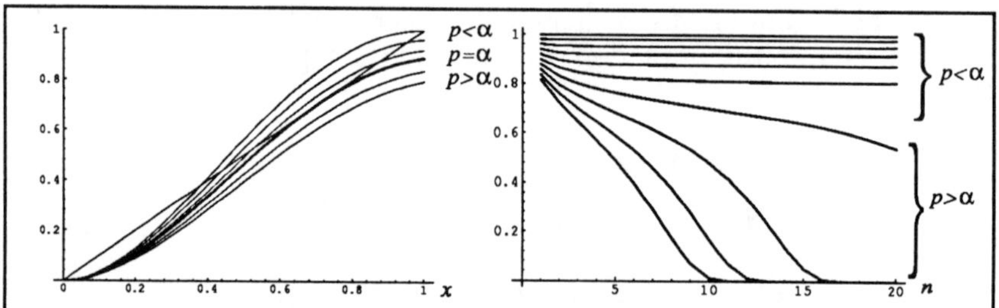

$G_0(x)$ for b=2 and p between 0 an 0.1. $F_{2n}(0)$ for p=0 to p=0.1, b=2.

If $p > \alpha$ then G_0 has only zero as fixed point. Thus $F_{2n}(0)$ converges to zero. We get similar results with $F_{2n+1}(0)$. It tends to a positive limit if an only if $p \in [0,\alpha]$. These facts can be shown using

$$F_{2n+1}(0) = 1 - \left(1 - (1-p)F_{2n}(0)\right)^b.$$

We still have to prove that $\alpha > 0$. For $p = 0$, we have

$$G_0(x) = \left(1 - (1-x)^b\right)^b, \; G_0(1) = 1, \; G_0'(1) = 0 < 1.$$

As G_0 is differentiable and $G_0'(1) = 0$, there exists $0 < y < 1$ such that $G_0(y) > y$. And by continuity of G_0 in p, there exists an $\varepsilon > 0$ such that for all $p < \varepsilon$, $G_0(y) > y$. This implies that for $p < \varepsilon$, G_0 has a fixed point on $(0,1)$ and then we have $\alpha > \varepsilon > 0$. Thus the first part of Theorem 1 is proved.

For general $b \geq 2$, we derive an upper bound for α. We have :

$$\begin{aligned} G_0(x) &= q^b\left(1-\left(1-qx\right)^b\right)^b \\ &\leq q^b(1-(1-qx)^b) \\ &\leq q^b(1-(1-bqx)) \\ &= bq^{b+1}x. \end{aligned}$$

Thus if $bq^{b+1} < 1$, $G_0(x) < x$ for all $x > 0$. This implies that it cannot have a non-zero fixed point and thus $p > \alpha$. Thus we have

$$\alpha \leq 1 - \sqrt[b+1]{\frac{1}{b}}.$$

This implies that α tends to zero when b tends to infinity. □

REMARK 1. For $b = 2$,

$$\alpha = 1 - \sqrt[3]{\left(\frac{27}{32}\right)} \approx 0.05506.$$

4. The main theorem of convergence. THEOREM 2. *For the incremental tree with bounded edge variable X, $\boldsymbol{E}V_n/n$ converges to a finite limit $\mathcal{V}(X)$ as n tends to infinity.*

The proof will be spread over the next two sections. Assume without loss of generality that for some finite μ, $0 \leq X \leq \mu$. It is helpful to introduce an associated tree T' in which we introduce new node values $V'(u)$. However, at every node, we force $V'(u) \geq V(u)$. First fix the integers $N \geq 1$ and $k \geq 1$. The leaves of T' have value zero. At any level i that is not a multiple of N, we follow the standard rules (2) as for an incremental tree with edge variable X. If i is a multiple of N, say $i = lN$, then we set for any node u at level i

$$W(u) = \begin{cases} \max_{v \in A_u} \{V'(v) + E(u,v)\} & \text{if } i \text{ is even} \\ \min_{v \in A_u} \{V'(v) + E(u,v)\} & \text{if } i \text{ is odd} \end{cases},$$

and,

$$V'(u) = \begin{cases} l\boldsymbol{E}V_N + (2l-1)k\mu & \text{if } W(u) \leq l\boldsymbol{E}V_N + (2l-1)k\mu \\ \infty & \text{if } W(u) > l\boldsymbol{E}V_N + (2l-1)k\mu \end{cases}.$$

Note that many nodes may have the value infinity. We call T' the (k, N) associated tree. Let V_n' be the random variable defined as the root value of such a tree with n levels of edges. Note that for all nodes at levels that are multiple of N, the values of $V'(u)$ are either ∞ or a given fixed finite value that is the same at that level. This device already hints at the behavior of incremental trees: all nodes at a given level have approximately equal values—no serious imbalances occur. The entire paper rests on three technical inequalities that deserve of a section of their own.

5. The fundamental inequalities. The first result states that it is "doubly exponentially" unlikely that the root of a Pearl tree with Bernoulli leaf values takes the value one, if the Bernoulli parameter is small enough.

LEMMA 1. *Let T be a b-ary Pearl tree with Bernoulli (q) leaf values, where*

$$q \leq \xi \stackrel{\text{def}}{=} \frac{1}{2} b^{-b} .$$

Then, regardless of whether we begin with either MIN *or* MAX *nodes, and regardless of the parity of n,*

$$\boldsymbol{P}\{V_n = 1\} \leq b 2^{-b^{\lfloor n/2 \rfloor}} .$$

Proof. V_n is maximal if we begin with a MAX level. For a leaf value V_0, we have $\boldsymbol{P}\{V_0 = 1\} = q$. Define $p_n = \boldsymbol{P}\{V_n = 1\}$. Then the following recursion holds: $p_0 = q$, and

$$p_{2n} = \left(1 - (1 - p_{2n-2})^b\right)^b \leq (b p_{2n-2})^b .$$

This yields

$$p_{2n} \leq b^{b^{n+1}} q^{b^n} .$$

For p_{2n+1} we get

$$p_{2n+1} \leq 1 - \left(1 - b^{b^{n+1}} q^{b^n}\right)^b \leq b b^{b^{n+1}} q^{b^n} .$$

Let q such that $b^b q \leq \frac{1}{2}$. Then

$$p_{2n} \leq 2^{-b^n}, \text{ and } p_{2n+1} \leq b 2^{-b^n},$$

and in general, regardless of whether we start with a MIN or a MAX level,

$$p_n \leq b 2^{-b^{\lfloor n/2 \rfloor}} .$$

□

The second technical result establishes quite simply that V_n is close to $\boldsymbol{E}V_n$ in all circumstances.

LEMMA 2. *Let $0 \le X \le \mu$ in a random incremental tree. For all $\varepsilon > 0$,*

$$\boldsymbol{P}\{|V_n - \boldsymbol{E}V_n| \ge \varepsilon\} \le 2e^{-2\varepsilon^2/n\mu^2} .$$

Proof. At the ith level of edges, starting from the topmost level, we find b^i independent edge values. These are collected in a random vector U_i. Clearly then, $V(u) = f(U_1, \ldots, U_n)$ for some function f. Furthermore, if U_i is replaced by a different vector U_i', $V(u)$ changes by at most μ. Thus, we can apply the McDiarmid's inequality (1989)[McDi89]: for all $\varepsilon > 0$,

$$\boldsymbol{P}\{|V_n - \boldsymbol{E}V_n| \ge \varepsilon\} \le 2e^{-2\varepsilon^2/n\mu^2} . \tag{5}$$

□

Our third inequality is fundamental in proving that V_n increases about linearly in n. It states that a node at level lN has a value not much larger than l times the value of a node at level N. The explicit non-asymptotic nature of the bound will be helpful as well.

LEMMA 3. *Let $0 \le X \le \mu$ in a random incremental tree. For N large enough and for all $l > 0$,*

$$\boldsymbol{P}\{V_{lN} \ge l\boldsymbol{E}V_N + (2l-1)k\mu\} \le b^{N-N^{2/3}} b\, 2^{-b^{\frac{N^{2/3}}{2}-1}} + 2e^{-2N^{1/3}} \stackrel{\text{def}}{=} R(b,N),$$

where $k = \lceil N^{2/3} \rceil$. Finally for all $\varepsilon > 0$ there exists an N such that for all $n > N$,

$$\boldsymbol{P}\{|V_n - l\boldsymbol{E}V_N| \ge (2l-1)k\mu + N\mu\} \le \varepsilon,$$

where $l = \lfloor n/N \rfloor$ and $k = \lceil N^{2/3} \rceil$.

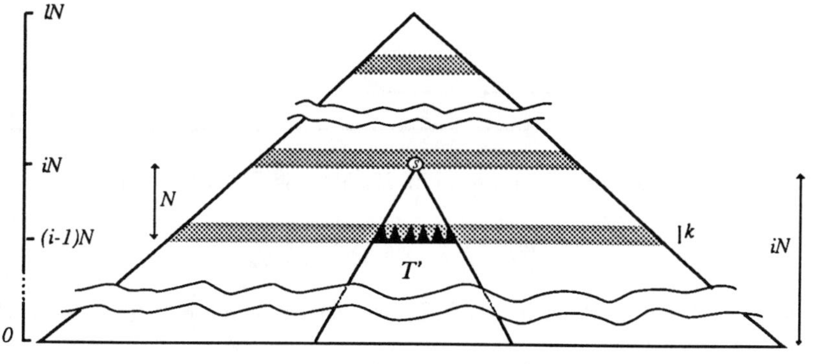

The construction

Proof. Assume that the tree has $n = lN$ levels. We consider the (k, N) associated tree with $k = \lceil N^{2/3} \rceil$. Let N be so large that $R(b, N) \le (1/2)b^{-b}$. We prove by induction that for such N and for all $i \ge 1$,

$$\boldsymbol{P}\{V'_{iN} = \infty\} \le R(b, N). \tag{6}$$

For $i = 1$, we obtain

$$\begin{aligned} \boldsymbol{P}\{V'_N = \infty\} &= \boldsymbol{P}\{V_N > \boldsymbol{E}V_N + k\mu\}, \\ &\le 2e^{-2(k\mu)^2/N\mu^2} \qquad \text{(according to (5))} \\ &\le 2e^{-2N^{1/3}} \\ &\le R(b, N). \end{aligned}$$

Now we assume that $\boldsymbol{P}\left\{V'_{(i-1)N} = \infty\right\} \le R(b, N)$. The nodes at level iN are i.i.d. distributed as V'_{iN}. Let T' be an associated tree with iN levels and s its root node. Then look at the $m < N$ levels of this tree from depth $N - m$ to depth N. This part consists of b^{N-m} m-level subtrees as shown in the next figure. Let T_m be one of these subtrees and let v_m be its root. Thus $V'(v_m)$ is distributed as $V'_{(i-1)N+m}$. The leaves w of T_m are nodes of T' at level $(i-1)N$. Thus their values $(V'(w))$ are bi-valued i.i.d. random variables distributed as $V'_{(i-1)N}$.

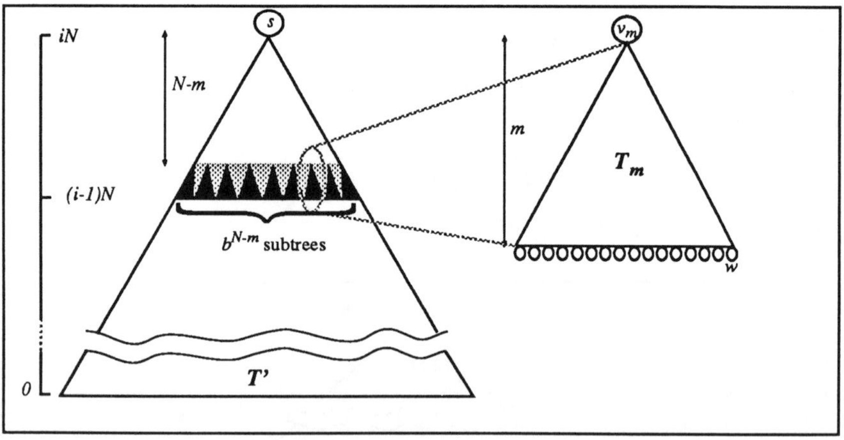

We assign to each leaf node w of T_m a value $V''(w)$ as follows:

$$V''(w) = \begin{cases} \infty & \text{if } V'(w) = \infty \\ 0 & \text{if } V'(w) < \infty \end{cases}.$$

And to each internal node u of T_m we assign a value $V''(u)$ using the MIN-MAX rules 1:

$$V''(u) = \begin{cases} \max_{v \in A_u} \{V''(v)\} & \text{if } u \text{ is a MAX node of } T' \\ \min_{v \in A_u} \{V''(v)\} & \text{if } u \text{ is a MIN node of } T' \end{cases}.$$

Then $V''(v_m)$ is distributed as the root of a m-level b-ary Pearl tree, where the leaves take value ∞ with probability $q = \boldsymbol{P}\{V''(w) = \infty\}$. The bottom level is a MIN or a MAX according to the parity of $(i-1)N$. Thus as $q = \boldsymbol{P}\{V''(w) = \infty\} = \boldsymbol{P}\left\{V'_{(i-1)N} = \infty\right\} \le R(b, N) \le (1/2)b^{-b}$, by Lemma 1 about Pearl trees,

$$\boldsymbol{P}\{V''(v_m) = \infty\} \le b2^{-b^{\lfloor m/2 \rfloor}}.$$

Let u be an internal node of T_m. As it is not at a level that is a multiple of N in T', $V'(u)$ is computed with the standard rules of the incremental model. Thus $V'(u)$ is infinity if and only if $V''(u)$ is infinity. If $V''(v_m) = 0$, then

$$V'(v_m) \le (i-1)\boldsymbol{E}V_N + (2i-3)k\mu + m\mu. \tag{7}$$

Thus,

$$\boldsymbol{P}\{V'(v_m) = \infty\} \le b2^{-b^{\lfloor m/2 \rfloor}}.$$

Furthermore we have

$$\begin{aligned} Q_m &\stackrel{\text{def}}{=} \boldsymbol{P}\{V'(v_m) = \infty \text{ for at least one node } v_m \text{ at depth } N-m \text{ from the top of } T'\} \\ &\le b^{N-m} b 2^{-b^{\lfloor m/2 \rfloor}}. \end{aligned}$$

Now we take $m = k = \lceil N^{2/3} \rceil$. If there is no infinity node at depth $N-k$ in T', then each $V'(v_m)$ is less than $(i-1)\boldsymbol{E}V_N + (2i-3)k\mu + k\mu$, and $V'(s)$ is stochastically less than $V'(v_m) + V_N$. Thus

$$\begin{aligned} \boldsymbol{P}\{V'_{iN} = \infty\} &= \boldsymbol{P}\{V'(s) > i\boldsymbol{E}V_N + (2i-1)k\mu\} \\ &\le Q_k + \boldsymbol{P}\{(i-1)\boldsymbol{E}V_N + (2i-3)k\mu + k\mu + V_N \\ &\qquad > i\boldsymbol{E}V_N + (2i-1)k\mu\} \\ &\le Q_k + \boldsymbol{P}\{V_N > \boldsymbol{E}V_N + k\mu\} \\ &\le b^{N-N^{2/3}} b\, 2^{-b^{\frac{N^{2/3}}{2}-1}} + 2e^{-2N^{1/3}} \\ &= R(b, N) \\ &\le \frac{1}{2} b^{-b}. \end{aligned}$$

Thus the induction proof of (6) is finished and we have

$$\boldsymbol{P}\{V_{lN} \ge l\boldsymbol{E}V_N + (2l-1)k\mu\} \le R(b, N).$$

Next, we are left with a minor cleanup to handle the case when n is not a multiple of N. Now we consider b-ary incremental trees with n levels, n not a multiple of N, and we set $l = \left\lfloor \frac{n}{N} \right\rfloor$. Using 7 with $m = n - lN < N$ we get

$$\boldsymbol{P}\{V_n \ge l\boldsymbol{E}V_N + (2l-1)k\mu + m\mu\} \le \boldsymbol{P}\{V''(v_m) = \infty\} \le b2^{-b^{\lfloor m/2 \rfloor}}.$$

Thus if $N > m \geq N^{1/4}$ we have,

$$\boldsymbol{P}\left\{V_n \geq l\boldsymbol{E}V_N + (2l-1)k\mu + N\mu\right\} \leq b2^{-b\lfloor N^{1/4}\rfloor/2}. \tag{8}$$

If $m \leq N^{1/4}$ the probability that $V_n \geq l\boldsymbol{E}V_N + (2l-1)k\mu + N\mu$ is less than the probability that there is at least an infinity node at level lN of the associated tree. Thus,

$$\begin{aligned}\boldsymbol{P}\left\{V_n \geq l\boldsymbol{E}V_N + (2l-1)k\mu + N\mu\right\} &\leq b^m R(b,N)\\ &\leq b^{N^{1/4}} R(b,N).\end{aligned} \tag{9}$$

Finally using (8) and (9) we have for all n,

$$\boldsymbol{P}\left\{V_n \geq l\boldsymbol{E}V_N + (2l-1)k\mu + N\mu\right\} \leq b2^{-b\left\lfloor N^{1/4}\right\rfloor} + b^{N^{1/4}} R(b,N).$$

The right hand side tends to zero when N tends to infinity. Thus, for all $\varepsilon > 0$, there exists an N such that for $n > N$,

$$\boldsymbol{P}\left\{V_n \geq l\boldsymbol{E}V_N + (2l-1)k\mu + N\mu\right\} \leq \varepsilon,$$

where $k = \lceil N^{2/3}\rceil$ and $l = \lfloor n/N \rfloor$. □

6. Convergence: proof of Theorem 2. We show that

$$\limsup_{n\to\infty} \frac{\boldsymbol{E}V_n}{n} \leq \liminf_{n\to\infty} \frac{\boldsymbol{E}V_n}{n}, \tag{10}$$

by showing that for given $\varepsilon > 0$,

$$\limsup_{n\to\infty} \frac{\boldsymbol{E}V_n}{n} \leq \frac{\boldsymbol{E}V_N}{N} + 3\varepsilon$$

for all N large enough. Then, by definition of the limit infimum we can find an N so large that

$$\frac{\boldsymbol{E}V_N}{N} \leq \liminf_{n\to\infty} \frac{\boldsymbol{E}V_n}{n} + \varepsilon,$$

so that we may conclude (10) by the arbitrary nature of ε.

We use the notation of the preceding part: u is the root of an n-level complete b-ary incremental tree with edge variable X, and $V'(u)$ is its value for the (k,N) associated tree. Define $l = \lfloor n/N \rfloor$. If $V'(u) \leq N\mu + l\boldsymbol{E}V_N + (2l-1)k\mu$ then

$$\begin{aligned}V(u) &\leq N\mu + \left\lfloor \frac{n}{N}\right\rfloor \boldsymbol{E}V_N + \left(2\left\lfloor\frac{n}{N}\right\rfloor - 1\right)k\mu\\ &\leq N\mu + \frac{n}{N}\boldsymbol{E}V_N + 2\frac{nk}{N}\mu,\end{aligned}$$

so that (recalling $k = \lceil N^{2/3} \rceil$),

$$\begin{aligned} \frac{V(u)}{n} &\le \frac{N}{n}\mu + \frac{\boldsymbol{E}V_N}{N} + 2\frac{k}{N}\mu \\ &\le \frac{N}{n}\mu + \frac{\boldsymbol{E}V_N}{N} + 2\left(\frac{N^{2/3}+1}{N}\right)\mu \\ &\le \frac{\boldsymbol{E}V_N}{N} + 2\varepsilon \qquad (11) \end{aligned}$$

for N large enough and $n \ge N/\mu\varepsilon$. Using Lemma 3, we can find N large enough such that

$$\boldsymbol{P}\{V_n \ge l\boldsymbol{E}V_N + (2l-1)k\mu + N\mu\} < \varepsilon$$

for all p and for $n > N$. Thus we have,

$$\begin{aligned} \frac{\boldsymbol{E}V_n}{n} &\le \boldsymbol{P}\left\{\frac{V_n}{n} > \frac{\boldsymbol{E}V_N}{N} + 2\varepsilon\right\} + \frac{\boldsymbol{E}V_N}{N} + 2\varepsilon \\ &\le \boldsymbol{P}\{V_n \le l\boldsymbol{E}V_N + (2l-1)k\mu + N\mu\} + \frac{\boldsymbol{E}V_N}{N} + 2\varepsilon \\ &\le \frac{\boldsymbol{E}V_N}{N} + 3\varepsilon \end{aligned}$$

for N large enough and $n \ge N/\mu\varepsilon$. This implies that

$$\limsup_{n\to\infty} \frac{\boldsymbol{E}V_n}{n} \le \frac{\boldsymbol{E}V_N}{N} + 3\varepsilon$$

as required. Thus $\boldsymbol{E}V_n/n$ has a limit $\mathcal{V}(X)$ when n tends to infinity. □

7. A law of large numbers. THEOREM 3. *If $X \in [0,\mu]$ and $\boldsymbol{P}\{X > 0\} > \alpha$, where α is defined in Theorem 1, we have*

$$\lim_{n\to\infty} \frac{\boldsymbol{E}V_n}{n} = \mathcal{V}(X) > 0$$

and

$$\lim_{n\to\infty} \frac{V_n}{\boldsymbol{E}V_n} = \lim_{n\to\infty} \frac{V_n}{n\mathcal{V}(X)} = 1$$

almost surely as n tends to infinity.

Proof. We first prove this lemma for Bernoulli trees with parameter p. Let u be a node of T in an n-level complete b-ary incremental tree with parameter p. We associate with u the value $V'(u)$ related to $V(u)$ by monotonicity: $V'(u) \le V(u)$. The idea is to cut the tree into pieces of N levels each, and for every second piece, we force all edge values to be zero. On those pieces, we use results about Pearl trees. The exact definition of

$V'(u)$ is given below. We denote by V'_n the value of the root of the n-level model. We will show that $\liminf_{n\to\infty} \boldsymbol{E}V'_n/n > 0$.

Let N be a large fixed positive integer. For all nodes u at level i, we determine $V'(u)$ from $V'(v), v \in A_u$ as follows for $l = \lceil n/(2N) \rceil$:

1) If $(2l-2)N < i < (2l-1)N$, then $V'(u)$ is determined from $V'(v), v \in A_u$ as in the incremental tree with parameter p.

2) If $i = (2l-1)N$, then first $W(u)$ is determined from $V'(v), v \in A_u$, by MIN-MAX rules as in the incremental tree with parameter p, and we set

$$V'(u) = \begin{cases} -\infty & \text{if } W(u) < l \\ l & \text{if } W(u) \geq l \end{cases}.$$

(Thus, at this level, $V'(u)$ is bi-valued!)

3) If $(2l-1)N < i \leq 2lN$, then the edge values are considered to be zero and thus $V'(u)$ is determined by the MIN-MAX rules

$$V'(u) = \begin{cases} \max_{v\in A_u} \{V'(v)\} & \text{if } u \text{ at even level in } T \\ \min_{v\in A_u} \{V'(v)\} & \text{if } u \text{ at odd level in } T \end{cases} \tag{12}$$

It is easy to verify by induction that $V'(u) \leq V(u)$ for every node. Now we prove by induction that if $p > \alpha$, for all $\varepsilon > 0$ we can find N such that for all integer $l > 0$ we have

$$\boldsymbol{P}\{V'_{lN} = -\infty\} \leq \min\left(\varepsilon, \frac{1}{2} b^{-b}\right). \tag{13}$$

For $l = 1$, this is true since

$$\boldsymbol{P}\{V'_N = -\infty\} = \boldsymbol{P}\{W_N = 0\} = F_N(0)$$

where F_N is the distribution function of the value of the root of an N-level incremental tree with parameter p, and $W_N = W(u)$ is the value of the root of this tree. (Recall that for $p > \alpha$, $F_N(0) \to 0$ as $N \to \infty$.) Thus we choose N so large that

$$F_N(0) < \frac{1}{2} \min\left(\varepsilon, \frac{1}{2} b^{-b}\right).$$

For the induction we have to distinguish between two cases. First we consider a node u at a level $2lN$. We assume (13) to be true for all $l' < 2l$. All nodes v at level $(2l-1)N$ have a value $V'(v)$ equal to l or $-\infty$. We consider the N-level subtree T_N rooted at the node u and in which the leaf values are the $V'(v)$ from level $(2l-1)N$ of T. Also $V'(u)$ is distributed as the root of a Pearl tree where the leaves have value l or $-\infty$. By the induction hypothesis the value $-\infty$ occurs with probability $q \leq (1/2)b^{-b}$. Thus by Lemma 1 we have,

$$\boldsymbol{P}\{V'(u) = -\infty\} \leq b2^{-b^{\lfloor N/2 \rfloor}} \leq \min\left(\varepsilon, \frac{1}{2} b^{-b}\right), \quad \text{for } N \text{ large enough.}$$

For all $m \geq 0$,

$$\text{(14)} \qquad \boldsymbol{P}\{V_{2lN+m} < l\} \leq \min\left(\varepsilon, \frac{1}{2}b^{-b}\right).$$

This concludes the first part. Let us now consider a node u at level $(2l - 1)N$. According to the hypothesis, at level $(2l-2)N$, there are nodes with value $(l-1)$ and nodes with value $-\infty$. The probability that at least one node v at level $(2l-2)N$ has value $V'(v) = -\infty$ is less than $b^N b 2^{-b^{\lfloor N/2 \rfloor}}$. If the b^N nodes at level $(2l-2)N$ have the value $l-1$, then $\boldsymbol{P}\{V'(u) = -\infty\} = \boldsymbol{P}\{W_N = 0\} = F_N(0)$. Thus if we choose N such that $b^{N+1}2^{-b^{\lfloor N/2 \rfloor}} \leq \frac{1}{2}\min(\varepsilon, \frac{1}{2}b^{-b})$,

$$\begin{aligned}
\boldsymbol{P}\{V'(u) = -\infty\} &\leq b^{N+1}2^{-b^{\lfloor N/2 \rfloor}} + F_N(0) \\
&< \frac{1}{2}\min\left(\varepsilon, \frac{1}{2}b^{-b}\right) + \frac{1}{2}\min\left(\varepsilon, \frac{1}{2}b^{-b}\right) \\
&= \min\left(\varepsilon, \frac{1}{2}b^{-b}\right).
\end{aligned}$$

Thus the induction is shown and we have for all integer l,

$$\boldsymbol{P}\{V'_{lN} = -\infty\} \leq \varepsilon.$$

Thus for all $l > 0$,

$$\boldsymbol{P}\{V_{2lN} \geq l\} \geq \boldsymbol{P}\{V'_{2lN} = l\} \geq 1 - \varepsilon.$$

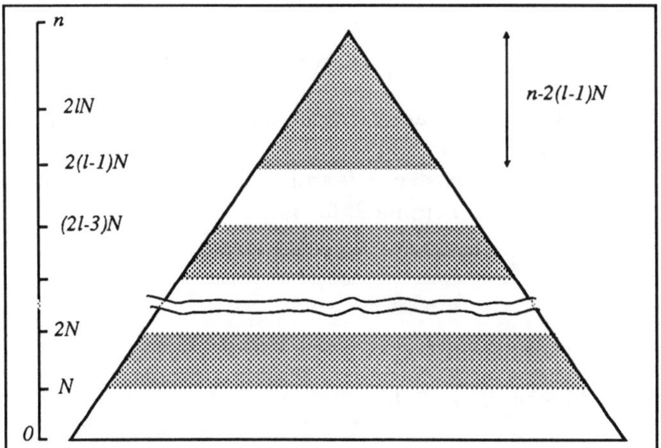

Inside the gray parts we assume the edge values to be zero.

We now generalize the result for incremental trees with a number of levels that is not a multiple of N. Let u be the root of T, an n-level b-ary incremental tree with parameter $p > \alpha$. Let $l = \lfloor n/2N \rfloor$. Using (14), with $m = n - 2lN$, we have

$$\boldsymbol{P}\{V(u) < l\} \leq \min\left(\varepsilon, \frac{1}{2}b^{-b}\right).$$

Thus,

$$\boldsymbol{E}V_n \geq l(1-\varepsilon).$$

As a consequence,

$$\frac{\boldsymbol{E}V_n}{n} \geq (1-\varepsilon)\left(\frac{n}{2N} - 1\right)\frac{1}{n} \geq \frac{1-\varepsilon}{2N} - o(1)\ ,$$

so that

$$\liminf_{n\to\infty} \frac{\boldsymbol{E}V_n}{n} \geq \frac{1-\varepsilon}{2N},$$

and finally

$$\lim_{n\to\infty} \frac{\boldsymbol{E}V_n}{n} > 0.$$

We now consider X to be a positive random variable bounded by μ such that

$$p \stackrel{\text{def}}{=} \boldsymbol{P}\{X > 0\} > \alpha\ .$$

Then there exists $\delta > 0$ such that

$$\boldsymbol{P}\{X \geq \delta\} > \alpha.$$

Then

$$\mathcal{V}(X) \geq \delta\mathcal{V}(p) > 0.$$

The theorem follows immediately. In particular, whenever X is continuous, $\boldsymbol{P}\{X > 0\} = 1$, and thus $\mathcal{V}(X) > 0$.

For all $p > \alpha$ there exist $c > 0$ and $n_0 > 1$ such that for all $n > n_0$, $\boldsymbol{E}V_n > cn$. According to Lemma 2, for any p, and any $\varepsilon > 0$,

$$\boldsymbol{P}\{|V_n - \boldsymbol{E}V_n| \geq \varepsilon\} \leq 2e^{-2\varepsilon^2/n\mu^2}.$$

Thus for $n > n_0$,

$$\begin{aligned}\boldsymbol{P}\left\{\left|\frac{V_n}{\boldsymbol{E}V_n} - 1\right| \geq \varepsilon\right\} &= \boldsymbol{P}\{|V_n - \boldsymbol{E}V_n| \geq \varepsilon\boldsymbol{E}V_n\} \\ &\leq \boldsymbol{P}\{|V_n - \boldsymbol{E}V_n| \geq \varepsilon cn\} \\ &\leq 2^{-2\varepsilon^2c^2n/\mu^2}.\end{aligned}$$

Thus by the Borel-Cantelli Lemma, $V_n/\boldsymbol{E}V_n \to 1$ when $n \to \infty$ almost surely. □

8. Robustness, continuity, and embedding. A natural question is to ask what happens for two incremental trees that are very much alike. Clearly, if the incremental model is to be widely accepted, it should have a certain robustness property with respect to small changes in the edge random variable X. In this section, a simple method is introduced for analyzing this sort of situation. For starters, we give the key technical result.

LEMMA 4. *Consider a complete b-ary tree with n levels of edges. With each edge e of this tree we associate a uniform* $[0,1]$ *random variable* U_e. *Denote by* $\mathcal{P}_n$ *the collection of all* b^n *paths from the root to the leaves. Let* $\varphi = \varphi(\varepsilon)$ *be defined as follows:*

$$\varphi = \inf \left\{ x : 1 > x \geq \varepsilon, b \left(\frac{1-\varepsilon}{1-x} \right)^{1-x} \left(\frac{\varepsilon}{x} \right)^x \leq 1 \right\}.$$

Then, for all $\varepsilon > 0$,

$$\limsup_{n \to \infty} \frac{1}{n} \boldsymbol{E} \left\{ \max_{P \in \mathcal{P}_n} \sum_{e \in P} I_{[U_e \leq \varepsilon]} \right\} \leq \varphi(\varepsilon).$$

Furthermore, $\varphi(\varepsilon) \to 0$ *as* $\varepsilon \to 0$. *For* $\varepsilon \geq 1/b$, $\varphi(\varepsilon) = 1$, *while for* $\varepsilon < 1/b$, $\varphi(\varepsilon) < 1$.

REMARK: BRANCHING RANDOM WALKS. We need the explicit bound of the previous Lemma in what follows. However, note that by the theory of maxima in branching random walks, it easily follows that

$$\lim_{n \to \infty} \frac{1}{n} \boldsymbol{E} \left\{ \max_{P \in \mathcal{P}_n} \sum_{e \in P} I_{[U_e \leq \varepsilon]} \right\} = \varphi(\varepsilon).$$

See for example the work by Hammersley [Ha74], Kingman [Ki75] or Biggins [Bi76,Bi77].

Proof. For $\varepsilon \geq 1/b$ the statement is trivial. We assume $\varepsilon < 1/b$. For every $P \in \mathcal{P}_n$, $B = \sum_{e \in P} I_{[U_e \leq \varepsilon]}$ is binomial (n, ε) distributed. Thus, by Bonferroni's inequality, for $1 > x \geq \varepsilon$,

$$\begin{aligned} \boldsymbol{P} \left\{ \max_{P \in \mathcal{P}_n} \sum_{e \in P} I_{[U_e \leq \varepsilon]} \geq xn \right\} &\leq \sum_{P \in \mathcal{P}_n} \boldsymbol{P}\{B \geq xn\} \\ &\leq b^n \left(\left(\frac{1-\varepsilon}{1-x} \right)^{1-x} \left(\frac{\varepsilon}{x} \right)^x \right)^n, \end{aligned}$$

where we use Chernoff's bound for the tail of a binomial distribution (see for example Hoeffding, [Hoeff63], Theorem 1). We denote

$$H(\varepsilon, x) = \left(\frac{1-\varepsilon}{1-x} \right)^{1-x} \left(\frac{\varepsilon}{x} \right)^x.$$

Thus $\varphi(\varepsilon)$ is the smallest solution greater than ε and smaller than 1 of

$$H(\varepsilon, x) = 1/b.$$

It is a simple analytical exercise to show that $H(\varepsilon, x)$ is monotonically decreasing from 1 at $x = \varepsilon$ to ε at $x = 1$ (see figure below).

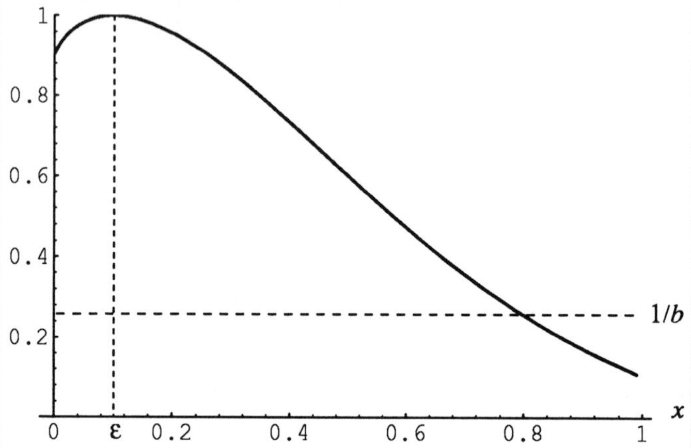

$H(\varepsilon{,}x)$, for ε=0.1

We see that $\varphi(\varepsilon)$ is well-defined and that for $\varepsilon < 1/b$, $\varepsilon < \varphi(\varepsilon) < 1$. Furthermore,

$$\varphi(\varepsilon) \to 0 \text{ as } \varepsilon \to 0$$

because

$$H(\varepsilon, f(\varepsilon)) \sim \varepsilon^{f(\varepsilon)} \to 0 \text{ as } \varepsilon \to 0$$

for any increasing function $f(\varepsilon)$ with $f(\varepsilon)\log(1/\varepsilon) \to \infty$, and $f(\varepsilon) \to 0$, as $\varepsilon \to 0$ ($f(\varepsilon) = 1/\sqrt{\log(1/\varepsilon)}$ will do). For ε small enough, $\varphi(\varepsilon) \le f(\varepsilon) \to 0$. We have for all $\delta > 0$,

$$\begin{aligned}\boldsymbol{E}\left\{\max_{P\in\mathcal{P}_n} \frac{\sum_{e\in P} I_{[U_e\le\varepsilon]}}{n}\right\} &\le \boldsymbol{P}\left\{\max_{P\in\mathcal{P}_n} \sum_{e\in P} I_{[U_e\le\varepsilon]} \ge (\varphi(\varepsilon)+\delta)n\right\} + \varphi(\varepsilon)+\delta \\ &\le o(1) + \varphi(\varepsilon) + \delta.\end{aligned}$$

By the arbitrary nature of δ, Lemma 4 follows. □

We can use this Lemma in a variety of ways.

EXAMPLE 1. Assume that we have two incremental trees with random edge variables X and Y respectively, where $\boldsymbol{P}\{X \ne Y\} = p$, and $0 \le X, Y \le \mu$.

The case of interest is when p is small. Then we may consider the complete b-ary tree with n levels of edges, in which we we give every edge e a value Z_e according to the rule

$$Z_e = \begin{cases} 0 & \text{if } X = Y \\ \mu & \text{if } X \neq Y \end{cases} .$$

Call $V_n(X)$ and $V_n(Y)$ the root values in both incremental trees. By Lemma 4,

$$|V_n(X) - V_n(Y)| \leq \max_{P \in \mathcal{P}_n} \sum_{e \in P} Z_e \leq (\varphi(p) + o(1))n\mu$$

almost surely. By a trivial argument,

$$|\mathcal{V}(X) - \mathcal{V}(Y)| \leq \varphi(p)\mu .$$

EXAMPLE 2. Assume that X and Y are Bernoulli edge variables with parameters p and q respectively. We may couple these on a common probability space. For example, we could consider a uniform $[0,1]$ random variable U associated with each edge in a complete b-ary tree. Set $X = I_{U \leq p}$ and $Y = I_{U \leq q}$. Clearly, $X \neq Y$ with probability $|p-q|$. Therefore, by Example 1,

$$|\mathcal{V}(X) - \mathcal{V}(Y)| \leq \varphi(|p-q|) .$$

The properties of φ insure that $\mathcal{V}(X)$ is a uniformly continuous function of p for Bernoulli trees.

EXAMPLE 3. For general random variables X and Y, having distribution functions F and G respectively, we may construct a common probability space based upon a uniform $[0,1]$ random variable U once again. Note that X is distributed as $F^{\text{inv}}(U)$ and Y as $G^{\text{inv}}(U)$. However, there are other kinds of couplings as well. Associate with each edge e in a complete b-ary tree a random variable Z_e, where

$$Z_e = \begin{cases} \mu & \text{if } |X-Y| > \delta ; \\ \delta & \text{if } |X-Y| \leq \delta , \end{cases}$$

where $\delta > 0$, and X and Y are coupled as above. A little thought shows that

$$\begin{aligned} |\mathcal{V}(X) - \mathcal{V}(Y)| &\leq \max_{P \in \mathcal{P}_n} \sum_{e \in P} Z_e + n\delta \\ &\leq n\mu(\varphi(\boldsymbol{P}\{|X-Y| > \delta\}) + o(1)) + n\delta \end{aligned}$$

almost surely, where we use Lemma 4. Therefore, collecting things,

$$|\mathcal{V}(X) - \mathcal{V}(Y)| \leq \inf_{\delta \geq 0} \left[\mu\varphi \left(\inf_{\text{all couplings of } X \text{ and } Y} \boldsymbol{P}\{|X-Y| > \delta\} \right) + \delta \right] .$$

This indeed shows the robustness of $\mathcal{V}(X)$ with respect to small changes in the distribution of X.

EXAMPLE 4. Given an arbitrary unbounded positive edge variable X with finite moment generating function ($\boldsymbol{E}\exp(tX) < \infty$ for some $t > 0$). Define a coupled random variable $Y = \min(X, \mu)$, where μ is large but fixed. The root values of the trees based upon Y and X may be linked in a simple manner. Using arguments not unlike those above, one can establish that $\boldsymbol{E}V_n/n \rightarrow \mathcal{V}(X) < \infty$. Thus, the boundedness condition in our main convergence theorem is not required after all.

9. Asymptotic behavior of the Bernoulli tree with parameter p. We first give a result stronger then Theorem 1 on the asymptotic behavior of V_n for Bernoulli trees when p is small. For more details, we refer to [Kam92].

LEMMA 5. *For $p \in [0, \alpha]$, there exist bona fide distribution functions F_∞ and H_∞ with finite expected values that put positive mass on all the nonnegative integers, such that*

$$\lim_{n\to\infty} F_{2n}(i) = F_\infty(i)$$

and

$$\lim_{n\to\infty} F_{2n+1}(i) = H_\infty(i) .$$

Furthermore, for $p > \alpha$, we have for all fixed $i \geq 0$,

$$\lim_{n\to\infty} F_{2n}(i) = \lim_{n\to\infty} F_{2n+1}(i) = 0 ,$$

and $V_n \rightarrow \infty$ almost surely when $n \rightarrow \infty$. Finally,

$$\alpha \leq 1 - b^{-1/(b+1)} \overset{b\to\infty}{\longrightarrow} 0.$$

We recall that if X is Bernoulli (p) and Y is Bernoulli (q), then

$$|\mathcal{V}(X) - \mathcal{V}(Y)| \leq \varphi(|p-q|) .$$

Then using Theorem 3, the following lemma is trivial.

LEMMA 6. *In a Bernoulli tree with parameter p, $\boldsymbol{E}V_n/n$ converges to a finite limit $\mathcal{V}(p)$, where $\mathcal{V}$ is a uniformly continuous function of p. Furthermore if $\alpha < p < 1-\alpha$ then $0 < \mathcal{V}(p) < 1$ and $V_n/\boldsymbol{E}V_n \rightarrow 1$ almost surely when $n \rightarrow \infty$.*

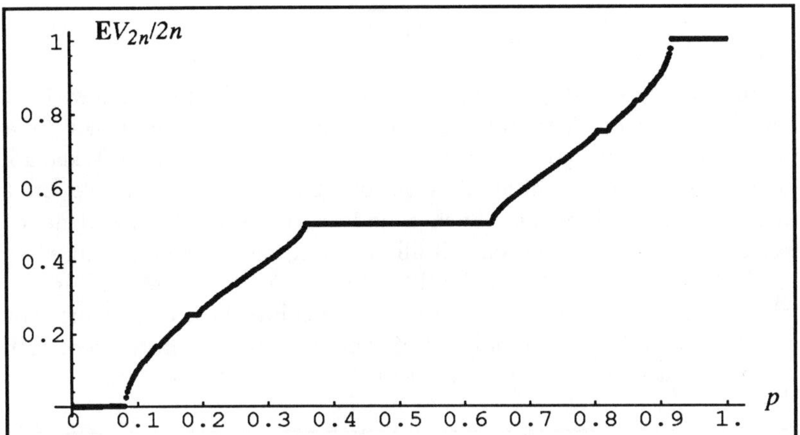

$\mathbf{E}V_{2n}/2n$ for b=10, n=200.

The above figure shows $\mathcal{V}(p)$ for $b = 10$, the flat parts close to 0 and 1 are explained by Theorem 1. Computations show that $\mathcal{V}(X)$ is flat around $p = 1/2$ as well whenever $b > 2$. In this respect, we offer the following theorem.

THEOREM 4. *For all b, $\mathcal{V}(1/2) = 1/2$ and for all p,*

$$\mathcal{V}(1-p) = 1 - \mathcal{V}(p).$$

Furthermore there exists $\beta \in (0, 1/2]$ such that

$$\mathcal{V}(p) \begin{cases} = \frac{1}{2} & \text{if } p \in [\beta, 1-\beta] \\ < \frac{1}{2} & \text{if } p \in [0, \beta) \\ > \frac{1}{2} & \text{if } p \in (1-\beta, 1] \end{cases} .$$

When $\beta < 1/2$, the range $[\beta, 1-\beta]$ is called the flat part around $p = 1/2$. For $p \in (0, 1/2)$, let $L(p, b)$ be the largest root of $1-(1-px^b)^b = x$ on $[0, 1]$. If $L(p, b) \neq 0$, then

$$\left\lfloor \frac{n}{2} \right\rfloor - \sqrt{n \log n} \leq \boldsymbol{E}V_n \leq \left\lfloor \frac{n}{2} \right\rfloor + \sqrt{n \log n}\ ,\ (n \geq \sqrt{2/L^b(p,b)})\ ,$$

$\beta \geq p$, and $V_n/n \to 1/2$ almost surely. Furthermore, β tends to zero as b tends to infinity. Thus the flat part exists and tends to the full range as $b \to \infty$. Finally, for $b \geq 8$, we have $0 < \beta < 1/2$.

Proof. We first show that

$$\left\lfloor \frac{n}{2} \right\rfloor - \sqrt{n \log n} \leq \boldsymbol{E} V_n .$$

The symmetric inequality is easily obtained by considering a tree with MAX nodes at the bottom (see [Kam92]). We consider a random $2n$-level b-ary incremental tree with parameter p. The nodes in the tree are marked good or bad. The leaves are all good. Consider a node at an odd level $2n+1$ with b children at level $2n$. Such a node corresponds to a MIN node in the tree. We mark it good only if all the children are good; otherwise, it is marked bad. For a node at level $2n$ (a MAX node) with b children, we mark it good if there exists at least one good child whose edge value is one. Thus, the root u is good if and only if there is a path from the root to the bottom level where all the MAX nodes provide at least one "1".

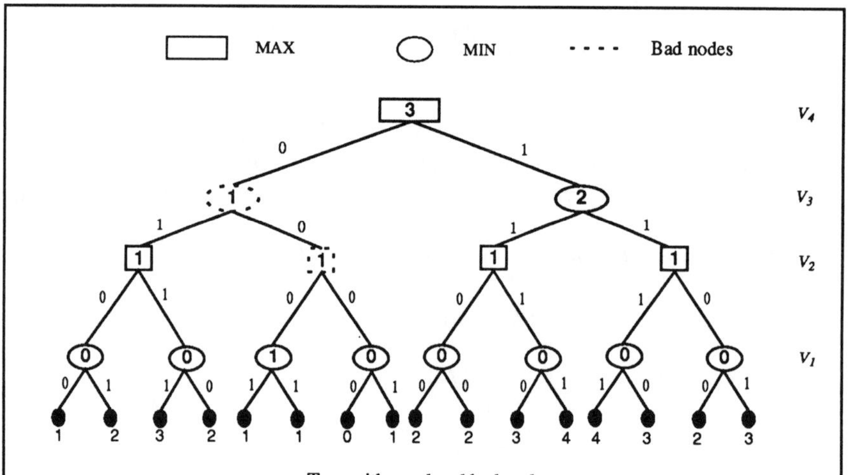

Tree with good and bad nodes.

If the value $V(u)$ of the root of the tree is good, then $V(u) \geq n$. Also, for a node v at level $2n+1$ we have $V(v) \geq n$. Let p_n denote the probability that a node at level n is marked as good. Then, by the previous discussion,

$$\boldsymbol{P}\{V_{2n} \geq n\} \geq p_n .$$

Furthermore, we have a simple recursion:

$$p_0 = 1 ,$$

and

$$p_{2n} = f(p_{2n-2}) ,$$

where

$$f(x) \stackrel{\text{def}}{=} 1 - \left(1 - px^b\right)^b .$$

The function f is continuous and increases monotonically from 0 to $f(1) = 1 - (1-p)^b$. We note therefore that p_{2i} decreases monotonically in i to a limit which is either zero or a positive number. The limit is the largest root on $[0, 1)$ of the equation $f(x) = x$. Let us call this limit $L(p, b)$. Thus, the following interesting inequalities are true:

$$\inf_n \mathbf{P}\{V_{2n} \geq n\} \geq L(p, b),$$
$$\inf_n \mathbf{P}\{V_{2n+1} \geq n\} \geq L^b(p, b).$$

Therefore,

$$\inf_n \boldsymbol{P}\left\{V_n \geq \left\lfloor \frac{n}{2} \right\rfloor\right\} \geq L^b(p, b).$$

Continuing this discussion, we consider the set of all p for which $L(p, b) > 0$. We know by McDiarmid's inequality that

$$\boldsymbol{P}\left\{|V_n - \boldsymbol{E}V_n| > \sqrt{n \log n}\right\} \leq \frac{2}{n^2}.$$

Therefore, if $2/n^2 < L(p, b)$, we see that

$$\boldsymbol{E}V_n \geq \left\lfloor \frac{n}{2} \right\rfloor - \sqrt{n \log n} .$$

For the remainder of the proof, we refer to [Kam92]. □

REMARK. Numerical computations show that the distribution of the root's value is even more concentrated than this Lemma shows.

10. Concluding remarks. Although we have discovered many properties of the limit $\mathcal{V}(X)$, we have been unable to provide a precise analytic formulation of this limit function. Many inequalities are available however. If $\mathcal{V}(p)$ denotes the limit for a Bernoulli tree with parameter p, we see that for any positive random variable X, since $X \geq aI_{X \geq a}$,

$$\mathcal{V}(X) \geq \sup_a a\mathcal{V}(\boldsymbol{P}\{X \geq a\}) .$$

(We hope the reader will permit this abusive notation.) In particular, if a is the median of a continuous random variable X, this inequality shows that

$$\mathcal{V}(X) \geq \frac{\text{median }(X)}{2} .$$

If $X \leq \mu$, we see that $\mathcal{V}(\mu - X) = \mu - \mathcal{V}(X) \geq a\mathcal{V}(\boldsymbol{P}\{\mu - X \geq a\})$. Thus,

$$\mathcal{V}(X) \leq \mu - \sup_a a\mathcal{V}(\boldsymbol{P}\{X \leq \mu - a\}) .$$

For example, when X is uniform on $[0,1]$, and we take $a = 1/2$, we obtain the inequalities

$$\frac{1}{4} \leq \mathcal{V}(X) \leq \frac{3}{4} .$$

This was of course predictable, as we know that $\mathcal{V}(X) = 1/2$ by a relatively simple symmetry argument.

Another inequality follows from $\mathcal{V}(1-\alpha) = 1$: we obtain for any positive X,

$$\mathcal{V}(X) \geq X_{(\alpha)} ,$$

where $X_{(\alpha)}$ is the α quantile of X.

Note that $\mathcal{V}(a + bX) = a + b\mathcal{V}(X)$. Assume that X is bounded and positive, and that Y is independent of X, bounded, and positive. Then, one would be tempted to infer that

$$\mathcal{V}(X+Y) = \mathcal{V}(X) + \mathcal{V}(Y) ,$$

where $\mathcal{V}(X+Y)$ refers to the incremental tree with edge values distributed as $X+Y$. This is clearly false: just take X and Y Bernoulli with parameter α. Then $X+Y$ is stochastically greater than a Bernoulli with parameter $1-(1-\alpha)^2 = 2\alpha - \alpha^2$. From our results,

$$\mathcal{V}(2\alpha - \alpha^2) > 0 ,$$

while

$$\mathcal{V}(X) = \mathcal{V}(Y) = 0 .$$

The lack of linearity seen throughout our analysis and experiments makes it difficult to get a good grip on the limit function except in special cases.

In the Bernoulli model, if the ones were represented in the minimax path in proportion to their frequency in the tree, we would have $\mathcal{V}(p) = p$. Clearly, we do not have this—the tree's behavior depends upon more than just frequencies or averages. Also, in general, we do not have $\mathcal{V}(X) = \boldsymbol{E}X$. The Bernoulli model shows that $\mathcal{V}(X)/\boldsymbol{E}X$ can be zero!

A comparison with the Pearl tree with the same distribution on the leaf values is helpful. In the Bernoulli model, we associate with the leaves i.i.d. binomial (n,p) random variables. Pearl's result implies that the root value V_n of his tree satisfies

$$\frac{V_n}{n} \to p = \boldsymbol{E}X$$

almost surely: the ones are proportionally represented in the minimax path.

For game tree searching strategies, the incremental model is very promising, as it incorporates different behaviors for different values of the edge parameters such as p in the Bernoulli model. The incremental trees are teeming with different sorts of life, perhaps modeling both very easy and very hard search problems for different choices of edge distributions. For p near zero, the zeroes overwhelm the ones, even more than their proportions would suggest.

The natural continuation of this study is the consideration of search heuristics. Here we note that algorithms that expand an edge have access to the edge's value. The purpose is to find strategies that expand few nodes (relatively speaking) in an incremental tree, yet lead to an n-level path of value close to V_n in some sense. For example, if we were to take a random child and if the opponent were infinitely smart (knowing the entire tree), we would end up with a path of total value less than about np (as this would be obtained against a random opponent), and less than about $n\mathcal{V}(p)$ (as we ourselves are not infinitely smart). The true path value thus concentrates around $nS(p)$, where $0 \leq S(p) \leq \min(p, \mathcal{V}(p))$. For more interesting heuristics based upon backtracking and pruning, the model promises to be exciting.

Finally, one may wonder what happens with random edge variables that have large infinite tails. This may occur for example for the normal distribution. More specifically, we may consider edge variables with a symmetric stable distribution with parameter $\alpha \in (0, 2]$ (these have characteristic function $\exp(-|t|^\alpha)$). Consider first a Pearl tree, in which we associate with each leaf independently a sum of n independent symmetric stable random variables, to be able to make a fair comparison with our model. We can prove that $V_n/\sqrt{n}$ oscillates on alternating levels between a positive value c and its negative counterpart $-c$, when the stable distribution is normal ($\alpha = 2$), and when new levels are added at the bottom of the tree. For the Cauchy distribution ($\alpha = 1$), $|V_n/n|$ tends to a positive constant c almost surely, while the sign of V_n alternates on different levels. For $\alpha < 1$, we have distributions with very big tails, and $|V_n/n| \to \infty$ almost surely. For these distributions, the swings in the oscillatory pendulum are unbearably big. In the incremental model, there are distributions for which $\limsup_{n\to\infty} V_n/n = \infty$ almost surely if we add new levels at the root.

REFERENCES

[Bi76] J. D. Biggins, "The first and last-birth problems for a multitype age-dependent branching process," *Advances in Applied Probability*, vol. 8, pp. 446-459, 1976.

[Bi77] J. D. Biggins "Chernoff's theorem in the branching random walk," *Journal of Applied Probability*, vol. 14, pp. 630-636, 1977.

[CM83] M.S. Campbell and T.A. Marsland, "A comparison of minimax tree search algorithms," in: *Artificial Intelligence*, 20: 347-367, 1983.

[CL91] E.G. Coffman, Jr. and G.S. Lueker, *Probabilistic Analysis of Packing and Partitioning Algorithms*, Wiley-Interscience Series, New York, 1991.

[Ha74] J. M. Hammersley, "Postulates for subadditive processes," *Annals of Probability*, vol. 1, pp. 652-680, 1974.

[Hoeff63] W. Hoeffding: "Probability inequalities for some series of bounded random variables," in: *Journal of the American Statistical Association*, 58:13-29,1963.

[Kam92] O. Kamoun, "A Probabilistc Min-Max Tree," M.Sc. Thesis, School of Computer Science, McGill University, Montreal, 1992.

[Ki75] J. F. C. Kingman, "Subadditive ergodic theory," *Annals of Probability*, vol. 1, pp. 883-909, 1973.

[KM75] D.E. Knuth and R.W. Moore, "An analysis of alpha-beta pruning," in: *Artificial Intelligence*, 6: 293-326, 1975.

[McDi89] C. McDiarmid, "On the method of bounded differences," in: *Surveys in Combinatorics 1989*, vol. 141, pp. 148–188, London Mathematical Society Lecture Notes Series, Cambridge University Press, Cambridge, 1989.

[Nau82a] D.S. Nau, "The last player theorem," in: *Artificial Intelligence*, 18: 53-65, 1982.

[Nau82b] D.S. Nau, "An investigation of the causes of pathology in games," in: *Artificial Intelligence*, 19: 257-278, 1982.

[Nau83] D.S. Nau, "Pathology on game trees revisited, and an alternative to minimaxing," in: *Artificial Intelligence*, 21: 221-244, 1983.

[Newb77] M. Newborn, "The efficiency of the alpha-beta search of trees with branch-dependent terminal node score," in: *Artificial Intelligence*, 8: 137-153, 1977.

[Pearl80] J. Pearl, "Asymptotic properties of Minimax trees and game-searching procedures," in: *Artificial Intelligence*, 14: 113-138, 1980.

[Pearl84] J. Pearl, *Heuristics*, Addison-Wesley Publishing Company, Reading, MA, 1984.

[Schw67] L. Schwartz, *Cours d'Analyse*, Hermann, Paris, 1967.

METRICS ON COMPOSITIONS AND COINCIDENCES AMONG RENEWAL SEQUENCES

PERSI DIACONIS*, SUSAN HOLMES†, SVANTE JANSON‡, STEVEN P. LALLEY§, AND ROBIN PEMANTLE¶

Abstract. We study several metrics on the space $C(m,n)$ of compositions of m into at most n parts. Understanding the geometry of these spaces leads to the study of the distribution of the distance between randomly chosen compositions. This in turn leads to some non-standard probability problems. One involves pinned Wiener processes. A second leads to the following renewal theory problem: Let $X_1, X_2, ...X_n; Y_1, Y_2, ...Y_n$ be positive integer valued random variables. Let $C_n = |\{i,j \le n : X_1 + ... + X_i = Y_1 + \cdots + Y_j\}|$ be the number of coincidences among the partial sums. We determine limiting approximations to the distribution of C_n. When X_i and Y_i are jointly independent and identically distributed; the limit is non normal. When $X_1 + \cdots + X_n = Y_1 + \cdots + Y_n$ is tied down (as in the application to compositions) the limit is normal. Our study was motivated by algorithms for careful approximation of the bootstrap.

1. Introduction. Let $C(m,n)$ be the set of compositions of m into at most n parts. Thus

$$(1.1)\quad C(m,\,n) = \{(h_1,\,h_2\cdots,\,h_n) : h_i \ge 0,\text{ integer},\ h_1 + \cdots + h_n = m\}$$

The familiar stars and bars argument shows $|C(m,n)| = \binom{m+n-1}{n-1}$.

For example, when $m = 4$, $n = 3$ the 15 compositions in $C(4,3)$ are

112	400	013	130	220
121	040	031	301	202
211	004	103	310	022

Compositions are a basic combinatorial object which arise in several statistical applications. For example, a class of m students given grades in $\{A, B, C, D, E\}$ give rise to a point in $C(m, 5)$. Aitchison [1] gives a comprehensive treatment of compositional data.

Our motivation for careful study of compositions arose from analysis of the statistical tool known as the bootstrap. This is based on repeated samples of of n items chosen with replacement from a list of n. Each sample can be associated to a point in $C(n,n)$, with h_i being the number of times item i appears in the sample. We were seeking "well distributed" points in

* Dept. of Mathematics, Harvard University, Cambridge, MA 02138.
† Unité De Biométrie, INRA-UMII-ENSA.M, 34060 Montpellier, France; and Dept. of Statistics, Sequoia Hall, Stanford University, Stanford, CA 94305.
‡ Dept. of Mathematics, Uppsala University, P.O. Box 480, S-75106 Uppsala, Sweden.
§ Dept. of Statistics, Purdue University, West Lafayette, IN 47907.
¶ Dept. of Mathematics, University of Wisconsin, Madison, WI 53706.

$C(n,n)$ this required natural motions of distance. See Diaconis and Holmes [7] for further discussion.

In Section 2 we introduce and study a number of metrics on $C(m,n)$. As explained below, this is closely related to the study of metrics on the space of probability measures. We study the size of the balls in a given metric by studying the following problem: Pick h, h' at random in $C(m,n)$. What is the distribution of $d(h,h')$? We give fairly complete answers on $C(n,n)$. The results lead to some non-standard probability problems.

The final two sections focus on some renewal theory problems arising from the study of our subset metric. Let $X_1, X_2, \cdots; Y_1, Y_2 \cdots$ be positive integer valued random variables. Let S_j^X and S_j^Y be the corresponding partial sums. Define the coincidence number

$$C_n = |\{i,\, j;\ 1 \le i,\, j \le n : S_i^X = S_j^Y\}| \tag{1.2}$$

That is, the number of common values between the sets

$$\{S_1^X,\, S_2^X,\, \cdots, S_n^X\} \text{ and } \{S_1^Y,\, S_2^Y,\, \cdots,\, S_n^Y\}\,.$$

We study the limit distribution of C_n under two distributional assumptions. We first treat the independent and identically distributed case. In Section 3 we prove

THEOREM 1.1. *Let $\{X_i\}_{i=1}^{\infty}$, $\{Y_j\}_{j=1}^{\infty}$ be jointly independent and identically distributed positive integer valued random variables. Suppose that g.c.d.$\{h : P\{X_1 = h\} > 0\} = 1$ and that X_1 has finite mean μ and finite positive variance σ^2. Then, as $n \to \infty$*

$$\frac{C_n - n\mu^{-1}}{\sqrt{n}} \Longrightarrow \min(Z_1,\, Z_2)$$

where the vector $(Z_1,\, Z_2)$ has a bivariate normal distribution with mean zero and non-degenerate (rank 2) covariance matrix. The limit is thus non-normal.

In Section 4 we prove

THEOREM 1.2. *Let $\{W_i\}_{i=1}^{n}$ be the cell counts when n balls are dropped into n boxes. Let $X_i = W_i + 1$. Let $\{Y_i\}_{i=1}^{n}$ be an independent copy of $\{X_i\}_{i=1}^{n}$. As $n \to \infty$, the C_n of (1.2) satisfies*

$$\frac{C_n - n/2}{\sqrt{n}} \Longrightarrow Z$$

where Z is normal with mean 0 and positive, finite variance.

The connection between Theorems 1.1 and 1.2 and the metrics of Section 2 is explained in the introduction to Section 3.

2. Metrics on compositions. Let $C(m,n)$ be the set of compositions of m into at most n parts. In this section we define and study several metrics on $C(m,n)$: total variation (Section 2.1), the subset metric (Section 2.2), and Vassershtein metric (Section 2.3). Basic properties and sampling distributions are developed. The sampling distributions are derived for both the uniform and multinomial distributions. The first gives a geometric feeling for the space. The second is natural for the bootstrap applications: the measure induced on $C(n,n)$ by bootstrap sampling is exactly the multinomial distribution of n balls dropped randomly into n boxes.

Compositions may be regarded as measures on $\{1,2,...,n\}$ with total mass m. This allows any distance on probabilities to be adapted to a metric on $C(m,n)$. Section 2.4 gives pointers to the relevant literature and some further examples.

Total variation emerges as our favorite metric. The others are developed because they have natural invariance properties or lead to interesting math problems.

2.1. Total variation. For $x, y \in C(m,n)$ define

$$d_{TV}(x,\,y) = \begin{cases} \text{minimum number of } \pm 1 \text{ switches} \\ \text{needed to bring } x \text{ to } y. \end{cases} \tag{2.1}$$

For example, take $m=4,\ n=3,\ x=(4,0,0),\ y=(1,1,2)$. We bring x to y by $400 \to 301 \to 202 \to 112$ so $d_{TV}(x,y)=3$.

The standard properties of the total variation distance between two probability measures (see e.g. Diaconis [6, Chapter 3]) can be translated to give the following equivalences.

LEMMA 2.1. *The total variation distance on $C(m,n)$ defined in (3.1) satisfies*

$$d_{TV}(x,\,y) = \frac{1}{2}\sum_{i=1}^{n} |x_i - y_i| = \frac{1}{2}\max_{|f_i|\le 1}\sum_{i=1}^{n} x_i f_i - y_i f_i = m - \sum_{i=1}^{n} \min(x_i, y_i)$$

Remark 2.2.

1. The first equality gives an easy way to calculate d_{TV}. The second equality shows that if two compositions are close then linear combinations of them are uniformly close. The third equality gives a statistically natural property of d_{TV}. In the bootstrap application, $\min(x_i,\,y_i)$ is the amount of overlap or redundancy between the two samples. For $m=n$, the distance is largest (equal to n) if the two compositions come from disjoint bootstrap replications.
2. The distance d_{TV} is invariant under coordinate permutations. On the other hand the number of points in a metric ball can depend on where the ball is centered. For example, when $m=4$ and $n=3$, there are two points at distance 1 from (004) and 6 points at

distance 1 from (112). Generic x have $n(n-1)$ points at distance 1 but for larger balls, all depends on how close x is to the corner of the simplex $C(m,n)$. This becomes more pronounced for larger n. For example, on $C(n,n)$, if Y has a multinomial distribution and $x = (n, 0, \cdots, 0)$, $d_{TV}(x, Y)$ is essentially constant at n while if $x = (1, 1, \cdots, 1)$ $d_{TV}(x,Y)$ has an approximate normal distribution centered at n/e. This can be proved using the argument of Lemma 2.3 which interpolates between these extremes by choosing x at random.

The next two lemmas give the approximate sampling distributions of $d_{TV}(X,Y)$ where X and Y are randomly chosen compositions. We derive limiting approximations when m and n are large with $m/n \to \lambda$. This is the domain of interest for bootstrap applications. The answers can be quite different in other zones (m small n large or vice-versa). Nowadays, one can easily simulate this distribution for any specific m, n of interest.

LEMMA 2.3. *Let X and Y be independently chosen from the multinomial distribution on $C(m, n)$. Then, for $m, n \nearrow \infty$ with $m/n \to \lambda$, $0 < \lambda < \infty$, $d_{TV}(X,Y)$ is approximately normally distributed with*

$$\textit{mean} \sim \frac{n}{2}\mu(\lambda), \quad \textit{var} \sim \frac{n}{4}\sigma^2(\lambda)$$

For $\mu(\lambda) = E|W - W'|$ *with W, W' independent Poisson* (λ) *and*

$$\sigma^2(\lambda) = [2\lambda - \mu(\lambda)^2]\{1 - 2[2\lambda^2 - \lambda\mu(\lambda)^2]^{-1}[E|W^2 - WW'| - \lambda\mu(\lambda)]^2\}$$

Proof. The means and the variances can be computed by elementary arguments or by using the conditioned limit arguments as below.

Without essential loss, take $\lambda = m/n$. Realize the multinomial variables X and Y as the conditional values of independent Poisson(λ) vectors X', Y', with $X_1' + \ldots + X_n' = Y_1' + \ldots + Y_n' = m$. Let normalized random variables be defined by:

$$\begin{aligned} A_n &= \tfrac{1}{\sqrt{n(2\lambda - \mu^2(\lambda))}} \textstyle\sum_{i=1}^n \{|X_i' - Y_i'| - \mu(\lambda)\} \\ B_n &= \tfrac{1}{\sqrt{n\lambda}} \textstyle\sum_{i=1}^n (X_i' - \lambda) \\ C_n &= \tfrac{1}{\sqrt{n\lambda}} \textstyle\sum_{i=1}^n (Y_i' - \lambda) \end{aligned}$$

Then the conditional law of A_n given $B_n = C_n = 0$ is the law of $(2d_{TV}(X,Y) - \mu(\lambda))/\sqrt{n(2\lambda - \mu^2(\lambda))}$. From the multivariate central limit theorem,

$$\begin{pmatrix} A_n \\ B_n \\ C_n \end{pmatrix} \longrightarrow \begin{pmatrix} A \\ B \\ C \end{pmatrix}$$

with $(A,B,C)^T$ trivariate normal having mean vector 0 and covariance

matrix $\begin{pmatrix} 1 & \rho & \rho \\ \rho & 1 & 0 \\ \rho & 0 & 1 \end{pmatrix}$ where $\rho = [2\lambda^2 - \lambda\mu(\lambda^2)]^{-\frac{1}{2}}[E|W^2 - WW'| - \lambda\mu(\lambda)]$.
Conditioned limit theory as in Holst [12] (cor. 3.6) implies that

$$\mathcal{L}(A_n|B_n = C_n = 0) \longrightarrow \mathcal{L}(A|B = C = 0)$$

Now, if a random normal vector Z is partitioned into Z_1 and Z_2, using standard notation, $\mathcal{L}(Z_1|Z_2 = z_2)$ is normal with mean $\mu_1 + \Sigma_{12}\Sigma_{22}^{-1}(z_2 - \mu_2)$ and covariance matrix $\Sigma_{11} - \Sigma_{12}\Sigma_{22}^{-1}\Sigma_{21}$. In the present case $\mathcal{L}(A|B = C = 0)$ is normal with mean 0 and variance $1 - 2\rho^2$. This yields the stated result after elementary rescaling. □

Remark 2.4. Ramasubban [19,20] studied $\Delta_r = E|W - W'|^r$. He gives $\mu(\lambda) = 2\lambda e^{-2\lambda}[I_0(2\lambda) + I_1(2\lambda)]$ with $I_n(x)$ the nth order modified Bessel function of the first kind. We compute

λ	.5	1	1.5	2	2.5	3
$\mu(\lambda)$	.674	1.048	1.319	1.543	1.738	1.912
$\sigma^2(\lambda)$	0.330	0.712	1.082	1.448	1.812	2.177

Feller [8] relates Poisson differences and Bessel functions.

Essentially the same proof, conditioning on geometric variables instead of Poisson variables, gives a limit theorem for the total variation distance under the uniform distribution on $C(m, n)$, in this case the variance has a closed form.

LEMMA 2.5. *Let X and Y be independently chosen from the uniform distribution on $C(m, n)$. Then, for $m, n \nearrow \infty$ with $m/n \to \lambda$, $0 < \lambda < \infty$, $d_{TV}(X, Y)$ is approximately normally distributed with*

$$mean \sim \frac{n}{2}\mu(\lambda)$$
$$var \sim \frac{n}{4}\sigma^2(\lambda)$$

For $\mu(\lambda) = E|W - W'| = \dfrac{2\lambda(1+\lambda)}{1+2\lambda}$ where W, W' are independent geometric $(\dfrac{1}{1+\lambda})$ variables $(P(W = j) = \theta(1-\theta)^j, 0 \le j < \infty, \theta = \frac{1}{1+\lambda})$ and

$$\begin{aligned}\sigma^2(\lambda) &= [2\lambda(\lambda+1) - \mu(\lambda)^2]\{1 - 2[2\lambda^2(\lambda+1)^2 \\ &\qquad -\lambda(\lambda+1)\mu^2(\lambda)]^{-1}[E|W^2 - WW'| - \lambda\mu(\lambda)]^2\} \\ &= 4\lambda^2(\lambda+1)^2(2\lambda^2 + 2\lambda + 1)/(2\lambda+1)^4\end{aligned}$$

Remark 2.6. We compute

λ	.5	1	1.5	2	2.5	3
$\mu(\lambda)$	0.75	1.333	1.875	2.4	2.92	3.43
$\sigma^2(\lambda)$	0.352	0.988	1.868	2.995	4.372	5.998

Thus, typical pairs X, Y tend to be further apart under the uniform as compared with the multinomial distribution.

2.2. Subset distance. Compositions are in 1–1 correspondence with subsets of size $n-1$ in a set of $m+n-1$ elements: arrange $1, 2, \cdots, m+n-1$ in a row and circle the elements in the subset. The associated composition has n parts corresponding to the number of elements between the circles. Thus, with $m = 4$, $n = 3$, the composition (400) corresponds to 1 2 3 4 ⑤ ⑥ while (013) corresponds to ①2 ③4 5 6. The composition $(k_1, k_2, \cdots, k_n)$ corresponds to the subset $\{k_1+1, k_1+k_2+2, \cdots, k_1+\cdots+k_{n-1}+n-1\}$. Write $s(x)$ for the subset corresponding to the composition x.

There is a natural metric on subsets which induces a metric on compositions. For $x, y \in C(m, n)$ define

$$d_s(x, y) = (n-1) - |s(x) \cap s(y)|$$

Thus for $x = (400)$, $y = (013)$; $s(x) = \{5, 6\}$, $s(y) = \{1, 3\}$, $d(x, y) = 2$.

The metric d_s depends on the ordering: $d_s(50000, 11111) = 2$, $d_s(05000, 11111) = 3$. On the other hand the metric d_s has the following invariance property: the number of points in the ball $\{y : d_s(x, y) \leq \omega\}$ does not depend on x. This follows from the invariance of the metric $(n-1) - |s \cap t|$ on subsets under the action of the permutation group S_{n+m-1}. The main reason for studying d_s is because of the frequent interplay between subsets and compositions in the combinatorial literature.

The following two results give the limiting distribution of $d_s(X, Y)$ under the two distributions on $C(m, n)$.

LEMMA 2.7. *Let X and Y be independently chosen from the uniform distribution on $C(m, n)$. Then, for $m, n \nearrow \infty$ with $m/n \to \lambda$, $0 < \lambda < \infty$, d_s has a hypergeometric distribution. It is approximately normally distributed with*

$$\begin{aligned} \textit{mean} &= (n-1) - \frac{(n-1)^2}{n+m-1} &\sim n\lambda/(1+\lambda) \\ \textit{var} &= \frac{(n-1)^2 m^2}{(m+n-1)^2(m+n-2)} &\sim n\lambda^2/(1+\lambda)^3 \end{aligned}$$

Proof. Using the correspondence, $d_s(X, Y)$ has the same distribution as $(n-1) - |S \cap T|$ where S and T are randomly chosen subsets of size $(n-1)$ from $\{1, 2, \cdots m+n-1\}$. By invariance, S may be fixed at

$\{1, 2, \cdots, n-1\}$. Now, the distribution of $|S \cap T|$ is hypergeometric and the result stated is classical. □

LEMMA 2.8. *Let X and Y be independently chosen from the multinomial distribution on $C(n, n)$. Then, as $n \nearrow \infty$, $d_s(X, Y)$ is approximately normally distributed with*

$$\begin{aligned} \textit{mean} &\sim n/2 \\ \textit{var} &\sim \sigma^2 n \quad \textit{for some } \sigma^2,\ 0 < \sigma^2 < \infty . \end{aligned}$$

Proof. Choosing $X = (X_1, X_2, \cdots, X_n) \in C(n, n)$ from the multinomial distribution amounts to dropping n balls into n boxes with X_i the number of balls in box i. Let Y be similarly distributed. Now $|s(X) \cap s(Y)| = C_{n-1}$ as defined in (1.2) applied to the variables $X_i + 1$. The result now follows from Theorem 1.2. See Section 4 below. □

Remark 2.9. Very similar arguments give results similar to Lemma 2.8 for general m, n.

2.3. Vassershtein distance. The final metric considered in detail is the analog of a standard metrization of the weak star topology. For $x, y \in C(m, n)$ define

$$d_V(x, y) = \begin{array}{l}\text{minimum number of adjacent } \pm 1 \\ \text{switches need to bring } x \text{ to } y .\end{array} \tag{2.2}$$

Thus $d_V(400, 112) = 5$ from $(400) \to (310) \to (301) \to (211) \to (202) \to (112)$. Rachev [18] discusses the history and literature for this distance on probability measures.

From results proved there we have the following equivalent versions:

LEMMA 2.10. *The Vassershtein distance d_V on $C(m, n)$ defined by (2.2) satisfies*

$$\begin{aligned} d_V(x, y) &= \textstyle\sum_{i=1}^{n} |x_i^+ - y_i^+|; & x_i^+ &= x_1 + \cdots + x_i \\ &= \max_{f \in Lip_1} |\textstyle\sum x_i f_i - y_i f_i|, & Lip_1 &= \{f_i \,|\, |f_i - f_{i+1}| \le 1 \quad , 1 \le i < n\} . \end{aligned}$$

Remark 2.11. Note that d_V depends on the coordinate ordering: $d_V(300, 030) = 3$, $d_V(300, 003) = 6$. However, in the bootstrap application, if the original sample values are ordered real numbers then adjacency has a natural meaning and the Vassershtein distance becomes interesting.

LEMMA 2.12. *Let X and Y be chosen independently from the uniform distribution on $C(m, n)$. Then, as $m, n \nearrow \infty$, with $m/n \to \lambda$, $d_V(X, Y)/(n^{\frac{3}{2}}(2\lambda(1+\lambda))^{\frac{1}{2}})$ converges in distribution to $\int_0^1 |B_0(t)| dt$ with $B_0(t)$ the standard Brownian bridge on $[0, 1]$.*

Under the multinomial distribution, $d_V(X, Y)/(n^{\frac{3}{2}}(2\lambda)^{\frac{1}{2}})$ *converges to the same limit.*

Proof. We give the proof for the multinomial distribution, the proof for the Uniform being similar.

Under the multinomial distribution

$$d_V(X, Y) \stackrel{d}{=} \sum_{i=1}^{n} |X_i^+ - Y_i^+|$$

with X_i^+ and Y_i^+ independent binomial $(m, \frac{i}{n})$. Thus, for $i = \theta n$, $0 < \theta < 1$ fixed, $(X_i^+ - Y_i^+)/\sqrt{n}$ has a normal limit with mean 0 and variance $2\theta(1-\theta)\lambda$. Checking the covariances, we see that the increments have the covariances of a Brownian bridge. To make the convergence argument rigorous, we may appeal to Billingsley [3, Theorem 24.2]. This asserts that if $\xi_1, \xi_2, \cdots, \xi_n$ are exchangeable random variables with sum zero, sum of squares tending to 1, and max tending to 0, then the associated random function converges to a Brownian bridge.

Multiplying and dividing the expression for $d_V(X, Y)$ by n we get a Riemann sum for the integral. The result follows from the continuous mapping theorem. □

Remark 2.13. Following work of Cifarelli and Regazzini [4], Shepp [22] and Rice [21] carried out a careful investigation of the law of $\int_0^1 |B_0(t)|dt$. Shepp gives an elegant derivation of the Fourier transform in terms of Airy functions and a recursion for moments. Rice managed to numerically invert the Fourier transform to give highly accurate percentage points and the following graph of the density.

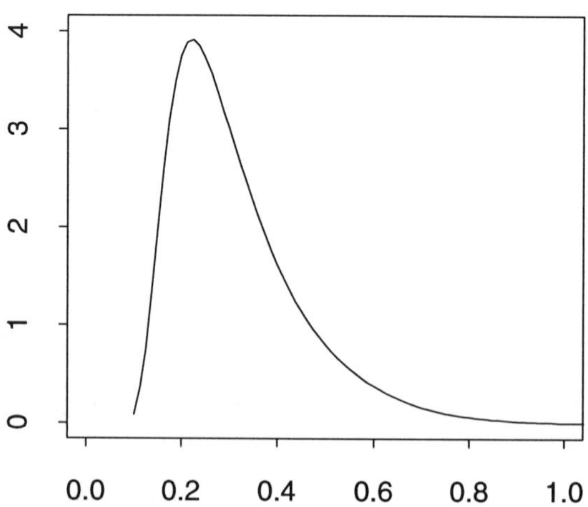

2.4. Other metrics. Compositions in $C(m, n)$ can be identified with probability vectors of length n by dividing by m. Thus any metric on probabilities can be carried over to compositions. Rachev [18] gives an encyclopedic survey of metrics on probabilities and Vegelius et al. [23] study measures of similarity between distributions. Two simple metrics are the Hellinger and ℓ_2 distances

$$d_{He}(x, y) = \sum_{i=1}^{n} (x_i^{\frac{1}{2}} - y_i^{\frac{1}{2}})^2 \qquad d_2^2(x, y) = \sum_{i=1}^{n} (x_i - y_i)^2$$

The first represents compositions as points on a sphere. The second has all of the advantages of Euclidian space. Each has an easily derived normal approximation under either the uniform or the multinomial distribution, using the techniques of Section 2.1.

We conclude by mentioning two further metrics which do not have natural versions on probabilities. The first is Hamming distance:

$$d_H(x, y) = |\{i : x_i \neq y_i\}|$$

This is invariant under permuting coordinates. It also has the following invariance property: $|\{y : d_H(x, y) \leq k\}|$ does not depend on x. Under both uniform distribution u and multinomial distribution m, $d_H(X, Y)$ has an approximate normal limiting distribution, when $m = n$, the means and variances are

$$E_u\{d_H(X, Y)\} = \frac{2}{3}n \qquad E_m\{d_H(X, Y)\} = \theta n,$$

$$1 - \theta = e^{-2} \sum_{j=0}^{\infty} \frac{1}{(j!)^2} \doteq .3085^+$$

$$\text{var}_u\{d_H(X, Y)\} \cong \tfrac{14}{81}n \qquad \text{var}_m\{d_H(X, Y)\} \cong \theta(1 - \theta)$$

$$-2 * [1 - \theta - e^{-2} \sum_k \frac{1}{(k-1)!k!}]^2$$

Our second metric may be called the childs metric

$$d_c(x, y) = \begin{array}{l}\text{miminum number of moves}\\ \text{required to bring X to Y}\end{array}$$

where a move takes $(x_1, \cdots, x_n) \to (x_1 \cdots, x_i - a, \cdots, x_j + a, \cdots, x_n)$ for any pair of coordinates $i \neq j$ and any integer a chosen so all coordinates are non-negative. Thus $d_c(50000, 02300) = 2$ from $(50000) \to (32000) \to (02300)$. If a composition is thought of as n piles of blocks, a move consists of picking any number of blocks from a pile and depositing them on any other pile. This metric has good invariance properties. Alas, R.L. Graham (personal communication) has shown that computing d_c is $\# - p$ complete. Indeed, given $x, y \in C(m, n)$ form $z_i = x_i - y_i$. Let t be the maximum number of blocks in a partition such that z_i, summed over each block, gives zero. The $d_c(x, y) = n - t$. However, deciding if $d_c(x, y) \le n - 2$ involves computing if some non-trivial subset sum of z_i is zero. This is a well known NP complete problem.

Final Remark 2.14. Metrics on combinatorial objects offer a rich area of study and application. See Critchlow [5] and Diaconis [6, Chapter 6] for examples in the space of permutations. We hope that some of the present results will be found similarly useful for analyzing compositional data.

3. Coincidences for independent renewal sequences. This section studies the number of coincidences between independent renewal sequences. Throughout we assume that $\{X_i\}_{i=1}^{\infty}$ and $\{Y_i\}_{i=1}^{\infty}$ are each independent and identically distributed random variables. With $\{X_i\}$ independent of $\{Y_i\}$. We do not assume X_i and Y_i have the same distribution. Assume further that all variables are strictly positive, integer valued, non-arithmetic (g.c.d. $\{h : P(X_1 = h) > 0\} = 1$), non degenerate (not almost surely equal to 1), and have finite second moments.

We let $S_i^X = X_1 + \cdots + X_i$ and S_i^Y be the partial sums. The object of study is the number of coincidences

$$C_n = \left|\{i, j \le n : S_i^X = S_j^Y\}\right| .$$

We will prove a limit theorem for C_n when X_i and Y_i have the same law in Section 3.1. We explain what happens when X_i and Y_i have different laws in Section 3.2, which also contains a review of relevant literature.

Our motivation came from studying the metric d_s explained in Section 2.2 above. Under the multinomial distribution on $C(n, n)$, this has the same distribution as the number of coincidences between the two sequences $\{W_1 + 1, W_1 + W_2 + 2, \cdots, W_1 + \cdots + W_{n-1} + (n-1)\}$ $\{Z_1 + 1, Z_1 + Z_2 + 2, \cdots, Z_1 + \cdots + Z_{n-1} + (n-1)\}$ where $\{W_i\}_{i=1}^{n}$ are the number of balls in box i when n balls are dropped into n boxes according to the multinomial distribution and $\{Z_i\}_{i=1}^{n}$ are independent, with the same distribution. Heuristically, W_i are approximately independent Poisson (1). The limit theory in this section was developed to study such coincidences. It turns out that the heuristic is wrong: the multinomial counts are tied down and this matters, the correct results for the balls in boxes case are explained in Section 4. The present results seem of independent interest. We found a rigorous developement under minimal conditions challenging.

We conclude this introduction with an overview of the argument. With notation as above, for the remainder of this introduction take X_i and Y_i with a common law. Define

$$R_X = \{S_1^X, S_2^X, \cdots\} \qquad R_Y = \{S_1^Y, S_2^Y, \cdots\}. \tag{3.1}$$

Let the common points of R_X and R_Y be $T_1 < T_2 < T_3 < \cdots$. There are infinitely many common points from the renewal theorem (condition on the set R_X and use the renewal theorem on the renewal process S_n^Y). The T_i form a renewal processes. Moreover, the X and Y excursions between successive T_i are independent and identically distributed. Consequently, if we define U_n, V_n by

$$S_{U_n}^X = S_{V_n}^Y = T_n \tag{3.2}$$

The successive (vector) increments of the two dimensional process

$$(U_n, V_n) \tag{3.3}$$

are *iid* so that (U_n, V_n) is a random walk which is strictly increasing in both coordinates. Theorem 1.1 of the introduction follows from a study of this walk. Indeed, C_n is just the number of points of the walk (U_j, V_j) that lie in the square $[0, n] \times [0, n]$. This is because each point of the walk corresponds to a coincidence of partial sums with (U_j, V_j) giving the times when the partial sums are equal. This coincidence is counted in C_n if and only if $U_i, V_j \leq n$.

We next explain (heuristically) why the limit law is the minimum of two correlated normal variables. The argument is based on a useful fact from renewal theory. Let $Z_1, Z_2, \ldots$ be non-negative independent and identically distributed, integer valued random variables with $E(Z_i) = \beta$, var $(Z_i) = \sigma^2$, $0 < \sigma^2 < \infty$. Let S_n^Z be the partial sum process. Define $W_h = |\{n : S_n^Z \leq h\}|$. Then, as $h \nearrow \infty$,

$$\frac{W_h - h\beta^{-1}}{\sqrt{h\beta^{-3}\sigma^2}} \Longrightarrow \xi$$

where ξ has a standard normal distribution. This is an immediate consequence of the ordinary central limit theorem since $P\{W_h < n\} = P\{S_n^Z > h\}$.

Now consider the renewal process (U_j, V_j) defined above. As noted, C_n is the number of points of this sequence in the square $[0, n] \times [0, n]$. This is clearly the same as the minimum of N_n^U, N_n^V where these are respectively the number of points in the renewal sequences U_j, V_j in $[0, n]$. Now Hunter [13] showed $(N_n^U - n\mu^{-1})/\sqrt{n}$, $(N_n^V - n\mu^{-1})/\sqrt{n}$ has a bivariate limiting normal distribution.

Thus C_n suitably normalized converges to the minimum of two normals.

To make the argument rigorous, we must study the distribution of (U_1, V_1). We show this has a non-degenerate covariance matrix in Section 3.2 which also contains further remarks and references.

3.1. Time between coincidences. Throughout we assume $\{X_i\}$, $\{Y_i\}$ satisfy the assumptions of the first paragraph of Section 3. We study the moments of T_n and (U_i, V_i) defined in (3.1, 3.2). We need a preliminary lemma for which we introduce notation.

Let $F(z) = \sum_{h=0}^{\infty} P_h z^h$, $|z| < 1$, be the probability generating function of the increments of a renewal process S_n. Let $U(z) = \sum_{h=0}^{\infty} u_h z^h$, where

$$u_h = \sum_{n=0}^{\infty} P\{S_n = h\} \tag{3.4}$$

is the renewal measure. Then $U(z) = 1/(1 - F(z))$.

LEMMA 3.1. *Assume $\mu = F'(1) < \infty$. Then, the increments have finite variance if and only if*

$$U(z) - \frac{1}{\mu(1-z)} = \sum_{h=0}^{\infty} (u_h - \frac{1}{\mu}) z^h$$

stays bounded as $z \to 1$. In this case, $U(z) - \frac{1}{\mu(1-z)} \to \frac{\sigma^2 + \mu^2 - \mu}{2\mu^2}$ with σ^2 the variance of the increment.

Proof. $U(z) - \frac{1}{\mu(1-z)} = \frac{1}{1-F(z)} - \frac{1}{\mu(1-z)} = \frac{\mu(1-z)-(1-F(z))}{\mu(1-z)(1-F(z))} = \frac{F(z)-1-(z-1)F'(1)}{\mu(1-z)(1-F(z))}$. Since $\frac{1-F(z)}{1-z} \to F'(1) = \mu$ as $z \nearrow 1$, $\sup_{0 \le z < 1} |U(z) - \frac{1}{\mu(1-z)}| < \infty$ if and only if $F(z) - 1 - (z-1)F'(1) = O\left((1-z)^2\right)$ as $z \nearrow 1$. Now

$$\frac{F(z) - 1 - (z-1)F'(1)}{(1-z)^2} = \int_0^1 F''(1 - t(1-z))(1-t)dt$$

By monotone convergence, the right hand side stays bounded as $z \to 1$ if and only if $F''(1-) < \infty$ if and only if the increments have finite variance. □

The next result shows that the increments of the renewal process T_i defined at (3.1, 3.2) have finite variance.

PROPOSITION 3.2. *Let $\{X_i\}_{i=1}^{\infty}$, $\{Y_i\}_{i=1}^{\infty}$, be independent and identically distributed non-arithmetic, integer valued random variables with finite means μ_X, μ_Y and finite, positive variances σ_X^2, σ_Y^2. Let $\{T_i\}_{i=1}^{\infty}$ be the intersection places (cf. 3.2) of the partial sum processes. Then, $\{T_i\}$ is a renewal process with iid increments $Z_i = T_{i+1} - T_i$ having $E(Z_i) = \mu_X \mu_Y$ and var $(Z_i) = \sigma^2(Z_i) < \infty$.*

Proof. Let u_h^Z be the renewal measure defined at (3.4). Clearly $\mu_h^Z = u_h^X u_h^Y$. Hence $u_h \to \mu_X^{-1} \mu_Y^{-1}$ as $h \nearrow \infty$. So $\mu_Z = \mu_X \mu_Y < \infty$. Now, with notation as in Lemma 3.1,

$$u_h^Z - \frac{1}{\mu_Z} = u_h^X u_h^Y - \frac{1}{\mu_X \mu_Y} = \frac{1}{\mu_Y}(u_h^X - \frac{1}{\mu_X}) + \frac{1}{\mu_X}(u_h^Y - \frac{1}{\mu_Y}) + (u_h^X - \frac{1}{\mu_X})(u_h^Y - \frac{1}{\mu_Y}).$$

thus

$$\begin{aligned} U^Z(z) - \frac{1}{\mu_Z(1-z)} &= \frac{1}{\mu_Y}\left[U^X(z) - \frac{1}{\mu_X(1-z)}\right] + \frac{1}{\mu_X}\left[U^Y(z) - \frac{1}{\mu_Y(1-z)}\right] \\ &\quad + \sum_{h=0}^{\infty} \left(u_h^X - \frac{1}{\mu_X}\right)\left(u_h^Y - \frac{1}{\mu_Y}\right) z^h . \end{aligned}$$

By Lemma 3.1, the first two terms on the right hand side stay bounded as $z \nearrow 1$. So, using Lemma 3.1 again, it suffices to show that the third term stays bounded.

Using the Cauchy-Schwartz inequality it suffices to show

$$\lim_{r \nearrow 1} \sum_{h=0}^{\infty} \left(u_h^X - \frac{1}{\mu_X} \right)^2 r^{2h} = \sum_{h=0}^{\infty} \left(\mu_h^X - \frac{1}{\mu_X} \right)^2 < \infty \tag{3.5}$$

and the same with Y replacing X. Using the Plancherel Theorem, the first sum in (3.5) equals

$$\frac{1}{2\pi} \int_0^{2\pi} \left| U^X(re^{it}) - \frac{1}{\mu_X(1-re^{it})} \right|^2 dt \leq \sup_{|z|<1} \left| U^X(z) - \frac{1}{\mu_X(1-z)} \right|^2$$

This last is finite since

$$U^X(z) - \frac{1}{\mu_X(1-z)} = \frac{1}{1-F(z)} - \frac{1}{\mu_X(1-z)}$$

is continuous on $|z| \leq 1$: The only singularity could be at $z = 1$ but the argument in Lemma 3.1 shows convergence for $z \to 1$ as long as $|z| \leq 1$. □

Remark 3.3. We do not know how to relate higher moments of X and Y to those of Z. We have shown that if X and Y have finite moment generating functions then Z does as well.

The next result of this section shows that the bivariate vector (U_1, W_1) of (3.2) has a finite, nondegenerate covariance matrix.

PROPOSITION 3.4. *Let $\{X_i\}_{i=1}^{\infty}$, $\{Y_i\}_{i=1}^{\infty}$ satisfy the assumptions of Proposition 3.2. Let (U_i, V_i) be the times of intersection of the two partial sum processes (cf. 3.2). Then (U_i, V_i) are iid with finite, rank two covariance matrix and $E(U_i) = \mu_X$, $E(V_i) = \mu_Y$.*

Proof. By Proposition 3.2, $ET_1^2 < \infty$. But $U_1 \leq T_1$ and $V_1 \leq T_1$. Thus the second moments are finite. For the means, condition on $\{Y_i\}_{i=1}^{\infty}$. Then, U_1 is a stopping time relative to $\{X_i\}_{i=1}^{\infty}$. Since $T_1 = \sum_{i=1}^{U_1} X_i$, $E(T_1|\{Y_i\}) = E(U_1|\{Y_i\})\mu_X$ by Walds Lemma (see e.g. Gut [10, Theorem I.5.3(i)]). Hence $E(T_1) = E(U_1)\mu_X$ but $E(T_1) = \mu_X\mu_Y$ by Proposition 3.2 so $E(U_1) = \mu_X$. A parallel argument gives $E(V_1) = \mu_Y$. □

The difficult part of the argument is showing that the covariance matrix of (U_1, V_1) is rank 2. If the covariance matrix is singular, then $aU_1 + bV_1 + c = 0$ a.s. for some a, b, c not all zero. We may take a, b, c integral. Now $a = b = 0$ implies $c = 0$ which is excluded. Hence without loss $a \neq 0$. We may further assume $a > 0$ and that a and b have no common divisors > 1. The argument proceeds in two cases:

Case 3.5. $b \geq 0$. Then, $aU_1 \leq -c$ so $U_1 \leq -c/a$ almost surely. This is impossible since $P\{U_1 > M\} > 0$ for any integer M by the following argument. We will assign values to $X_1, X_2, \cdots, X_M$ and $Y_1, Y_2, \cdots, Y_n$ (for some n) which are taken with positive probabilities and are such that $S_i^X \neq S_j^Y$ for all $i \leq M$, $j \leq n$ and $S_M^X < S_n^Y$. Thus $S_i^X \neq S_j^Y$ for all $i \leq M, j \leq n$ and so $U_1 > M$. To do the assignment, start with any X_1 and $Y_1 \neq X_1$. If say $Y_1 < X_1$, choose Y_2 such that $S_2^Y \neq S_1^X = X_1$. If $S_2^Y < S_1^X$, Choose Y_3 but if $S_2^Y > S_1^X$, choose X_2. Continue, always choosing the next value in the sequence with the smallest sum. This is always possible since there are at least two allowed values for each variable and only one of these is excluded by the construction. The construction stops when X_{M+1} is to be chosen.

Case 3.6. $b < 0$. Let $\beta = -b > 0$. The solutions to the diophantine equation $aU - \beta V + c = 0$ are of the form $U = u_h = u_0 + \beta h$, $V = v_h = v_0 + ah$ for some u_0, $v_0 > 0$ and $h = 0, 1, 2, \cdots$. Consider the sequence $\{S_{u_h}^X - S_{v_h}^Y\}_{h \geq 0}$. This is a random walk (starting at $S_{u_0}^X - S_{v_0}^Y$) with increments distributed as $S_\beta^X - S_a^Y$. Since the starting point and the increments take on at least three possible values with positive probability, there exists, for each $M > 0$, a possible realization of $\{S_{u_h}^X - S_{v_h}^Y\}_{h=0}^n$ for some $n \geq 0$, with all terms $\neq 0$ and $|S_{u_n}^X - S_{v_n}^Y| > M$. This corresponds to a realization of $\{X_i\}_{i=1}^{u_n}$ and $\{Y_i\}_{i=1}^{v_n}$ such that $(U_1, V_1) \neq (u_h, v_h)$ for every $h \leq n$. Thus, by the assumption $aU_1 + bV_1 + c = 0$, $U_1 \geq u_{n+1}$ and $V_1 \geq v_{n+1}$ for a.e. continuation of $\{X_i\}$, $\{Y_i\}$. If M is chosen large enough, every integer larger than M is a sum of possible values of X_i and a sum of possible values of Y_i. Hence if say $S_{u_n}^X - S_{v_n}^Y < -M$, there are possible continuations $\{X_i\}_{i=u_n+1}^{u_n+b}$, with $b \geq 1$, and $X_{u_n+1} + \cdots + X_{u_n+b} = S_{v_n}^Y - S_{u_n}^X$. This gives a realization $\{X_i\}_{i=1}^{u_n+b}$, $\{Y_i\}_{i=1}^{v_n}$ with $S_{u_n+b}^X = S_{v_n}^Y$, and thus $V_1 \leq v_n$ which is a contradiction.

Remark 3.7. It is the nondegeneracy of the covariance matrix which gives a non-normal limit in Theorem 1.1. We have no real hold on any of the elements of the covariance matrix.

Proof of Theorem 1.1. Suppose now that $\{X_i\}$, $\{Y_j\}$ have common distributions. From Proposition 3.4, the bivariate random walk (U_i, V_i) has $E(U_1) = E(V_1) = \mu = E(X_1)$ and finite rank two covariance matrix

$$\Sigma = \begin{pmatrix} \sigma^2 & \rho\sigma^2 \\ \rho\sigma^2 & \sigma^2 \end{pmatrix}$$

Hunter [13] shows

$$\lim_{n\to\infty} P\left\{\frac{N_n^U - n\mu^{-1}}{\sigma(n/\mu^3)^{\frac{1}{2}}} \leq \alpha, \quad \frac{N_n^V - n\mu^{-1}}{\sigma(n/\mu^3)^{\frac{1}{2}}} \leq \beta\right\} = \Phi_\rho(\alpha, \beta)$$

with Φ_ρ a bivariate normal distribution with mean 0, variances 1, and correlation ρ.

From this, $(C_n - n\mu^{-1})/\sqrt{n}$ converges to $\min(Z_1, Z_2)$ where (Z_1, Z_2)

is bivariate normal with mean 0 and covariance matrix

$$\begin{pmatrix} \sigma^2\mu^{-3} & \rho\sigma^2\mu^{-3} \\ \rho\sigma^2\mu^{-3} & \sigma^2\mu^{-3} \end{pmatrix}$$

□

Remark 3.8. Under the assumptions of Theorem 1.1, standard renewal theory (e.g. [10, Theorem II.5.1]) shows $\frac{1}{n}N_n^U \to \mu^{-1}$ a.s. and $\frac{1}{n}N_n^V \to \mu^{-1}$ a.s. hence $\frac{1}{n}C_n \to \mu^{-1}$ a.s.

3.2. Remarks and related literature.

1. Consider the number of coincidences under the conditions of Theorem 1.1 when X_1 and Y_1 have different distributions. Proposition 3.4 and Hunter's Central Limit Theorem [13] can be used to show that if the means differ then C_n has a normal limit. If the means are equal, then C_n has a non-normal limit as above.
2. Under the assumptions of Theorem 1.1 the number of coincidences among the partial sums of j *iid* random walks is distributed as the minimum of a j-variate normal (with suitable norming).
3. Things change radically if the increments are allowed to take negative values. Then, the sequence of partial sums will tend to fill out an interval and stochastic fluctuations take place at the fringes.
4. Proposition 3.2 of Section 3 is closely related to the regenerative phenomena studied by Kendall and Kingman [14] [16]. They study the Abelian semi-group of renewal sequences $\{u_n\}_{n=1}^{\infty}$ under the coordinate wise product. A masterful summary of this work is in Kingman [16]. Fristed [9] gives recent developments. A survey of its extensions to the theory of delphic semi-groups can be found in the work of Kendall and Harding [15]. This includes a survey of the work of Rollo Davidson which has connections to several of the authors of the present paper.

   Specialize the set up of Proposition 4.2 to the case where X_1 and Y_1 have the same distribution. There is then a map from measures to measures (from the Law of X_1 to the Law of Z_1) this corresponds to squaring the renewal measure: $u_n^Z = (u_n^X)^2$. Evidently this map is one to one and continuous. To see that it is not onto, we observe that if $P(Y = 1) = P(Y = 2) = \frac{1}{2}$, the corresponding renewal sequence does not have a square-root. Kendall [14] classified the infinitely divisible renewal sequences, showing they form a convex set with a countable set of extreme points.
5. The problem of coincidences among renewal sequences arises in the analysis of coupling arguments (see e.g. Lindvall [17]). Here the focus is on the occurrence of the first coincidence.

4. Coincidences with tied down sequences. This section gives a proof of Theorem 1.2, restated here for the readers convenience. Consider n balls, placed uniformly and independently in n boxes. For $1 \le j \le n$, let S_j equal j plus the number of balls in boxes $1, \ldots, j$. Let $\{S_j^X : 1 \le j \le n\}$ and $\{S_j^Y : 1 \le j \le n\}$ each be jointly distributed as $\{S_j : 1 \le j \le n\}$ and be independent of each other. Define

$$C_n = C_n\left(\{S_j^X\}, \{S_j^Y\}\right) = \#\left\{(i, j) : S_i^X = S_j^Y\right\} \tag{4.1}$$

to be the number of coincidences among the partial sums.

THEOREM 4.1.

$$\frac{C_n - (n/2)}{\sqrt{n}} \overset{\mathcal{D}}{\Rightarrow} Z \quad as\ n \to \infty, \tag{4.2}$$

where Z is normal with zero mean and positive finite variance.

The steps in proving Theorem 4.1 are:

(1) reduce to a one-sided local Central Limit Theorem (CLT);
(2) Poissonize;
(3) use the Poisson representation to embed in a renewal problem;
(4) apply a known local CLT to the renewal problem to prove the one-sided local CLT.

We begin with step 1, a reduction to Theorem 4.1′.

THEOREM 4.1.′ There exist $\sigma > 0$ and ϕ satisfying $\sup_k \phi(n, k) \to 0$ as $n \to \infty$, such that for all integers n and k,

$$\mathbf{P}(C_n = k) \ge \frac{1}{\sigma\sqrt{2\pi n}}\left(e^{-(n-2k)^2/(2n\sigma^2)} - \phi(n, k)\right). \tag{4.3}$$

To see that this implies Theorem 4.1, let $f : \mathbf{R} \to [0, 1]$ be any continuous function, and let $W_n = (C_n - n/2)/\sqrt{n}$. Then

$$\begin{aligned}
\liminf_n \mathbf{E}f(W_n) &\ge \lim_{A\to\infty} \liminf_{n\to\infty} \sum_{k:|k-n/2|<A\sqrt{n}} \frac{1}{\sigma\sqrt{2\pi n}} e^{-(n-2k)^2/(2n\sigma^2)} \\
&\qquad f\left((k - n/2)/\sqrt{n}\right) \\
&= \lim_{A\to\infty} \int_{-A}^{A} f(x)dG(x) \\
&= \int f(x)dG(x),
\end{aligned}$$

where G is the distribution of $\sigma/2$ times a standard normal. Replacing f by $1-f$ shows that $\mathbf{E}f(W_n) \to \int f(x)dG(x)$ for all bounded continuous f, establishing the implication.

For step 2, we construct a version of the process $\{S_j\}$ from a Poisson point process.

PROPOSITION 4.2. *Let $\{N_t : t \geq 0\}$ be a rate 1 Poisson point process. Then the conditional law of the process $\{N_j + j : j = 1, 2, 3, \ldots\}$ given $N_n = n$ is the same as the law of the process $\{S_j\}$ above.*

Proof. One way to assign n balls uniformly and independently into n boxes is to generate n IID random variables $\{T_j\}$, uniform on the real interval $[0, n)$, and then to place ball j in box $\lfloor T_j \rfloor$. The values of $\{S_j\}$ arising in this way will be unaffected if the sequence $T_1, \ldots, T_n$ is replaced by its ascending rearrangement, $T_{(1)}, \ldots, T_{(n)}$. But the times of the Poisson process $\{N_t\}$ conditioned on $N_n = n$ are distributed as the order statistics of n independent draws from $[0, n)$, and therefore $\{N_j + j\} \stackrel{\mathcal{D}}{=} \{S_j\}$. □

Step 3 is to construct a version of C_n via the Poisson representation. Let $\{N_t^X\}$ and $\{N_t^Y\}$ be independent rate 1 Poisson processes. Let $T_0 = 0$ and inductively define T_n to be the least integer $k > T_{n-1}$ such that

$$N_i^X + i = N_j^Y + j = k \quad \text{for some integers } i \text{ and } j\,.$$

In other words, T_1, $T_2, \ldots$ are the joint renewal times of the independent renewal processes $\{N_j^X + j\}$ and $\{N_j^Y + j\}$, whose increments are each 1 plus a Poisson of mean 1. Let $R_0^X = R_0^Y = 0$ and define R_k^X (respectively R_k^Y) to be the integer i for which $N_i^X + i = T_k$ (respectively $N_i^Y + i = T_k$). Then the triples $\{(U_k, V_k, Z_k) : k \geq 1\}$ defined by

$$(4.4) \qquad (U_k, V_k, Z_k) = \left(R_k^X - R_{k-1}^X,\ R_k^Y - R_{k-1}^Y,\ T_k - T_{k-1}\right)$$

are IID. In words, the blocks between joint renewals are IID, and these blocks contain the information: number of renewals in first process, number of renewals in second process, total time from last joint renewal.

This is an example of the situation studied in Section 3. In particular, Propositions 3.2 and 3.4 show that U_k, V_k and Z_k have finite variances and $EU_k = EV_k = E(N_1 + 1) = 2$ and $EZ_k = 4$.

Let $\tilde{C}_n = \sup\{k : T_k \leq 2n\}$ denote the number of joint renewals before time $2n$. Let G denote the event $\{N_n^X = N_n^Y = n\}$. Proposition 4.2 implies that the conditional law of $\{N_i^X + i,\ N_j^Y + j : 1 \leq i,\ j \leq n\}$ given G is the same as the law of $\{S_i^X,\ S_j^Y : 1 \leq i,\ j \leq n\}$. Observe that on the intersection of G and $\{\tilde{C}_n = k\}$, one has $R_k^X = R_k^Y = n$. Denoting

$$p_{n,k} = \mathbf{P}(R_k^X = R_k^Y = n,\ \ T_k = 2n)\,,$$

we then have

$$(4.5) \qquad \mathbf{P}(C_n = k) = \mathbf{P}(\tilde{C}_n = k | G) = \frac{p_{n,k}}{\sum_j p_{n,j}}\,,$$

Finally, step 4 consists of applying a local CLT to $\{(U_j, V_j, Z_j)\}$ to derive Theorem 4.1′ from (4.5). This is done in three substeps. First we show that

$$(4.6)\quad p_{n,k} = ck^{-3/2}\exp\left(-\frac{Q(n-2k,\ n-2k,\ 2n-4k)}{2k}\right) + g(n, k),$$

where Q is the quadratic form obtained by inverting the covariance matrix of the triple (U_1, V_1, Z_1) and

$$\lim_{k\to\infty}\sup_n k^{3/2}\left(1+\frac{(n-2k)^2}{k}\right)|g(n, k)| = 0.$$

Secondly we show that

$$(4.7)\qquad p_{n,k} = c2^{3/2}n^{-3/2}\exp\left(-\frac{(n-2k)^2}{2n\sigma^2}\right) + h(n, k),$$

where

$$(4.8)\qquad \lim_{k\to\infty}\sup_n k^{3/2}\left(1+\frac{(n-2k)^2}{k}\right)|h(n, k)| = 0.$$

Thirdly, we show that (4.7), (4.8) and (4.5) imply Theorem 4.1′.

Step 4a is an application of the local CLT from Bhattacharya and Rao [2, Theorem 22.1, Corollary 22.3]. The conclusion (4.6) follows immediately once the following hypotheses are verified:

(*i*) $\mathbf{E}(U_1, V_1, Z_1) = (2, 2, 4)$;
(*ii*) $\mathbf{E}(U_1^2 + V_1^2 + Z_1^2) < \infty$;
(*iii*) $\mathbf{P}(Z_1 > t) < C_1 e^{-C_2 t}$ for some $C_1, C_2 > 0$;
(*iv*) (U_1, V_1, Z_1) generates a truly 3-dimensional lattice.

Of these, (*iv*) is obvious, (*iii*) is a consequence of an easy coupling argument, and (*i*) and (*ii*) are consequences of a remark above.

Step 4b, the derivation of (4.7) from (4.6), begins with the easy observation that $Q(x, x, 2x)$ is nondegenerate, and can be written as $x^2/(2\sigma^2)$ for some positive finite σ. [For if this fails, summing (4.6) in k contradicts $\lim_{n\to\infty}\mathbf{P}(R_n^X = R_n^Y = n) \to 0$.] Letting (4.7) define the quantity $h(n, k)$, it remains to establish (4.8). In other words, we must show that

$$(4.9)\quad \mathcal{E}(n, k) := \left|(2k)^{-3/2}\exp\left(-\frac{(n-2k)^2}{4k\sigma^2}\right) - n^{-3/2}\exp\left(-\frac{(n-2k)^2}{2n\sigma^2}\right)\right|$$

satisfies $\sup_n \mathcal{E}(n, k)k^{3/2}(1+(n-2k)^2/k) \to 0$. Fix k, and consider first the case $|n-2k| > k^{0.55}$. The exponential terms in (4.9) are then at most $e^{-ck^{0.1}}$, which establishes (4.8). On the other hand, when $|n-2k| \le k^{0.55}$, then the mean value theorem gives

$$|n^{-3/2} - (2k)^{-3/2}| \le Ck^{-5/2}|n-2k| \le Ck^{-1.95}.$$

Also, for sufficiently large k,

$$\left|\exp\left(-\tfrac{(n-2k)^2}{2n\sigma^2}\right) - \exp\left(-\tfrac{(n-2k)^2}{4k\sigma^2}\right)\right| \quad \leq \quad \left|\tfrac{(n-2k)^2}{2n\sigma^2} - \tfrac{(n-2k)^2}{4k\sigma^2}\right| = \tfrac{|n-2k|^3}{4nk\sigma^2}$$

which gives a contribution to (4.8) of

$$(1 + \frac{(n-2k)^2}{k})\frac{|n-2k|^3}{4nk\sigma^2} \leq C\frac{|n-2k|^3}{k^2} + C\frac{|n-2k|^5}{k^3}$$

This tends to 0 if $|n-2k| < k^\delta$, for any δ strictly less than 0.6, this yields (4.8) in the case $|n-2k| \leq k^{0.55}$.

The final substep, 4c, is accomplished by showing

$$\text{(4.10)} \qquad \lim_{A\to\infty} \limsup_{n\to\infty} \frac{\sum_{k:|k-n/2|>A\sqrt{n}} p_{n,k}}{\sum_k p_{n,k}} = 0\,.$$

For this shows that given $\epsilon > 0$, we can choose A such that for sufficiently large n, all but mass ϵ of the conditional law of $\tilde{C}_n$ given G is contained in the range $[-A\sqrt{n},\, A\sqrt{n}]$, from which is is immediate that (4.7) and (4.5) imply Theorem 4.1′.

Firstly, from (4.7),

$$\sum_k p_{n,k} \geq \sum_{k:|k-n/2|\leq\sqrt{n}} p_{n,k} \geq 2\sqrt{n}\left[cn^{-3/2}\exp\left(-2/(\sigma^2)\right) - \sup_k h(n,\, k)\right],$$

and this is at least $c'n^{-1}$ for some constant c' and sufficiently large n. Secondly, let $c_0 = \sup_{n,k} |h(n,\, k)k^{3/2}(1 + (n-2k)^2/k)|$. Thus $|h(n,\, k)| < c_0 k^{-1/2}(n-2k)^{-2}$. Estimating the numerator of (4.10) gives

$$\sum_{k:|k-n/2|>A\sqrt{n}} p_{n,k}$$

$$= \sum_{k:|k-n/2|>A\sqrt{n}} cn^{-3/2}\exp\left(-\frac{(n-2k)^2}{2n\sigma^2}\right) + h(n,\, k)$$

$$= \sum_{k:|k-n/2|>A\sqrt{n}} cn^{-3/2}\exp\left(-\frac{(n-2k)^2}{2n\sigma^2}\right) + \sum_{k<n/3} h(n,\, k) +$$

$$\sum_{k\geq n/3,\, |k-n/2|>A\sqrt{n}} h(n,\, k)\,.$$

The first of these terms is $2n^{-1}(1-\Phi(2A/\sigma))$ plus an error term going to zero as $n\to\infty$ (uniformly in A, though we don't need the uniformity). The second term is at most

$$c_0 \sum_{k<n/3} k^{-1/2}(n-2k)^{-2} \leq 9c_0 n^{-3/2}\,.$$

And the third term is at most

$$2c_0 \sum_{j>A\sqrt{n}} (n/3)^{-1/2}(2j)^{-2} \leq \frac{c_0}{An}.$$

These three estimates establish (4.10), thus finishing step 4 and the proof of Theorem 4.1′. □

Acknowledgement: This work was carried out in the halls of the IMA workshop during the session on Monte Carlo and Markov chains. We thank the staff of the IMA for its terrific, supportive environment. We thank Bert Fristedt, Allan Gut, Greg Lawler, Jim Pitman and Larry Shepp for their help.

REFERENCES

[1] J. Aitchison, *The Statistical Analysis of Compositional Data*, Chapman and Hall, New York 1986.

[2] R.N. Bhattacharya and R.R. Rao, *Normal Approximation and Asymptotic Expansions*, Wiley, New York 1976.

[3] P. Billingsley, *Convergence of Probability Measures*, Wiley, New York 1968.

[4] D. Cifarelli and E. Regazzini, *On the asymptotic distribution of a statistic arising in testing the homogeneity of two samples*, Giornole Degli Economisie (1975), pp. 233–249.

[5] D. Critchlow, *Metric methods for analyzing partially ranked data* in *Lecture Notes in Statistics No. 34*, Springer-Verlag, Berlin 1985.

[6] P. Diaconis, *Group Representations in Probability and Statistics*, Institute of Mathematical Statistics, Hayward, California 1988.

[7] P. Diaconis and S. Holmes, *Gray codes for randomization procedures*, Statistics and Computing, no 4, 1994.

[8] W. Feller, *An Introduction to Probability Theory and its Applications Vol II*, (second edition) Wiley, New York 1971.

[9] B. Fristedt, *The central limit problem, for infinite products of, and Lévy processes of renewal sequences*, Z. Wahr. Verw. Gebiete **58** (1981), pp. 479–507.

[10] A. Gut, *Stopped Random Walks*, Springer-Verlag, Berlin 1987.

[11] L. Holst, *Two conditional limit theorems with applications*, Ann. Stat. **7** (1979), pp. 551–557.

[12] L. Holst, *Some conditional limit theorems in exponential families*, Ann. Prob. **9** (1981), pp. 818–830.

[13] J. Hunter, *Renewal theory in two dimensions: asymptotic results*, Adv. Appl. Prob. **6** (1974), pp. 546–562.

[14] D. Kendall, *Renewal sequences and their arithmetic* in *Symp. on Probability Methods in Analysis*, Springer Lecture Notes in Math. **31** (1967), pp. 147–175.

[15] D. Kendall and E. Harding, *Stochastic Analysis*, Wiley, London 1973.

[16] J. Kingman, *Regenerative Phenomena*, Wiley, London 1972.

[17] T. Lindvall, *Coupling Methods*, Cambridge University Press, Cambridge 1993.

[18] S. Rachev, *The Monge-Kantorovich mass transference problem and its stochastic applications*, Theor. Prob. Appl. **29** (1986), pp. 647–676.

[19] T. Ramasubban, *The mean difference and the mean deviation of some discontinuous distributions*, Biometrika **45** (1958), pp. 549–556.

[20] T. Ramasubban, *The generalized mean differences of the binomial and Poisson distributions*, Biometrika **46** (1959), pp. 223–229.

[21] S. RICE, *The integral of the absolute value of the pinned Wiener processes—calculation of the probability density by numerical integration*, Ann. Prob. **10** (1982), pp. 240–243.

[22] L. SHEPP, *On the integral of the absolute value of the pinned Wiener processes*, Ann. Prob. **10** (1982), pp. 234–239. (Acknowledgement of priority Ann. Prob. **19**, p. 1397.)

[23] J. VEGELIUS, S. JANSON and F. JOHANSSON, *Measures of similarity between distributions*, Quality and Quantity **20** (1986), pp. 437–441.

THE NO LONG ODD CYCLE THEOREM FOR COMPLETELY POSITIVE MATRICES

JOHN H. DREW* AND CHARLES R. JOHNSON*

Abstract. We present a self-contained proof of the following fact. Let an undirected graph G be given. Every symmetric matrix A, with graph G, that is both entry-wise nonnegative and positive semidefinite can be written as $A = BB^T$ with B entry-wise nonnegative if and only if G has no odd cycle of length 5 or more. In the process, we determine the worst case for the minimum number of columns in B in the representation of such an A.

An n-by-n matrix $A = (a_{ij})$ is called *completely positive* if A may be written

$$A = BB^T$$

in which B is n-by-m and entry-wise nonnegative. We write $A \in CP$, or CP_n if it is necessary to indicate the dimension. Though they arise in a variety of ways [H], there is yet no definitive test for a matrix to be completely positive. Recent work has also related CP matrices to exchangeable probability distributions on finite sample spaces [D]. A straightforward observation is that the definition of complete positivity could equivalently be stated as

$$A = b_1 b_1^T + b_2 b_2^T + \cdots + b_m b_m^T,$$

in which $b_1, b_2, \ldots, b_m$ are entry-wise nonnegative column vectors of n components: the vector b_i corresponds to the i-th column of the matrix B. This observation makes it clear that CP_n is a cone in $M_n(R)$. If A is completely positive, the *minmum* value for m for which there is such a representaion is denoted as $\#(A)$.

It is clearly necessary for complete positivity that A be symmetric positive semidefinite and entry-wise nonnegative. Such a matrix is called *doubly nonnegative*, and we write $A \in DN$ or DN_n. It is clear that DN_n is also a cone, the intersection of the cone of positive semidefinite matrices and the cone of entry-wise nonnegative matrices.

If $A = (a_{ij})$ is an n-by-n symmetric matrix, the undirected graph $G = G(A)$ is, as usual, the graph on n vertices in which there is an edge $\{i, j\}$ if and only if $a_{ij} \neq 0$. Standard graph theoretic background may be found in a variety of texts. Matrix theoretic background may be found in [HJ1] and [HJ2]. We use the notation $G = (V, E)$ to identify the vertex set

---

* Department of Mathematics, The College of William and Mary, Williamsburg, Virginia 23187-8795(The work of Charles R. Johnson was supported in part by National Science Foundation grant DMS 92-00899 and by the Office of Naval Research contract N00014-90-J-1739).

V and edge set E of the graph G. A *cycle* of length k in the graph G is a sequence of edges: $\{p_1,p_2\},\{p_2,p_3\},\cdots,\{p_{k-1},p_k\},\{p_k,p_1\}$ of G, in which $p_i \neq p_j$ if $i \neq j$.

Generally, DN is a larger set than CP. An example of an $A \in DN_5$ such that $A \notin CP_5$ is

$$A = \begin{bmatrix} 4 & 2 & 0 & 0 & 2 \\ 2 & 4 & 2 & 0 & 0 \\ 0 & 2 & 4 & 3 & 0 \\ 0 & 0 & 3 & 4 & 2 \\ 2 & 0 & 0 & 2 & 4 \end{bmatrix}.$$

Details of this example have been verified in [H], and an alternate brief explanation may be found in [DJL]. The value $n = 5$ is the smallest for which there is such an example, and

$G(A) =$ 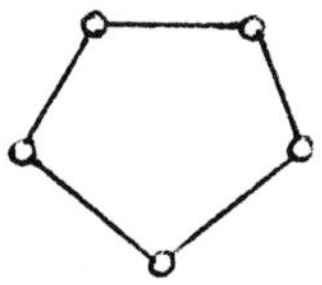 .

It is natural to ask, then, what are all the graphs G such that, whenever $A \in DN$ and $G(A) = G$, then $A \in CP$. Such graphs have been called *completely positive*, and they represent one class of patterns for which there is a definite check for complete positivity, namely just check the obvious necessary condition of double nonnegativity. (Another class of patterns, for which there is a different definitive check, has been described in [DJL].)

Surprisingly, there are many graphs that are completely positive, and a lovely characterization (see theorem 1 below) has recently been given in the series of papers [BG], [BH], [KB]], and [B] which, in addition to these four papers, relies upon the result of [MM] or [GW], dealing with the cases $n = 2,3,4$. Our purpose here is to give a self-contained and rather different proof of this "no long odd cycle" theorem. In the process, we are able to comment precisely on the possibilities for $\#(A)$. For simplicity, we call a graph a *no long odd cycle* (NLOC) graph if it has no cycles of odd length 5 or more.

Before stating the main results and proving them via a sequence of lemmas, we make several simple observations that will be useful. Both classes, DN_n and CP_n, are closed under permutation similarity and under positive diagonal congruence. Because of the former, a matrix under consideration may be taken to be in any form that permutation similarity allows or a graph may be taken to be in any form that isomorphism allows. The latter allows us to assume the diagonal entries of a matrix under consideration are all 1's, if convenient. Both classes are closed under addition of nonnegative diagonal matrices. Both classes are also closed under direct summation: $A \oplus B \in CP$(resp. DN) if and only if $A, B \in CP$(resp. DN). Moreover, if $A, B \in CP$, then $\#(A \oplus B) = \#(A) + \#(B)$. Finally, if $C \in M_n$ is nonsingular and $C^{-1} \geq 0$, then $C^T AC \in CP_n$ implies $A \in CP_n$.

The two main results are the following. As mentioned earlier, theorem 1 is known through a sequence of six earlier papers. Theorem 2 however, along with other specific comments about $\#(A)$, is new.

THEOREM 1. A graph is completely positive if and only if it is a NLOC graph.

THEOREM 2. If G is a connected NLOC graph and A is CP, with $G(A) = G$, then

$$\#(A) \leq \max\{|E(G)|, \text{rank}A\}.$$

In particular, if G is any NLOC graph and A is CP, with $G(A) = G$, then $\#(A) \leq [\frac{n^2}{4}]$, for $n \geq 4$.

The proofs of the two theorems are in parallel via a sequence of lemmas. The first two, one purely combinatorial, the other matrix theoretic, allow the proofs to focus upon a small set of special situations. Recall that a vertex v is a *cutpoint* of a graph G if the degree of v is ≤ 1 or if the removal of v from G leaves a graph G' with more connected components than G.

The strategy of proof is as follows. For sufficiency of the NLOC condition, we note that (1) after removal of cutpoints, there are just three basic types of NLOC graphs. (2) Matrix theoretically, cutpoints may be removed from a graph without disturbing the CP property and with a predictable effect upon the $\#(\cdot)$ function. (3) For each of the basic NLOC graphs, $CP = DN$, and the $\#(\cdot)$ function may be precisely estimated. Finally, (4) for necessity, if G is an odd length cycle on 5 or more vertices, we provide an example of an $A \in DN$, with $G(A) = G$ and $A \notin CP$; such examples may be perturbed to produce examples for arbitrary non-NLOC graphs.

A *basic* NLOC graph is one with no cutpoints. We denote by T_n the special graph on n vertices that consists of one or more triangles articulated along an edge. For example,

$T_6 =$ 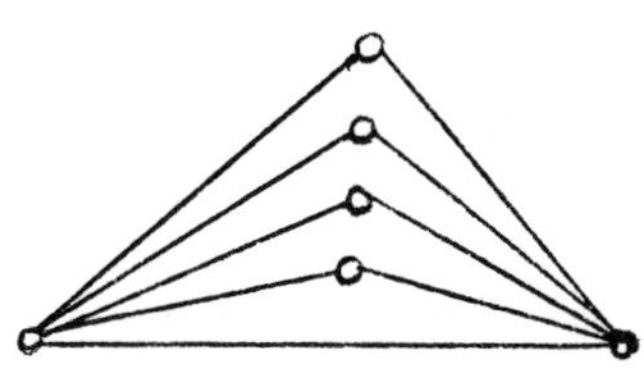 .

As usual, K_n denotes the complete graph on n vertices.

LEMMA 1. A connected, basic NLOC graph is either

a) bipartite;

b) $T_n, n \geq 3$; or

c) K_4.

Proof. Let G be a connected NLOC graph with no cutpoints. If G contains only even length cycles, then G is bipartite, as is well known. Alternatively, G contains at least one 3-cycle, H. If $H = G$, then $G = T_3$, and we are done. If $H \neq G$, imagine successively adding vertices (and edges) until we have G. Because of the "no cutpoint" assumption, there must be a vertex v_1 of G, not in H, that is adjacent to some vertex of H and lies on a cycle C of G that includes the edge of adjacency and an edge of H.

Consider the subgraph H' of G induced by the vertices of C together with those of H. If the length of C is even and H intersects C in just one edge, we would have an odd cycle of length at least 5 in G. If the length of C is even and at least 6 and H shares two edges with C, we would again have an odd cycle of length at least 5 in G. If the length of C is odd, it must be 3. We conclude that H' is either T_4 or K_4. If H' is K_4, the "no cutpoint" assumption implies $G = K_4$, and we are done. If H' is T_4 and $G \neq T_4$, again pick a vertex v_2 of G not in H', adjacent to a vertex of H' and lying on a cycle C' that includes an edge of H' and the edge of adjacency. Since v_2 is adjacent to one of the vertices of one of the triangles of $H' = T_4$, we argue as before to find that the subgraph of G induced by $H' \cup C'$ is either T_5 or contains a 5 cycle, which cannot occur. If $H' \cup C'$ is T_5 and $G \neq T_5$, again pick a vertex v_3 of G not in $H' \cup C'$, adjacent to a vertex of $H' \cup C'$ and lying on a cycle C'' that includes an edge of $H' \cup C'$ and the edge of adjacency. Since the subgraph of G induced by $H' \cup C' \cup C''$ is either

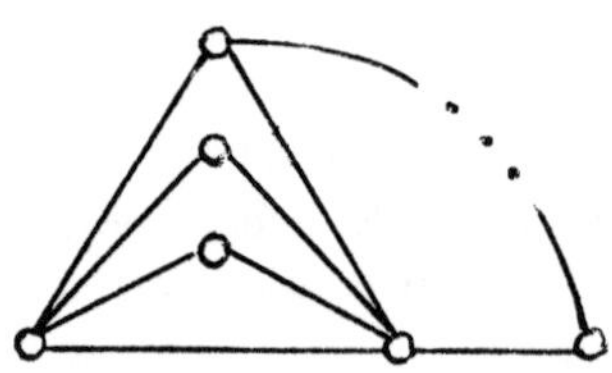

or

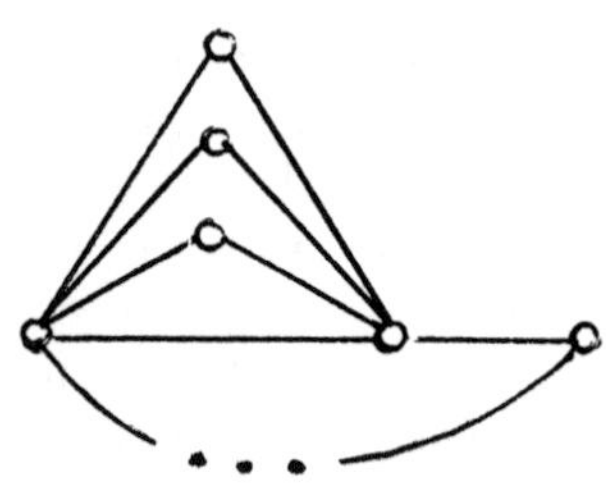

and G contains no long odd cycles, we conclude that $H' \cup C' \cup C''$ is T_6. Continuing to bring in further vertices as available, shows that G is T_n for some $n \geq 3$. □

In order to understand the effect of cutpoints matrix theoretically, we need the following lemma, which has been implicitly used for some time in the completion literature and which is most easily seen using Schur complements. Recall that the *Schur complement* [HJ1] of the nonsingular principal submatrix A_{11} in the square matrix

$$A = \begin{bmatrix} A_{11} & A_{12} \\ A_{21} & A_{22} \end{bmatrix}$$

is $A_{22} - A_{21}A_{11}^{-1}A_{12}$, which we denote as (A/A_{11}). More generally, if $\alpha \subseteq N = \{1, 2, \cdots, n\}$, the Schur complement of the nonsingular principal submatrix $A[\alpha]$ in the n-by-n matrix A is $A[\alpha^c] - A[\alpha^c, \alpha]A[\alpha]^{-1}A[\alpha, \alpha^c]$, denoted $(A/A[\alpha])$. Here, we use the standard notation $A[\alpha, \beta]$ to denote the submatrix of A lying in the rows $\alpha \subseteq N$ and columns $\beta \subseteq N$; $A[\alpha]$ is an abbreviation for the principal submatrix $A[\alpha, \alpha]$. Recall that $\text{rank}A = \text{rank}A[\alpha] + \text{rank}(A/A[\alpha])$ and that if $A[\alpha]$ is positive definite, the Hermitian matrix A is positive semidefinite if and only if $(A/A[\alpha])$ is.

LEMMA 2. Let

$$A = \begin{bmatrix} A_{11} & A_{12} & 0 \\ A_{12}^* & A_{22} & A_{23} \\ 0 & A_{23}^* & A_{33} \end{bmatrix}$$

be an n-by-n Hermitian matrix with A_{11}, A_{22}, and A_{33} square. Then A is positive semidefinite if and only if there exist Hermitian matrices A'_{22} and A''_{22} such that $A_{22} = A'_{22} + A''_{22}$ and such that

$$\hat{A}_{11} = \begin{bmatrix} A_{11} & A_{12} \\ A_{12}^* & A'_{22} \end{bmatrix} \text{ and } \hat{A}_{33} = \begin{bmatrix} A''_{22} & A_{23} \\ A_{23}^* & A_{33} \end{bmatrix}$$

are positive semidefinite.

Proof. ($\Leftarrow$) This follows directly from the fact that the sum of positive semidefinite matrices is positive semidefinite. ($\Rightarrow$) First consider the case in which A is positive definite. Then both A_{11} and the Schur complement of A_{11} in A,

$$(A/A_{11}) = \begin{bmatrix} A_{22} - A_{12}^*A_{11}^{-1}A_{12} & A_{23} \\ A_{23}^* & A_{33} \end{bmatrix},$$

are positive definite [HJ1]. Since (A/A_{11}) is positive definite, the Schur complement of A_{33} in (A/A_{11}),

$$((A/A_{11})/A_{33}) = A_{22} - A_{12}^*A_{11}^{-1}A_{12} - A_{23}A_{33}^{-1}A_{23}^*,$$

is positive definite. Let $A'_{22} = \frac{1}{2}((A/A_{11})/A_{33}) + A_{12}^*A_{11}^{-1}A_{12}$ and $A''_{22} = \frac{1}{2}((A/A_{11})/A_{33}) + A_{23}A_{33}^{-1}A_{23}^*$. Since $A_{12}^*A_{11}^{-1}A_{12}$ is positive definite, so is A'_{22}. Similarly A''_{22} is positive definite. $\hat{A}_{11}$ is positive definite since A_{11} is positive definite and A_{11}'s Schur complement in $\hat{A}_{11}$ is $A'_{22} - A_{12}^*A_{11}^{-1}A_{12} = \frac{1}{2}((A/A_{11})/A_{33})$ which is positive definite. Similarly $\hat{A}_{33}$ is positive definite as required.

In the case that A is positive semidefinite, but not positive definite, consider the matrix $A(\epsilon) = A + \epsilon I$. For any $\epsilon > 0$, $A(\epsilon)$ is positive definite and, as has been proven, yields two positive definite matrices, $\hat{A}_{11}(\epsilon)$ and $\hat{A}_{33}(\epsilon)$. Letting ϵ approach 0 produces the desired result. □

We note that for a symmetric matrix A, $G(A)$ has a cutpoint if and only if A (up to permutation similarity) may be partitioned as in lemma 2 with A_{22} a 1-by-1 matrix.

COROLLARY 1. Suppose that A is as in lemma 2 and that $A_{22} = a_{22}$ is 1-by-1. Then $A \in DN$ if and only if a_{22} may be written as $a_{22} = a'_{22} + a''_{22}$ so that $\hat{A}_{11}$, $\hat{A}_{33} \in DN$.

Proof. ($\Leftarrow$) The sum of doubly nonnegative matrices is doubly nonnegative. ($\Rightarrow$) By lemma 2, $\hat{A}_{11}$ and $\hat{A}_{33}$ are positive semidefinite. Thus $a'_{22} \geq 0$ and $a''_{22} \geq 0$, implying that $\hat{A}_{11}$ and $\hat{A}_{33}$ are both entrywise nonnegative. □

We observe that the decomposition $a_{22} = a'_{22} + a''_{22}$ may be chosen so that either $\hat{A}_{11}$ or $\hat{A}_{33}$ is singular. In particular, if $\hat{A}_{11}$ is not singular, we may decrease a'_{22} until it is. Because $\hat{A}_{11}$ is positive semidefinite, the singularity must occur for some $a'_{22} \geq 0$, thus maintaining $\hat{A}_{11}$'s double nonnegativity. Of course, $\hat{A}_{33}$ remains doubly nonnegative when a''_{22} is increased a corresponding amount.

COROLLARY 2. Suppose that A is as in lemma 2 and that A_{22} is 1-by-1. Then $A \in CP$ if and only if a_{22} may be written $a_{22} = a'_{22} + a''_{22}$ so that $\hat{A}_{11}$ and $\hat{A}_{33}$ are CP. Moreover, if $A \in CP$, then $\#(A) = \min\{\#(\hat{A}_{11}) + \#(\hat{A}_{33})\}$, in which the minimum is taken over all possible decompositions $a_{22} = a'_{22} + a''_{22}$ leading to CP matrices $\hat{A}_{11}$ and $\hat{A}_{33}$.

Proof. Although the first part of corollary 2 can be proven directly from lemma 2, we provide instead an independent proof. Suppose, without loss of generality, that

$$A[\alpha] = \begin{bmatrix} A_{11} & A_{12} \\ A_{12}^T & A_{22} \end{bmatrix} \text{ and } A[\beta] = \begin{bmatrix} A_{22} & A_{23} \\ A_{23}^T & A_{33} \end{bmatrix},$$

in which $\alpha = \{1, \cdots, k\}$, $\beta = \{k, \cdots, n\}$, and that k is the cutpoint. ($\Rightarrow$) Since A is CP, $A = \sum b_i b_i^T$. Because of the zero blocks in A, the support of each b_i lies wholly in α or wholly in β. Let $\hat{A}_{11}$ be the sum of the $b_i b_i^T$ over those i for which the support of b_i is a subset of α. Let $\hat{A}_{33}$ be the sum using β. (If support$(b_i) = \{k\}$, assign $b_i b_i^T$ to, say, $\hat{A}_{11}$.) Clearly $\hat{A}_{11}$ and $\hat{A}_{33}$ are CP. ($\Leftarrow$) This implication follows from the fact that the sum of CP matrices is CP.

If $A \in CP$, the initial part of this proof provides information about $\#(A)$. We observe that any pair of CP matrices $\hat{A}_{11}$ and $\hat{A}_{33}$ have individual representatiions of the type $\sum b_i b_i^T$ which may essentially be added together to form a representation for A. Also, any representation $\sum b_i b_i^T$

for A may be divided into two parts to form a pair of CP matrices $\hat{A}_{11}$ and $\hat{A}_{33}$. Ambiguity occurs only if some b_i has support $\{k\}$, in which case $b_i b_i^T$ may be divided arbitrarily between $\hat{A}_{11}$ and $\hat{A}_{33}$. It follows that $\#(A) = \min\{\#(\hat{A}_{11}) + \#(\hat{A}_{33})\}$, where the minimum is taken over all decompositions $a_{22} = a'_{22} + a''_{22}$ leading to CP matrices $\hat{A}_{11}$ and $\hat{A}_{33}$. □

Via lemma 1 and the corollaries to lemma 2, it will be seen that it suffices to study the graphs mentioned in lemma 1 in order to prove theorems 1 and 2. In the next few lemmas we discuss the case of bipartite graphs. Recall that an n-by-n matrix $A = (a_{ij})$ is (weakly) *diagonally dominant* if

$$|a_{ii}| \geq \sum_{j \neq i} |a_{ij}|, \qquad i = 1, \cdots, n.$$

LEMMA 3. If $A \in DN$ and $G(A)$ is bipartite, then there is a positive diagonal matrix D such that DAD is diagonally dominant.

Proof. After permutation similarity and positive diagonal congruence (which alter neither the hypothesis nor the conclusion), we may assume that A has the form

$$A = \begin{bmatrix} I_{n_1} & A_{12} \\ A_{12}^T & I_{n_2} \end{bmatrix}.$$

Let $F = -I_{n_1} \oplus I_{n_2}$. Then

$$F^T A F = FAF = \begin{bmatrix} I_{n_1} & -A_{12} \\ -A_{12}^T & I_{n_2} \end{bmatrix}$$

is also positive semidefinite and, thus (because all off-diagonal entries are nonpositive), an M-matrix [HJ2]. It follows [HJ2] that there is a positive diagonal matrix D such that $D(FAF)D$ is diagonally dominant. As the entries of $DFAFD$ equal those of DAD in absolute value, DAD is also diagonally dominant. □

We may now verify

LEMMA 4. If $A \in DN$ and $G(A)$ is bipartite, then $A \in CP$.

Proof. We may assume, without loss of generality, that A is irreducible. If $A = (a_{ij})$ is singular, then we may assume, by lemma 3, that

$$\sum_{j \neq i} a_{ij} = a_{ii}, \qquad i = 1, \cdots, n.$$

(If A is diagonally dominant and irreducible with strict inequality in any row, A is nonsingular [HJ1].) We may then write

$$A = \sum a_{ij}(e_i + e_j)(e_i + e_j)^T,$$

in which e_i is the i-th unit vector and the sum is over pairs i, j such that $i < j$ and $a_{ij} > 0$. If $A = (a_{ij}) \in M_n$ is nonsingular, we provide an overly-detailed discussion in order to facilitate the proof of lemma 5. Consider two possibilities. If $G(A)$ is a tree, write $A = A_1 + \alpha e_1 e_1^T$, in which $\alpha > 0$ and $A_1 \in DN$ is singular. The matrix A_1 may be treated as before to write A in terms of $\alpha e_1 e_1^T$ and $|E(G(A))| = n - 1$ other rank one nonnegative matrices, a total of $n = \text{rank}A$. If $G(A)$ is not a tree, $G(A)$ must have an edge $\{i, j\}$ whose removal leaves the graph connected. Write A as

$$A = A_1 + R$$

in which $R = xe_ie_i^t + ye_je_j^t + a_{ij}(e_ie_j^T + e_je_i^T)$ and $x, y > 0$ are chosen so that $A_1 \in DN$ is, again, singular and rank $R = 1$. (It is easy to check that this may be done.) Now, A_1 is irreducible and may be written in terms of $|E(G(A_1))| = |E(G(A))| - 1$ nonnegative rank one matrices, as before, and then A may be written in terms of $|E(G(A))|$ rank one matrices, where $n = \text{rank}A \leq |E(G(A))|$, which completes the proof. □

A careful examination of the proof of lemma 4 reveals how to determine $\#(A)$ in the case of a bipartite graph.

LEMMA 5. If $A \in CP$ and $G(A)$ is bipartite and connected, then

$$\#(A) = \max\{|E(G(A))|, \text{rank}A\}.$$

Proof. That $\#(A) \leq \max\{|E(G(A))|, \text{rank}A\}$ is clear from the proof of lemma 4. Clearly $\#(A) \geq \text{rank}A$, always. We complete the proof by showing that since $G(A)$ is bipartite, $\#(A) \geq |E(G(A))|$ also. Note that the graph of any rank one matrix bb^T is a clique of cardinality equal to that of the support. However, as no cancellation may occur in a representation of A as a sum of rank one nonnegative matrices and a maximal clique in a connected bipartite graph is an edge, there must be at least one rank one matrix for each edge of $G(A)$, i.e. $\#(A) \geq |E(G(A))|$, as claimed. □

Of course, $\text{rank}(A) \leq |E(G(A))| + 1$ whenever $G(A)$ is connected (i.e. A irreducible) and the only case in which $|E(G(A))|$ does not attain the maximum in lemma 5 is the case in which $G(A)$ is a tree and A is nonsingular. In this case $\#(A) = \text{rank}(A) = |E(G(A))| + 1$. If A is a direct sum (i.e. $G(A)$ is bipartite but not connected), then rank A could be greater than $|E(G)|$ by as much as the number of connected components (i.e. if all the components were trees). In any event, as $|E(G)| \leq \left[\frac{n^2}{4}\right]$ if $G(A)$ is a bipartite graph on $n \geq 3$ vertices, it follows from lemma 5 that $\#(A) \leq \left[\frac{n^2}{4}\right]$ if $G(A)$ is bipartite and A is completely positive and n-by-n, $n \geq 4$.

For $n \leq 4$, the general situation is fairly simple.

LEMMA 6. If $A \in DN_n$, $n \leq 3$, then $A \in CP_n$. Furthermore, if $A \in CP_n$, $n \leq 3$, then

$$\#(A) = \text{rank}A.$$

Proof. We may assume throughout that $a_{ii} = 1$, for $i = 1, ..., n$. The case $n = 1$ is clear. When $n = 2, A = BB^T$ where $B = \begin{bmatrix} 1 & 0 \\ a_{12} & \sqrt{\det A} \end{bmatrix}$. When $n = 3$, if $a_{12} = a_{13} = a_{23} = 1$, then $A = BB^T$, where $B = A/\sqrt{3}$. Otherwise, because CP is invariant under permutation similarity, we may assume $a_{12} < 1$ and $a_{23} \geq a_{12}a_{13}$. Letting $b = (1, a_{12}, a_{13})^T$,

$$A = bb^T + \begin{bmatrix} 0 & 0 & 0 \\ 0 & 1 - a_{12}^2 & a_{23} - a_{12}a_{13} \\ 0 & a_{23} - a_{12}a_{13} & 1 - a_{13}^2 \end{bmatrix} = bb^T + \left[\begin{array}{c|cc} 0 & 0 & 0 \\ \hline 0 & F & \\ 0 & & \end{array}\right].$$

Since A is positive semidefinite and $[a_{11}]$ is positive definite, the Schur complement of $[a_{11}]$ in A must be positive semidefinite. Thus F is positive semidefinite. Because F is 2-by-2 and in DN, $F \in CP$ and hence $A \in CP$.

Clearly, when $n = 2, \#A = \text{rank}A$. When $n = 3$, it follows from the above representation of A that $\#(A) \leq 3$. Thus, if rank $A = 3, \#A = \text{rank}A$. If rank $A = 2$, F is singular, so $\#(F) = 1$ and $\#(A) = 2$. Thus $\#(A) = \text{rank}A$. □

Our treatment of the case K_4 and T_n, $n \geq 4$, have common technical elements, for which we need the following observation.

LEMMA 7. If $0 < x < 1, a > 0, b > 0$ and $\frac{x}{1+\sqrt{1-x^2}} < \min\{a/b, b/a\}$, then

$$\begin{bmatrix} 1 & x \\ x & 1 \end{bmatrix}^{-1/2} \begin{bmatrix} a \\ b \end{bmatrix} > 0.$$

Proof. Let $\begin{bmatrix} 1 & x \\ x & 1 \end{bmatrix}^{-\frac{1}{2}} = \begin{bmatrix} \alpha & -\beta \\ -\beta & \alpha \end{bmatrix}$. Then $\alpha^2 + \beta^2 = \frac{1}{1-x^2}$ and $2\alpha\beta = \frac{x}{1-x^2}$. Solving for α and β yields $\frac{\beta}{\alpha} = \frac{x}{1+\sqrt{1-x^2}}$. Assuming $\beta/\alpha < min\{a/b, b/a\}$,

$$\begin{bmatrix} 1 & x \\ x & 1 \end{bmatrix}^{-\frac{1}{2}} \begin{bmatrix} a \\ b \end{bmatrix} = \begin{bmatrix} \alpha & -\beta \\ -\beta & \alpha \end{bmatrix} \begin{bmatrix} a \\ b \end{bmatrix} = \begin{bmatrix} \alpha a - \beta b \\ \alpha b - \beta a \end{bmatrix} = \alpha \begin{bmatrix} a - (\beta/\alpha)b \\ b - (\beta/\alpha)a \end{bmatrix} > 0.$$

□

We may next treat the case of K_4, which lacks the useful structure of a bipartite graph or of T_4. Previous proofs have appeared, for example, in [MM] and [GW]. Our proof exploits some of the features of the former.

LEMMA 8. If $A \in DN_4$ and $G(A) = K_4$, then $A \in CP_4$. Moreover, if $A \in CP_4$, then $\#(A) \leq 4$.

Proof. We assume that $a_{ii} = 1$, for $i = 1, ..., 4$. There are two cases to consider. First, suppose $A - a_i a_i^T \geq 0$ for some column a_i of A. For convenience, assume $i = 1$.

$$A = a_i a_i^T + \left[\begin{array}{c|ccc} 0 & 0 & 0 & 0 \\ \hline 0 & & & \\ 0 & & F & \\ 0 & & & \end{array}\right]$$

where F is the Schur complement of $[a_{11}]$ in A. Since both A and $[a_{11}]$ are positive semidefinite, so is F. Because F is 3-by-3 and in DN, lemma 6 implies $F \in CP$ and $\#(F) \leq 3$. Hence $A \in CP$ and $\#A \leq 4$.

Second, suppose $A - a_i a_i^T \geq 0$ does not hold for any column of A. An examination of the resulting sets of inequalities, combined with the double nonnegativity of A and the fact that CP is closed under permutation similarity, allows us to assume that $a_{12} < a_{13}a_{23}$, $a_{12} < a_{14}a_{24}$, $a_{34} < a_{13}a_{14}$, and $a_{34} < a_{23}a_{24}$. Partition A into 2-by-2 submatrices $A_{ij}, 1 \leq i, j \leq 2$, and define $D = A_{11}^{-\frac{1}{2}} \oplus I_2$. Letting $B = D^T AD$, we have

$$B = \begin{bmatrix} I_2 & A_{11}^{-\frac{1}{2}} A_{12} \\ (A_{11}^{-\frac{1}{2}} A_{12})^T & A_{22} \end{bmatrix}.$$

Lemma 7 implies $A_{11}^{-\frac{1}{2}} A_{12} > 0$, since $\frac{a_{12}}{1+\sqrt{1-a_{12}^2}} \leq a_{12} \leq min\{\frac{a_{13}}{a_{23}}, \frac{a_{23}}{a_{13}}, \frac{a_{14}}{a_{24}}, \frac{a_{24}}{a_{14}}\}$. Thus $B \geq 0$ and B is positive semidefinite, so $B \in DN$.

If $B - b_i b_i^T \geq 0$ for some column b_i of B, then B, and hence A, is CP by the same argument used for A above. Also $\#(A) = \#(B) \leq 4$. If not, we may assume by permutation similarity that $b_{34} = 0$, $b_{12} < b_{13}b_{23}$ and $b_{12} < b_{14}b_{24}$. Partitioning B into 2-by-2 submatrices and letting $H = B_{11}^{-\frac{1}{2}} \oplus I_2$, then

$$C = H^T BH = \begin{bmatrix} I_2 & B_{11}^{-\frac{1}{2}} B_{12} \\ (B_{11}^{-\frac{1}{2}} B_{12})^T & I_2 \end{bmatrix}$$

is DN. Since $G(C)$ is bipartite, lemma 4 implies that C, and hence A, is CP. In this case, $\#(A) = \#(B) = \#(C) \leq 4$ by lemma 5. □

Our treatment of the case of T_n is rather similar.

LEMMA 9. If $A \in DN_n$ and $G(A) = T_n$, then $A \in CP_n$. Moreover, if $A \in CP_n$, $n > 3$, and $G(A) = T_n$, then $\#(A) \leq |E(G(A))| - 1$.

Proof. For $n = 3$, the result follows directly from lemma 6. We may assume $n \geq 4$ and that A has the form

$$A = \left[\begin{array}{cc|c} 1 & a_{12} & A_{12} \\ a_{12} & 1 & \\ \hline A_{12}^T & & I_{n-2} \end{array}\right]$$

with $a_{12} > 0$ and $A_{12} > 0$. There are two cases to consider. First, if $a_{12} \geq a_{1k}a_{2k}$ for some $k \geq 3$, we can reduce row k and column k of A to all zeros by subtracting a suitable rank 1 matrix from A, leaving a DN matrix whose graph is T_{n-1}, and then proceed by induction. More specifically, we may assume $k = 3$ and let $\tilde{A} = A - bb^T$ where $b^T = (a_{13}, a_{23}, 1, 0, ..., 0)$.

$$\tilde{A} = \left[\begin{array}{ccc|c} 1 - a_{13}^2 & a_{12} - a_{13}a_{23} & 0 & \\ a_{12} - a_{13}a_{23} & 1 - a_{23}^2 & 0 & \tilde{A}_{12} \\ 0 & 0 & 0 & \\ \hline & \tilde{A}_{12}^T & & I_{n-3} \end{array}\right] \geq 0$$

Note that the Schur complement of I_{n-3} in $\tilde{A}$ is positive semidefinite because, if we remove its last row and column of zeros, we are left with the Schur complement of I_{n-2} in A, which is positive semidefinite. Thus $\tilde{A}$ is positive semidefinite, so $\tilde{A} \in DN$. Note also that $\#(A) \leq \#(\tilde{A}) + 1$ and $|E(G(A))| = |E(G(\tilde{A}))| + 2$, which is consistent with an inductive proof of the bound for $\#(A)$.

In the second case $a_{12} < a_{1k}a_{2k}$ for $k = 3, ..., n$. Let

$$A_{11} = \begin{bmatrix} 1 & a_{12} \\ a_{12} & 1 \end{bmatrix} \text{ and } D = \begin{bmatrix} A_{11}^{-\frac{1}{2}} & 0 \\ 0 & I_{n-2} \end{bmatrix}.$$

It will be shown that $DAD \in DN$ and $G(DAD)$ is bipartite. Then by lemma 4, $DAD \in CP$, and since $D^{-1} \geq 0, A \in CP$. It remains to show $DAD \in DN$ and $G(DAD)$ is bipartite. $G(DAD)$ is bipartite since

$$DAD = \begin{bmatrix} I_2 & A_{11}^{-\frac{1}{2}} A_{12} \\ A_{12}^T A_{11}^{-\frac{1}{2}} & I_{n-2} \end{bmatrix}.$$

DAD is positive semidefinite since A is positive semidefinite. $A_{11}^{-\frac{1}{2}} A_{12} > 0$ since we may apply lemma 7 to the columns of A_{12}. Thus $DAD \in DN$. This procedure shows that $\#(A)$ is no greater than $\#(DAD)$. Since $n > 3$, $G(DAD)$ is not a tree, and we have $\#(DAD) = |E(G(DAD))|$ from the comments following the proof of lemma 5. Because $|E(G(DAD))| = |E(G(A))| - 1$, we conclude $\#(A) \leq |E(G(A))| - 1$. □

We note that if $A \in CP_n, n > 3$, and $G(A) = T_n$, then $n - 2 \leq \#(A) \leq 2n - 4 = |E(T_n)| - 1 \leq \left[\frac{n^2}{4}\right]$. The graph of a rank 1 matrix is a clique, and the largest clique in T_n is K_3. As no cancellation is possible in the representation of a completely positive matrix as a sum of nonnegative rank one matrices, this explains the lower bound. The upper bound is lemma 9, and this upper bound is attainable for $n \geq 4$, as is the lower bound. Of course, any value between the minimum and maximum is possible.

The preceding lemmas form the basis for a proof of the sufficiency in theorem 1 and of theorem 2. For the necessity in theorem 1, we need the following observation.

LEMMA 10. If a graph G has an odd length cycle of length ≥ 5, then there exists a matrix $A \in DN$ such that $G(A) = G$ and $A \notin CP$.

Proof. Let $n = |V(G)|$ and let C be a 0-1 matrix whose graph is an odd cycle of length $m \geq 5$, with $G(C) \subseteq G$. The matrix C can be taken to be the m-by-m circulant matrix whose first row $(0, 1, 0, ..., 0, 1)$ has exactly two nonzero entries. By the Perron-Frobenius theorem, $\rho(C) = 2$. Since all eigenvalues of C lie in the interval $[-2, 2]$, all the eigenvalues of $B = C + 2I_m$ lie in $[0, 4]$. But 0 cannot be an eigenvalue of B because B is invertible. (B^{-1} is a circulant matrix whose first row is $\frac{1}{4}(m, -(m-2), (m-4), -(m-6), ..., -(-1)^{[\frac{m}{2}]}3, (-1)^{[\frac{m}{2}]}, (-1)^{[\frac{m}{2}]}, -(-1)^{[\frac{m}{2}]}3, ..., (m-4), -(m-2))$.) Thus B is positive definite. Since the eigenvalues of B depend continuously on the entries of B, the $(1, m)$ and $(m, 1)$ entries may be increased by some $\epsilon > 0$ without sacrificing positive definiteness. In this way we create a new matrix $\tilde{B} \in DN$, with $\tilde{B} \notin CP$, as verified below. If $\tilde{B} \in CP$, then $\tilde{B} = \sum b_i b_i^T$. Since $G(\tilde{B})$ contains no 3-cliques, each b_i can have at most two nonzero entries, say x_i and y_i. Because $x_i^2 + y_i^2 \geq 2x_i y_i$, the sum of the diagonal entries of each $b_i b_i^T$ cannot be smaller than the sum of the off-diagonal entries. Clearly $\tilde{B}$ inherits this property, but $2m < 2m + 2\epsilon$, a contradiction.

It follows easily that $A = \tilde{B} \oplus I_{n-m}$ is positive definite, $A \notin CP$ and $G(A) \subseteq G$. Since CP is a closed set, appropriate 0 elements of A can be increased slightly to create a matrix $\tilde{A}$ with $G(\tilde{A}) = G$ and $\tilde{A} \notin CP$. Small increases maintain positive definiteness, so $\tilde{A} \in DN$ as required. □

Proof of theorems 1 and 2. Lemma 10 shows that the NLOC condition is necessary in theorem 1. It suffices to prove sufficiency in theorem 1 for connected graphs since CP is closed under direct summation. Also, it suffices to prove theorem 2 for graphs that are not trees because if $G(A)$ is a tree, $\#(A) \leq \operatorname{rank} A$ by lemma 5 and $\operatorname{rank} A \leq n \leq \left[\frac{n^2}{4}\right]$ for $n \geq 4$. We will use induction to prove theorems 1 and 2 in parallel, augmenting theorem 2 by the statement that $\#(A) \leq |E(G(A))|$, except when A is nonsingular and $G(A)$ is a tree, in which case $\#(A) = \operatorname{rank} A = |E(G(A))| + 1$. Theorems 1 and 2 are true for $1 \leq n \leq 4$ by lemmas 4, 5, 6, 8 and 9. We now assume that both theorems are true for $n = 1, 2, \cdots, k \geq 4$ and show that both are true for $n = k + 1$.

Let $A \in DN_{k+1}$, $k \geq 4$, and let $G(A)$ be a connected NLOC graph. If $G(A)$ has no cutpoints, then it is a basic NLOC graph and, by lemma 1, is bipartite or T_{k+1}. By lemmas 4, 5 and 9, $A \in CP$ and $\#(A) \leq \max\{|E(G(A))|, \operatorname{rank} A\} \leq \left[\frac{(k+1)^2}{4}\right]$. If $G(A)$ has at least one cutpoint, then there are matrices $\hat{A}_{11} \in DN_{n_1}$ and $\hat{A}_{33} \in DN_{n_2}$, in which $n_1 + n_2 = k + 1$, as in corollary 1. Since $G(\hat{A}_{11})$ and $G(\hat{A}_{33})$ are connected NLOC graphs and n_1, $n_2 \leq k$, we have by the induction hypothesis that $\hat{A}_{11}$ and $\hat{A}_{33}$ are in CP and thus $A \in CP$.

By assumption, $G(A)$ is not a tree so $G(\hat{A}_{11})$ and $G(\hat{A}_{33})$ cannot both be trees. If neither is a tree, we have, by the augmented form of theorem 2, that $\#(\hat{A}_{11}) \leq |E(G(A_{11}))|$ and $\#(\hat{A}_{33}) \leq |E(G(A_{33}))|$ and then, by corollary 2, that $\#(A) \leq \#(\hat{A}_{11}) + \#(\hat{A}_{33}) \leq |E(G(A))|$. If one of them is a tree, say $G(\hat{A}_{11})$, this same result may be achieved since we may require, by the observation following corollary 1, that $\hat{A}_{11}$ is singular so that, by the augmented form of theorem 2, $\#(\hat{A}_{11}) \leq |E(G(A_{11}))|$. Thus, in any case, $\#(A) \leq \max\{|E(G(A))\}, \mathrm{rank} A\}$.

We now show $\#(A) \leq \left[\frac{(k+1)^2}{4}\right]$, first considering the case in which $G(A)$ is connected. If $n_1 \geq 4$ and $n_2 \geq 4$, then $\#(A) \leq \#(\hat{A}_{11}) + \#(\hat{A}_{33}) \leq \left[\frac{n_1^2}{4}\right] + \left[\frac{n_2^2}{4}\right] \leq \left[\frac{(n_1+n_2-1)^2}{4}\right] = \left[\frac{(k+1)^2}{4}\right]$, in which the last inequality is routine to verify. If exactly one of the n_i's is less than 4, say $n_1 < 4$, we may require, by the observation following corollary 1, that $\hat{A}_{11}$ is singular so that, by lemma 6, $\#(\hat{A}_{11}) = \mathrm{rank}\hat{A}_{11} \leq \left[\frac{n_1^2}{4}\right]$ and then, as before, $\#(A) \leq \left[\frac{(k+1)^2}{4}\right]$. Finally, if $n_1 = n_2 = 3$, then $k+1 = 5$ and, by lemma 6, $\#(A) \leq \#(\hat{A}_{11}) + \#(\hat{A}_{33}) \leq 3 + 3 = \left[\frac{(k+1)^2}{4}\right]$.

In the case that $A \in DN_{k+1}$, $k \geq 4$, and $G(A)$ is an NLOC graph, but is not connected, we may write $A = A_1 \oplus A_2$ in which $A_1,\ A_2 \in DN$ and $G(A_1),\ G(A_2)$ are NLOC graphs. Applying the induction hypothesis and lemma 6 to $\#(A) = \#(A_1) + \#(A_2)$ shows that $\#(A) \leq \left[\frac{(k+1)^2}{4}\right]$. □

REFERENCES

[B] A. Berman, Completely positive graphs in "Combinatorial and Graph-Theoretical Problems in Linear Algebra", The IMA Volumes in Mathematics and its Applications, Volume 50, Springer-Verlag, New York, 1993, pp. 229-233.

[BG] A. Berman and R. Grone, Bipartite completely positive matrices, Proc. Cambridge Philos Soc. 103(1988), pp. 269-276.

[BH] A. Berman and D. Hershkowitz, Combinatorial results on completely positive matrices, Linear Algebra and its Applications 95(1987), pp. 111-125.

[D] P. Diaconis, Talk at IMA Workshop, November, 1993.

[DJL] J. Drew, C.R. Johnson and R. Loewy, Completely positive matrices associated with M-matrices, to appear in Linear and Multilinear Algebra (1994).

[GW] L.J. Gray and D.G. Wilson, Nonnegative factorization of positive semidefinite nonnegative matrices, Linear Algebra and its Applications 31(1980), pp. 119-127.

[H] M. Hall Jr., Combinatorial Theory, Blaisdell, Lexington, 1967, 2nd ed. 1986.

[HJ1] R. Horn and C.R. Johnson, Matrix Analysis, Cambridge University Press, Cambridge, 1985.

[HJ2] R. Horn and C.R. Johnson, Topics in Matrix Analysis, Cambridge University Press, Cambridge, 1991.

[KB] N. Kogan and A. Berman, Characterization of completely positive graphs, Discrete Math, 114 (1993), pp. 297-304.

[MM] J.E. Maxfield and H. Minc, On the equation $X'X = A$, Proc. Edinburgh Math. Soc. 13(1962), pp. 125-129.

A NOTE ON TRIANGLE-FREE GRAPHS

PAUL ERDŐS*, SVANTE JANSON†, TOMASZ ŁUCZAK‡, AND JOEL SPENCER§

1. Introduction. There is a natural, if imprecise, notion that the requirement of trianglefreeness on a graph G forces it into a bipartite-like form. In an extreme case, if G has n vertices and $e = \lfloor \frac{n^2}{4} \rfloor$ edges then Turán's Theorem gives that it must be the bipartite graph $K_{\lfloor n/2 \rfloor, \lceil n/2 \rceil}$. Our concern here is what happens when e is smaller, must G still exhibit bipartite-like behavior. Very roughly, our answer is: Yes, if $e \gg n^{3/2}$, No otherwise.

We define $B(G)$ to be the maximal number of edges over all *induced* bipartite subgraphs of G. We define $f(n,e)$ to be $\min B(G)$ over all G with n vertices and e edges. We define $g(e)$ to be $\min B(G)$ over all G with e edges so that $g(e) = \min_n f(n,e)$.

THEOREM 1.

$$c_1 e^{1/3} \le g(e) \le c_2 e^{1/3} \ln^2 e$$

THEOREM 2. For $e < c_5 n^{3/2}$

$$c_1 e^{1/3} \le f(n,e) \le c_2 e^{1/3} \ln^2 e$$

while for $c_5 n^{3/2} < e < n^2/4$

$$c_3 e^3 n^{-4} \le f(n,e) \le c_4 e^3 n^{-4} \ln^2 n$$

Here $c_1, c_2 \ldots$ represent absolute positive constant whose optimal values we do not attempt to compute. The polylogarithmic factor between lower and upper bounds would be very nice to eliminate but we have not been able to do so.

2. In a paper of Erdős.... The upper bound of Theorem 1 is based on a 1961 paper by the senior author [1]. There it is shown that the random graph G on n vertices and $n^{3/2} A^{-1/2}$ edges (A a sufficiently large constant) almost surely contains a trianglefree graph G_1 with no independent set of size $x = A n^{1/2} \ln n$. In modern language the random $G(n,p)$ with edge

* Mathematics Institute, Realtanoda 13, Budapest, Hungary.

† Department of Mathematics, Uppsala University, P.O. Box 480, S-751 06 Uppsala, Sweden.

‡ Department of Mathematics and Computer Science, Emory University, Atlanta, GA 30322.

§ Department of Computer Science, Courant Institute, New York University, 251 Mercer Street, New York, NY 10012.

probability $p = A^{-1/2}n^{-1/2}$ almost surely has this property. Let $B^+(G_1)$ be the maximal number of edges of G_1 on a subset of $2x$ vertices. Then $B(G_1) \leq B^+(G_1)$ almost surely and $B^+(G_1) \leq B^+(G)$ tautologically. Now

$$\Pr[B^+(G) \geq \alpha] \leq \binom{n}{2x}\binom{\binom{2x}{2}}{\alpha}p^\alpha \leq n^{2x}\left(\frac{2ex^2p}{\alpha}\right)^\alpha = o(1)$$

for $\alpha = cn^{1/2}\ln^2 n$. That is, $B(G_1) = O(n^{1/2}\ln^2 n)$ almost surely. Thus there exist G_1 with n vertices, $e = O(n^{3/2})$ edges with

$$B(G_1) = O(n^{1/2}\ln^2 n) = O(e^{1/3}\ln^2 e)$$

as claimed.

3. Two proofs of a lower bound. Here we give two arguments that $g(e) = \Omega(e^{1/3})$.

For the first proof we show

$$g(e) \geq \min_{1 \leq d \leq e} d + f(e - d(d^2+1))$$

where we interpret $f(a)$ as zero if a is negative. The conclusion $g(e) = \Omega(e^{1/3})$ follows from elementary analysis. Let G be a graph with e edges and let d be the maximal degree $\deg(x)$ over $x \in V(G)$. Let $N(x)$ denote the set of neighbors of x, an independent set as G is trianglefree. Consider the restriction of G to $x \cup N(x)$, $B(G) \geq \deg(x) = d$. Let $N^2(x)$ denote the vertices at distance two from x. Let H be G with vertices $x \cup N(x) \cup N^2(x)$ deleted. At most $1+d+d(d-1) = d^2+1$ vertices and hence at most $d(d^2+1)$ edges have been deleted so that H has an induced bipartite $A \times B$ with $f(e - d(d^2+1))$ edges. But then $(A \cup \{x\}) \times (B \cup N(x))$ is an induced bipartite graph with $d + f(e - d(d^2+1))$ edges.

The second proof is probabilistic. Take $c = .1$ for definiteness and we wish to show $g(e) \geq ce^{1/3}$. We can assume $d \leq ce^{1/3}$ since, as already stated, $B(G) \geq d$. Pick a random subset S of the vertices by

$$\Pr[x \in S] = p = ce^{-2/3}$$

From S delete all pairs of points at distance one or two, yielding a random set S^-. Let T be the set of neighbors of points of S^-. From T delete all y, y' for which there is a path $xyy'x'$ of length 3 with $x, x' \in S^-$. Then G restricted to $S^- \cup T^-$ is an induced bipartite graph – indeed, the union of stars centered at the $x \in S^-$. Let X be the number of edges in this random restriction. We claim

$$E[X] = \Omega(e^{1/3})$$

Let's concentrate on a single edge $\{x, y\}$ of G. With probability $p(1-p) = \Omega(e^{-2/3})$ we select x but not y in S. Now if we don't have $x \in S^-, y \in T^-$

we must have some $x' \neq x, y$ at distance one or two from x or y. There are at most $2(d^2+1)$ potential such x' so the probability some such $x' \in S$ is at most $2(d^2+1)p$ which is certainly less than .01 by our choice of c. With probability at least .99 this does not occur. Thus the probability that $x \in S^-, y \in T^-$ is still $\Omega(e^{-2/3})$. The bound on $E[X]$ follows by the linearity of expectation. Some choice must reach the expectation so there exist S, T and hence S^-, T^- with $X \geq c_2 e^{1/3}$ as desired.

4. n vertices, e edges. Here we show Theorem 2. Suppose $e \leq c_5 n^{3/2}$. Clearly $f(n,e) \geq g(e) = \Omega(e^{1/3})$. The construction of Theorem 1 has $e = \Theta(n^{3/2})$. For $e \leq c_5 n^{3/2}$ there is such a constuction with $n' = \Theta(e^{3/2}) \leq n$ vertices. Adding isolated vertices if necessary gives a graph G with n vertices, e edges and $B(G) = O(e^{1/3}\ln^2 e)$.

Now suppose $e > c_5 n^{3/2}$ and we show $f(n,e) = O(e^3 n^{-4} \ln^2 n)$. As $f(n,e) \leq \binom{n}{2}$ tautologically we can further assume $e = o(n^2)$. Set $m = e^2 n^{-3}$ and split the n vertices in disjoint sets I_i, $1 \leq i \leq \frac{n}{m}$, each of size m. There is a graph G_0 on vertex set $1, \ldots, \frac{n}{m}$ with $\Theta((n/m)^{3/2})$ edges and $B(G_1) = O((n/m)^{1/2}\ln^2(n/m))$. Define G on n vertices by blowing up point i into set I_i. That is, the vertex set is the union of the I_i and $x \in I_i, y \in I_j$ are joined if and only if i, j are adjacent in G_0. Then G is trianglefree and has $\Theta(m^2(n/m)^{3/2}) = \Theta(e)$ edges with

$$B(G) = O(m^2(n/m)^{1/2}\ln^2(n/m)) = O(e^3 n^{-4}\ln^2(n/m)) = O(e^3 n^{-4}\ln^2 n)$$

as desired.

Finally, we turn to the lower bound. Set $d = 2e/n$, the average degree. Remove sequentially vertices of degree at most, say, $d/10$. This can remove at most $nd/10 = e/5$ edges so one is left with an induced graph G' on $n' \leq n$ vertices and at least $.8e$ edges, all degrees at least $d/10$. Now, restricting our attention to G' we find an induced subgraph of the form $N(x) \times N(y)$. Note that since G' is trianglefree this graph will be an induced bipartite graph. (If the sets intersect we consider it as $N(x) \times N(y) - N(x)$.) Set $f(x,y)$ equal to the number of edges in $N(x) \times N(y)$. Observe that $f(x,y)$ is precisely the number of paths of length three from x to y since each path $xaby$ corresponds to a unique edge ab. Thus $\sum f(x,y)$ gives the number of ordered 3-paths in G'. But this number is at least $(1.6e)(.1d-1)^2 = \Omega(ed^2)$ since we can choose the ordered edge ab is at least $1.6e$ ways and then neighbors x of a and y of b. Therefore $\sum f(x,y) = \Omega(ed^2)$. We deduce that for the average x, y and hence for some x, y

$$f(x,y) = \Omega(ed^2/(n')^2) = \Omega(ed^2/n^2) = \Omega(e^3 n^{-4})$$

as desired.

REFERENCES

[1] P. Erdős, *Graph Theory and Probability II*, Canad. J. Math **13** (1961), 346–352.

INTERSECTIONS AND LIMITS OF REGENERATIVE SETS

BERT FRISTEDT*

1. Introduction. A regenerative set M is a random set of real numbers that has a type of Markov property, not necessarily at arbitrary times like the set of instances following independent exponentially distributed waiting times, but only at those times belonging to M. Early uses of the term 'regenerative set' appear in [Meyer 1970] and [Maisonneuve and Morando 1970] — and, earlier, Tackás [1956] used 'regeneration point' for a member of a certain type of regenerative set. As clarification for some readers but possibly prematurely for others, I mention that Feller [1968], in Chapter XIII, used the term 'recurrent event', but this term has been largely discarded because a regenerative set is not what is usually called an event. Hoffman-Jørgensen [1969] uses the term 'Markov set', whereas Krylov and Yushkevich [1965] use 'Markov random set' for a stochastic process related to a regenerative set. Breiman [1968] and Kingman [1972] focus on the indicator function of M, using the terms 'renewal process' and 'regenerative phenomenon', respectively. We will stick to 'regenerative set'.

When the regenerative set M is assumed to be a subset of the set $\mathbf{Z}^+$ of nonnegative integers or of the set $\mathbf{R}^+$ of nonnegative real numbers it will be assumed that $0 \in M$ with probability one. On the other hand, when M is assumed to be a subset of the set $\mathbf{Z}$ of integers or of the set $\mathbf{R}$ of real numbers, no particular number will play a special role, but rather stationarity will be assumed. (Taksar [1987, Remark (6.17)] notes that stationarity is not as special an assumption as one might think.) We identify any particular one of the four cases by a phrase such as 'regenerative set in $\mathbf{Z}$'.

Section 2 treats the definition of 'regenerative set'. In Section 3 we identify a natural assumption under which intersections of regenerative sets are regenerative sets. The assumption is satisfied when the regenerative sets are independent. The focus in Section 3 is on conditions for a countably infinite intersection to be trivial, but such conditions are obtained in only two of the four settings. In the $\mathbf{R}^+$- and $\mathbf{R}$-settings, the intersection of just two independent regenerative sets is not fully understood. In Section 4, convergence concepts are introduced. Section 5 identifies infinitely divisible regenerative sets (for the operation of intersection), but in only two of the four settings do I know that all the infinitely divisible regenerative sets are encompassed by discussion in this article. Section 6 treats the convergence problem for triangular arrays with the operation of intersection. The story is complete for the $\mathbf{Z}^+$ and $\mathbf{Z}$-settings and virtually vacuous, waiting for resolution, in the other two cases.

---

* School of Mathematics, University of Minnesota, 127 Vincent Hall, 206 Church Street, Minneapolis, MN 55455.

The preceding paragraph indicates that there is much to be done in the $\mathbf{R}^+$- and $\mathbf{R}$-settings. However, much is known about a certain class of regenerative sets in $\mathbf{R}^+$, namely, those which have positive Lebesgue measure. [Kendall 1968] and [Kingman 1972] are two references and any literature about 'p-functions' is likely to be relevant.

In theory each of the following sections has four subsections, one for each of the four settings. However, the fourth subsection, reserved for the $\mathbf{R}$-setting, is non-existent in some sections because I have nothing significant to say beyond that which follows immediately by combining ideas from the second and third subsections in a natural manner. The second and third subsections of a section may be read in either order. And a reader may choose quite a different path from the one I have chosen — say, by reading the introduction and first subsection within every section before attacking any second, third, or fourth subsections.

2. Definitions. The definitions we give, specialized to the various settings, are similar to that given by Hoffmann-Jørgensen [1969]. Section 0 of [Fitzsimmons, Fristedt, and Maisonneuve 1985] is also relevant. There have been other defintions that are close to equivalent to each other and to the definitions we will use. Near the end of his paper Horowitz [1972] discusses some disparities among definitions. Papers approaching the concept of a regenerative set various via stochastic processes including some that are linear in pieces with jumps between the pieces are [Krylov and Yushkevich 1965], [Meyer 1970], [Maisonneuve 1971], [Horowitz 1972], and [Maissonneuve 1983].

2.1. $\mathbf{Z}^+$-setting: definitions. The conventions that ∞ is larger than any integer and that $\inf \emptyset = \infty$ are in force. The symbol $\overline{\mathbf{Z}}^+$ denotes the set obtained by adjoining ∞ to $\mathbf{Z}^+$. Let $(\Omega, \mathcal{F}, \mathcal{P})$ denote a probability space and $(\mathcal{F}_\sqcup : \sqcup \in \mathbf{Z}^+)$ a filtration of sub-σ-fields of $\mathcal{F}$. Thus, $\mathcal{F}_\sqcup \subseteq \mathcal{F}_{\sqcup+\infty}$ for each t. We set $\mathcal{F}_\infty$ equal to the smallest σ-field containing $\mathcal{F}_\sqcup$ for every t. A $\overline{\mathbf{Z}}^+$-valued random variable T is a *stopping time* with respect to the filtration $(\mathcal{F}_\sqcup : \sqcup \in \mathbf{Z}^+)$ if $\{\omega : T(\omega) \leq t\} \in \mathcal{F}_\sqcup$ for all $t \in \mathbf{Z}^+$. The σ-field $\mathcal{F}_T$ of 'information available before or at the stopping time T' is defined by

$$\mathcal{F}_\mathcal{T} = \{\mathcal{A} : \mathcal{A} \cap \{\omega : \mathcal{T}(\omega) \leq \sqcup\} \in \mathcal{F}_\sqcup \,,\ \sqcup \in \overline{\mathbf{Z}}^+\} \,.$$

A *regenerative set* with respect to a filtration $(\mathcal{F}_\sqcup : \sqcup \in \mathbf{Z}^+)$ is a function

$$M : \Omega \to 2^{\mathbf{Z}^+}$$

having certain properties. One property is that M be *progressively measurable*, that is, for each $t \in \mathbf{Z}^+$,

$$\{(s, \omega) : s \in M(\omega) \cap \{0, 1, 2, \ldots, t\}\} \in \mathcal{B}_\sqcup \times \mathcal{F}_\sqcup \,, \tag{2.1}$$

where $\mathcal{B}_\sqcup$ denotes the σ-field consisting of all subsets of $\{0, 1, 2, \ldots, t\}$. This progressive measurability property implies and is also implied by

$$\{\omega : t \in M(\omega)\} \in \mathcal{F}_\sqcup \tag{2.2}$$

for every t. (It is (2.1) rather than (2.2) that we use in our definition because that is the one that carries over to regenerative sets in $\mathbf{R}^+$.) Another consequence of (2.1) is that the following function is well defined:

$$(C, t) \rightsquigarrow P(\{\omega : M(\omega) \cap \{0, 1, \ldots, t\} = C\}) \ , \quad C \subseteq \mathbf{Z}^+, t \in \mathbf{Z}^+ \ ,$$

called the *distribution function* of the random set M. We often omit 'ω' in notation, and sometimes the braces used to indicate a set. Thus the distribution function of a progressively measurable random set M in $\mathbf{Z}^+$ is the function

$$(C, t) \rightsquigarrow P(M \cap \{0, 1, \ldots, t\} = C) \ , \quad C \subseteq \mathbf{Z}^+, t \in \mathbf{Z}^+ \ .$$

(The values of the distribution function at pairs (C, t) for which $C \subseteq \{0, 1, \ldots, t\}$ are the interesting values; the other values equal 0.) We will touch on but not focus on the issue of what other, possibly simpler, functions could equally well be given the name 'distribution function', in the sense that they would both determine and be determined by the function to which we have attached this name.

For T a stopping time that satisfies

$$T(\omega) \in M(\omega) \cup \{\infty\} \ \text{ for almost every } \omega \in \Omega \ , \tag{2.3}$$

set

$$(M - T)^+(\omega) = \{s - T(\omega) : s \in M(\omega),\ s \geq T(\omega)\} \ , \quad \omega \in \{T < \infty\}$$

For M to be called a regenerative set it must have the *regeneration property* — namely, that the random set $(M - T)^+$ and the σ-field $\mathcal{F}_T$ be conditionally independent given the event $\{T < \infty\}$ and, moreover, that the conditional distribution function of $(M - T)^+$ given the event $\{T < \infty\}$ not depend on the particular stopping time T. Thus, the regeneration property is that

$$\begin{aligned} & P(\{\omega : (M - T)^+(\omega) \cap \{0, 1, \ldots, t\} = C\} \cap A \mid \{T < \infty\}) \\ (2.4) = \ & P((M - T)^+ \cap \{0, 1, \ldots, t\} = C \mid \{T < \infty\}) \, P(A \mid \{T < \infty\}) \end{aligned}$$

for every $C \subseteq \mathbf{Z}^+$, $t \in \mathbf{Z}^+$, and $A \in \mathcal{F}_T$; and, moreover, that the first factor on the right depend only on (C, t), not on T. (We regard (2.4) as being satisfied vacuously if $P(T = \infty) = 1$.)

A *regenerative set* adapted to a filtration $(\mathcal{F}_\sqcup : \sqcup \in \mathbf{Z}^+)$ is a random set M having the progressive measurability and regeneration properties described above and also satisfying $P(0 \in M) = 1$. Consideration of the stopping time $T \equiv 0$ then shows, for a regenerative set M in $\mathbf{Z}^+$ and any stopping time T satisfying (2.3), that the conditional distribution function of $(M-T)^+$ given $\{T < \infty\}$ equals the distribution function of M.

Sometimes we speak of a regenerative set in $\mathbf{Z}^+$ without mentioning the filtration to which the regenerative set is adapted.

Let $(S_0 = 0, S_1, S_2, \ldots)$ be an increasing random walk in $\overline{\mathbf{Z}}^+$, defined on some probability space $(\Omega, \mathcal{F}, \mathcal{P})$ and adapted to some filtration $(\mathcal{G}_\backslash : \backslash \in \mathbf{Z}^+)$. Suppose further that $S_{n+1}(\omega) \neq S_n(\omega)$ for every n and ω for which $S_n(\omega) < \infty$. For $t = 0, 1, 2, \ldots$, define stopping times

$$\varsigma_t(\omega) = \sup\{n : S_n(\omega) \leq t\} \ .$$

It is straightforward to show that the set-valued function

$$\omega \rightsquigarrow \{S_n(\omega) : n = 0, 1, 2, \ldots\} \setminus \{\infty\}$$

is a regenerative set adapted to the filtration $(\mathcal{G}_{\varsigma_\sqcup} : \sqcup \in \mathbf{Z}^+)$.

Conversely, let M be a regenerative set adapted to a filtration $(\mathcal{F}_\sqcup : \sqcup \in \mathbf{Z}^+)$. Let $S_0 = 0$ and for $n \geq 0$, let

$$S_{n+1} = \begin{cases} \inf\{t > S_n : t \in M\} & \text{if} \quad S_n < \infty \\ \infty & \text{if} \quad S_n = \infty \end{cases}$$

and $\mathcal{G}_\backslash = \mathcal{F}_{\mathcal{S}_\backslash}$. Then $(S_n : n \in \mathbf{Z}^+)$ is a random walk in $\overline{\mathbf{Z}}^+$ and is adapted to the filtration $(\mathcal{G}_\backslash : \backslash \in \mathbf{Z}^+)$. Moreover, if one starts with a regenerative set and finds a corresponding random walk as in this paragraph and then finds the regenerative set corresponding to this random walk as in the preceding paragraph one obtains the regenerative set with which one started, but the same is not necessarily true of the filtrations.

We have seen that we may study regenerative sets in $\mathbf{Z}^+$ by studying random walks in $\overline{\mathbf{Z}}^+$ that are strictly increasing when finite, and conversely. Since the members of a regenerative set can be identified with the finite values of the corresponding random walk it is natural to call the step distribution $F = (f_1, f_2, \ldots, f_\infty)$ of the random walk the *gap distribution* of the regenerative set. (The term *interarrival distribution* is appropriate for the gap distribution of a regenerative set when certain applications are involved.)

For a regenerative set M in $\mathbf{Z}^+$ we define its *potential measure* U by

$$U(C) = E(\#(C \cap M)) \ , \quad C \subseteq \mathbf{Z}^+ \ ,$$

where # denotes cardinality. Its derivative

$$U' = (u_0, u_1, u_2, \ldots)$$

with respect to counting measure necessarily exists and is called the *potential derivative* of M and also of the corresponding random walk. Since the random variable $\#(\{t\} \cap M)$ takes only the values 0 and 1, we have

$$u_t = E(\#(\{t\} \cap M)) = P(t \in M) . \tag{2.5}$$

It is well known and easily proved that there is a one-to-one correspondence between the set of gap distributions and the set of potential derivatives: the potential derivative $(u_0, u_1, \ldots)$ can be obtained recursively from the corresponding gap distribution $(f_1, f_2, \ldots, f_\infty)$:

$$u_0 = 1 , \tag{2.6}$$

$$u_t = \sum_{n=1}^{t} f_n u_{t-n} , \quad 1 \leq t < \infty . \tag{2.7}$$

Since $u_0 = 1$ we can solve for f_n:

$$f_n = u_n - \sum_{q=1}^{n-1} f_q u_{n-q} , \quad 1 \leq n < \infty , \tag{2.8}$$

$$f_\infty = 1 - \sum_{n=1}^{\infty} f_n = \frac{1}{\sum_{t=0}^{\infty} u_t} . \tag{2.9}$$

These last formulas show explicitly how the potential derivative of a regenerative set in $\mathbf{Z}^+$ determines the gap distribution, but they do not indicate how to recognize whether a particular sequence of numbers in $[0, 1]$ is the potential derivative for some regenerative set. There seems to be no simple known criterion for a sequence to be a potential derivative.

EXAMPLE 2.1. The set of times when a simple symmetric random walk starting at 0 in $\mathbf{Z}$ is at 0 is a regenerative set in $\mathbf{Z}^+$. Its gap distribution and potential density are identified in Feller [1968, Sections III.2-III.3]:

$$f_n = \begin{cases} \frac{1}{2^n} \frac{(n-2)!\, n}{(n/2)!\,(n/2)!} & \text{for} \quad n = 2, 4, 6, \ldots \\ 0 & \text{for} \quad n = 1, 3, 5, \ldots, \infty \end{cases}$$

and

$$u_t = \begin{cases} \frac{1}{2^t} \frac{t!}{(t/2)!\,(t/2)!} & \text{for} \quad t = 0, 2, 4, \ldots \\ 0 & \text{for} \quad t = 1, 3, 5, \ldots \end{cases} .$$

The distribution function of a regenerative set M in $\mathbf{Z}^+$ can be written in terms of its gap distribution F. For any nonempty finite subset C of $\mathbf{Z}^+$ written as a finite sequence

$$C = \{c_0 < c_1 < \cdots < c_n\} ,$$

$$(2.10)\quad P(M \cap \{0,1,\ldots,c_n\} = C) = \begin{cases} \prod_{m=1}^{n} f_{c_m - c_{m-1}} & \text{if} \quad c_0 = 0 \\ 0 & \text{if} \quad c_0 \neq 0 \,. \end{cases}$$

Since the distribution function of M is expressed so simply in terms of the gap distribution, we think of the gap distribution as being that which identifies a given renewal sequence — that is, what one has in hand if one is 'given a renewal sequence'. One would thus like theorems that involve F. On the other hand, it is often the potential derivative that is most easily related to issues about which one is concerned. (For real-valued random variables a similar situation arises; certain questions have quick answers in terms of characteristic functions but one wants answers in terms of distribution functions.)

The *distribution function of an ordered pair* (M_1, M_2) of regenerative sets in $\mathbf{Z}$ is the function

$$(C_1, C_2, t) \quad \rightsquigarrow \quad P(M_1 \cap \{0,1,\ldots,t\} = C_1,\ M_2 \cap \{0,1,\ldots,t\} = C_2)\,,$$
$$C_1, C_2 \subseteq \mathbf{Z}^+,\ t \in \mathbf{Z}^+ \,.$$

This definition can be generalized to any finite number of regenerative sets in the obvious manner.

Let $\{M_k : k = 1, 2, \ldots\}$ be a finite or infinite sequence of regenerative sets adapted to a common filtration $(\mathcal{F}_\sqcup : \sqcup \in \mathbf{Z}^+)$. We say that this sequence is *jointly regenerative* if, for every finite subset K of the index set and every stopping time T that satisfies

$$T(\omega) \in \bigcap_{k \in K} M_k(\omega) \cup \{\infty\}$$

for almost every ω, $((M_k - T)^+ : k \in K)$ and $\mathcal{F}_T$ are conditionally independent given $\{T < \infty\}$ and the conditional distribution function of $((M_k - T)^+ : k \in K)$ given $\{T < \infty\}$ does not depend on the particular stopping time T.

2.2. $\mathbf{R}^+$-setting: definitions. In the $\mathbf{R}^+$-setting a filtration takes the form $(\mathcal{F}_\sqcup : \sqcup \in \mathbf{R}^+)$ Although it is not a sequence of σ-fields, the definition of filtration does entail $\mathcal{F}_\int \subseteq \mathcal{F}_\sqcup$ for $s \leq t$; and, moreover, we assume that the filtration is right-continuous, that is, for each $s \in \mathbf{R}^+$,

$$\mathcal{F}_\int = \bigcap_{\sqcup > \int} \mathcal{F}_\sqcup \,.$$

The definition of a stopping time T and the notations $\mathcal{F}_T$, $\mathcal{F}_\infty$, and $\overline{\mathbf{R}}^+$ are analogous to notations of the preceding subsection.

For the $\mathbf{R}^+$-setting, Hoffmann-Jørgensen [1969] defines 'strong Markov set' for which, with a change described in [Fitzsimmons, Fristedt, and Maisonneuve, 1985], we will use the term 'regenerative set'. A *regenerative set* adapted to a filtration $(\mathcal{F}_\sqcup : \sqcup \in \mathbf{R}^+)$ in a probability space

$(\Omega, \mathcal{F}, \mathcal{P})$ is a closed-set-valued function

$$M : \Omega \to 2^{\mathbf{R}^+}$$

that is *progressively measurable*, has the *regeneration property*, and satisfies $P(0 \in M) = 1$. In this setting (2.1) still applies as the definition of *progressive measurability* with $\mathcal{B}_{\sqcup}$ now denoting the Borel σ-field of the interval $[0, t]$. The *regeneration property* described in the preceding subsection carries over to the present setting provided that *distribution function* is defined appropriately, as in the following paragraph.

We define a *modified Hausdorff distance* between two closed sets C_1 and C_2 of $\mathbf{R}^+$ as

$$\begin{aligned} \rho(C_1, C_2) &= \frac{1}{2} \sup_{x_1 \in C_1} \inf_{x_2 \in C_2} |\arctan x_1 - \arctan x_2| \\ &\quad + \frac{1}{2} \sup_{x_2 \in C_2} \inf_{x_1 \in C_1} |\arctan x_1 - \arctan x_2| \, . \end{aligned} \tag{2.11}$$

The adjective 'modified' refers to the fact that arctan rather than the identity function is used. The implication of this modification is that two sets might be close together in this metric even if one of them, but not the other, has members that are outside, maybe far outside, some large interval. The *distribution function* of a progressively measurable set M is the function

$$(C, r) \rightsquigarrow P(\rho(M, C) \leq r) \; , \quad C \subseteq \mathbf{R}^+, \; C \text{ closed}, \; r \geq 0 \, .$$

In Example 2.2 below it is mentioned that the set of times when a Brownian motion starting at 0 is at 0 is a regenerative set M in $\mathbf{R}^+$. For any $t > 0$, $P(t \in M) = 0$. Thus, even for this important example, it would not have been possible to state the regeneration property in terms of fixed times t rather than stopping times, replacing the condition $\{T < \infty\}$ by the condition $\{t \in M\}$. Maisonneuve [1983] describes what can be done by first working with fixed times and then turning to stopping times, but to do so he treats a stochastic process associated with the random set, somewhat in the spirit of [Krylov and Yushkevich 1965].

Let (γ, ν) satisfy

$$\gamma + \int_{(0,\infty]} (x \wedge 1) \, \nu(dx) = 1 \; , \tag{2.12}$$

where $\gamma \geq 0$, ν is a measure on $(0, \infty]$, and $(x \wedge 1)$ denotes the minimum of x and 1. Let Q be a Poisson point process in $(0, \infty) \times (0, \infty]$ with intensity $\lambda \times \nu$, where λ denotes Lebesgue measure. Thus, for any Borel subset B of $(0, \infty) \times (0, \infty]$, $\#(Q \cap B)$ has a Poisson distribution with mean $\lambda \times \nu$, with the convention that a Poisson random variable with infinite mean equals ∞ with probability one. Moreover, these cardinalities for disjoint sets B are independent.

Let

$$X(\tau) = \gamma\tau + \sum_{\{(\sigma,x)\in Q:\sigma\le\tau\}} x\,, \qquad 0 \le \tau < \infty\,. \tag{2.13}$$

The stochastic process X is (not necessarily strictly) increasing and has stationary independent increments. Any right-continuous increasing process X that has stationary and independent increments, that is not identically 0, and that satisfies $X(0) = 0$ is called a *subordinator* (or *increasing Lévy process*) and has the same distribution as $\tau \rightsquigarrow X(c\tau)$ for some $c \in (0, \infty)$ and some X constructed as just described. (The constant c is needed because we have insisted on a particular value for the left side of (2.12), rather than merely requiring that it be both positive and finite.) (In the context of subordinators the measure μ is called the *Lévy measure* and the parameter γ is called the *drift*. In accordance with the goal of focusing on regenerative sets rather than corresponding processes, we will introduce alternative terms below.)

For X a subordinator, it is trivial that 0 belongs to

$$L = \{t < \infty : X(\tau) = t \text{ for some } \tau\} \tag{2.14}$$

and it can be shown that L satisfies the progressive measurability and regeneration properties. However, L may not be closed; in case $\gamma > 0$ or $\nu(0, \infty] = \infty$, L does not contain $X(\tau-)$ when $X(\tau-) < X(\tau)$. The closure M of L can be obtained from L by adjoining at most countably many points. It can be shown that any stopping time that belongs to $M \cup \{\infty\}$ with probability one also belongs to $L \cup \{\infty\}$ with probability one, from which it follows that M has the regenerative property. Therefore, M is a regenerative set in $\mathbf{R}^+$

Conversely, any regenerative set in $\mathbf{R}^+$ has the same distribution as the closure of the finite portion of the image of some subordinator for which (2.12) holds. We have insisted on the condition (2.12) so that there is, in so far as distributions are concerned, a one-to-one correspondence between regenerative sets in $\mathbf{R}^+$ and subordinators under discussion, namely those satisfying (2.12). Thus, the distribution of a regenerative set in $\mathbf{R}^+$ is characterized by a pair (γ, ν) satisfying (2.12).

We have not mentioned the filtrations explicitly in the correspondence just described because the story for them parallels the similar story for filtrations in connection with the correspondence between regenerative sets in $\mathbf{Z}^+$ and strictly increasing random walks.

For (γ, ν), corresponding to a regenerative set M in $\mathbf{R}^+$, we will call γ the *thickness describer* of M, ν the *gap describer* of M, and (γ, ν) the *describer* of M. The *potential measure* U of M, with corresponding subordinator X, is given by

$$U(B) = E(\lambda(\{\tau : X(\tau) \in B\}))\,, \quad B \subseteq \mathbf{R}^+,\ B \text{ Borel}\,.$$

If U is absolutely continuous with respect to λ, then any version U' of its derivative with respect to λ is the *potential derivative* of M. It can be shown that the potential derivative exists if $\gamma > 0$ and that the potential derivative does not exist if $\gamma = 0$ and $\nu(0,\infty] < \infty$. One reason that our story for regenerative sets in $\mathbf{R}^+$ will not be as complete as that for regenerative sets in $\mathbf{Z}^+$ is that there exists ν for which $\nu(0,\infty] = \infty$ and the potential derivative corresponding to $(0,\nu)$ does not exist. (Kallenberg [1981, Section 5] shows that a gap describer ν that pairs with a thickness describer equal to 0 to give a potential measure that is not absolutely continuous must, in a sense, be close to finite.)

Theory for the case of positive thickness describer is often identified by the term p-function, where $p(t)$ denotes the probability that t belongs to the corresponding regenerative set. See [Kingman 1972] for instance. In this restricted setting the distribution of a regenerative set is characterized by its p-function. The following example falls outside of the realm of p-function theory.

EXAMPLE 2.2. The set of times M when Brownian motion starting at 0 in $\mathbf{R}$ is at 0 is a regenerative set in $\mathbf{R}^+$. Its describer, as identified by Lévy [1968, Section 50] or, alternatively, by Itô and McKean, Jr. [1965, Section 2.2], is given by

$$\gamma = 0 \quad \text{and} \quad \frac{d\nu}{d\lambda}(x) = \frac{1}{4}x^{-3/2}\,, \quad 0 < x < \infty\,, \text{ with } \nu\{\infty\} = 0\,,$$

where λ denotes Lebesgue measure. The corresponding potential measure can be calculated using Laplace transforms:

$$\frac{dU}{d\lambda}(t) = \frac{2}{\pi}t^{-1/2}\,, \quad 0 \le t < \infty\,.$$

Although not important for our discussion, the subordinator corresponding to M is called the *strictly stable subordinator* of *index* $\frac{1}{2}$.

There are strictly stable subordinators of all positive indices no larger than 1. The one of *index* 1 arises from $\gamma = 1$ and $\nu = 0$. Those of other indices are treated in the next example.

EXAMPLE 2.3. Let $b \in (0,1)$,

$$\gamma = 0\,, \text{ and } \frac{d\nu}{d\lambda}(x) = \frac{1}{b(1-b)}x^{-1-b}\,, \quad 0 < x < \infty\,, \text{ with } \nu\{\infty\} = 0\,,$$

where λ denotes Lebesgue measure. The corresponding potential measures can be calculated using Laplace transforms:

$$\frac{dU}{d\lambda}(t) = \frac{1}{\Gamma(b)\,\Gamma(2-b)}t^{-(1-b)}\,, \quad 0 \le t < \infty\,,$$

where Γ denotes the gamma function. There are, of course, a corresponding regenerative set and a corresponding subordinator, the latter of which is the *strictly stable subordinator* of *index* b.

A necessary and sufficient condition for a regenerative set M in $\mathbf{R}^+$ to be a random subset of $\mathbf{Z}^+$ is that its thickness describer be 0 and its gap describer be supported by $\{1, 2, \ldots, \infty\}$, in which case M can also be regarded as a regenerative set in $\mathbf{Z}^+$.

The *distribution function of an ordered pair* (M_1, M_2) of regenerative sets in $\mathbf{R}^+$ is the function

$$(C_1, r_1, C_2, r_2) \rightsquigarrow P(\rho(M_1, C_1) \leq r_1, \rho(M_2, C_2) \leq r_2) .$$

Generalization to any finite number of regenerative sets raises no new issues, and the definition of 'jointly regenerative' is analogous to that used for regenerative sets in $\mathbf{Z}^+$.

2.3. Z-setting: definitions. In $\mathbf{Z}$ we only treat *stationary regenerative sets.* The condition $0 \in M$ in force for regenerative sets in $\mathbf{Z}^+$ is replaced here by the condition that the distribution of $M - t$ is the same as the distribution of M for each $t \in \mathbf{Z}$. Of course, for M to be a regenerative set in $\mathbf{Z}$ it must be progressively measurable with respect to a filtration $(\mathcal{F}_{\sqcup} : \sqcup \in \mathbf{Z})$ and have the regeneration property. Since the only regenerative sets in $\mathbf{Z}$ that we will treat are stationary we will usually drop the adjective 'stationary'. The *distribution* of a regenerative set M in $\mathbf{Z}$ is the function

$$(C, t) \rightsquigarrow P(M \cap \{-t, \ldots, -1, 0, 1, \ldots, t\} = C) , \quad C \subseteq \mathbf{Z} .$$

Let M be a regenerative set in $\mathbf{Z}$ such that $P(M = \emptyset) = 0$. Just as for regenerative sets in $\mathbf{Z}^+$, there is a corresponding *gap distribution*

$$F = (f_1, f_2, \ldots, f_\infty)$$

that determines and is determined by the distribution of M. However, as proved in [Feller 1968, XV.7(g)], only those gap distributions F having finite mean, and, in particular, $f_\infty = 0$, can arise in this way. Also proved there is the converse: for any distribution F having finite mean, there is a regenerative set M in $\mathbf{Z}$ such that $P(M = \emptyset) = 0$ and F is the gap distribution of M. The set that equals $\emptyset$ with probability one is a regenerative set in $\mathbf{Z}$. The distribution of any other regenerative set M in $\mathbf{Z}$ is determined by its gap distribution F, which has finite mean, and a real number $\alpha \in (0, 1]$, where $\alpha = P(M \neq \emptyset)$. We call α the *nonempty probability* and use $\alpha = 0$ to identify the random set ${}_\infty M$ that equals $\emptyset$ with probability one, but, of course, in this case there is no gap distribution (or, equivalently, any gap distribution will do).

Because of stationarity, $P(t \in M)$ does not depend on $t \in \mathbf{Z}$. It is straightforward to calculate the value of this constant:

$$P(t \in M) = \frac{\alpha}{\sum_{n=1}^\infty n f_n} . \tag{2.15}$$

Let $\alpha \in [0,1]$ and a distribution F on $\mathbf{Z}^+$ having finite mean be given. To construct a regenerative set M in $\mathbf{Z}$ with nonempty probability α and gap distribution F, we first set $M = \emptyset$ with probability $1-\alpha$ and with probability α we construct an infinite random set as follows. We choose, in an appropriate random manner, $t_- < 0 \le t_+$ to be the largest negative and smallest nonnegative members of M. Then $M \cap \mathbf{Z}^+$ will then be the image of a random walk starting at t_+ with step distribution F. The set $M \setminus \mathbf{Z}^+$ will be the image of a random walk starting at t_- with step distribution that assigns probability f_n to the negative integer $-n$. The two random walks are to be independent of each other and depend on t_- and t_+ only through their starting points.

It remains to describe an appropriate method for choosing the interval $I = (t_-, t_+]$, the unique random interval such that $t_- \in M$, $t_+ \in M$, $0 \in I$, and no member of I except t_+ belongs to M. The length of I is not a typical gap. When we specify that $0 \in I$ we bias our choice of I in favor of longer intervals. The length N of I should be chosen in a size-biased manner:

$$P(N=n) = \frac{nf_n}{\sum_{m=1}^{\infty} mf_m} .$$

Then, once N is chosen, the position of I should be chosen on an equiprobable basis from among the N possible choices.

The definitions of *distributions of ordered pairs* and of *jointly regenerative* carry over in a natural way from the corresponding definitions for regenerative sets in $\mathbf{Z}^+$.

2.4. R-setting: definitions. The story for this setting is obtained by combining appropriate pieces of the stories from the preceding two subsections. As described by Taksar [1980, (1.2)], an additional condition on (γ, ν) is needed for it to be the *describer* of a *stationary regenerative set* in $\mathbf{R}$. It must satisfy

$$\gamma + \int_{(0,\infty]} x\, \nu(dx) < \infty , \tag{2.16}$$

which, of course, entails $\nu\{\infty\} = 0$. Taksar [1980, (1.3)] showed that

$$P(t \in M) = \frac{\alpha\gamma}{\gamma + \int_{(0,\infty)} x\, \nu(dx)} . \tag{2.17}$$

(This formula would have a simpler appearance if one were to replace the normalizing condition (2.12) by the condition that the left side of (2.16) equal 1. However, such a replacement would give a more complicated appearance to other parts of our story, such as Proposition 4.7.) As in the $\mathbf{Z}$-setting, we will usually not include the adjective 'stationary' in the $\mathbf{R}$-setting, since in this setting no other type of regenerative set will be treated. Even if the thickness describer of a regenerative set $M \neq {}_\infty M$ in $\mathbf{R}$ equals 0 and its gap describer is supported by $\{1, 2, \ldots, \infty\}$, M cannot be regarded as a regenerative set in $\mathbf{Z}$ because the stationary requirement in $\mathbf{R}$ entails $P(M \subseteq \mathbf{Z}) = 0$.

Let $\alpha \in [0,1]$, $\gamma \in [0,1]$, and a measure ν on $(0,\infty)$ satisfying (2.16) and (2.12) be given. To construct a corresponding regenerative set M in $\mathbf{Z}$ with nonempty probability α and describer (γ, ν), we first set $M = \emptyset$ with probability $1-\alpha$ and with probability α we construct a nonempty set as follows. We choose, in an appropriate random manner, $t_- \leq 0 \leq t_+$ to be the largest nonpositive and smallest nonnegative members of M. The set $M \cap \mathbf{R}^+$ will then be the image of a subordinator corresponding to (γ, ν) starting at t_+. The set $M \cap \mathbf{R}^-$ will be the negative of the image of a subordinator that corresponds to (γ, ν) and starts at $-t_-$ (where $\mathbf{R}^-$ denotes the set of nonpositive real numbers). The two subordinators are to be independent of each other and depend on t_- and t_+ only through their starting points.

It remains to describe an appropriate method for choosing the interval $I = [t_-, t_+]$, the unique random interval such that $t_- \in M$, $t_+ \in M$, $0 \in I$, no member of I except t_- and t_+ belong to M, and $t_- = t_+ = 0$ if $0 \in M$. From (2.17) we obtain the requirement

$$P(t_- = t_+ = 0 \mid M \neq \emptyset) = \frac{\gamma}{\gamma + \int_{(0,\infty)} x\, \nu(dx)}\,.$$

This equals $P(X = 0)$, where X denotes the length of I. Away from 0 size-biasing shows that the distribution of X is absolutely continuous with respect to ν with

$$\frac{P(X \in dx \mid M \neq \emptyset)}{\nu(dx)} = \frac{x}{\gamma + \int_{(0,\infty)} x\, \nu(dx)}\,.$$

Once X is chosen and it happens to be greater than 0, then the right endpoint of position of I should be placed in the interval $(0, X)$ according to the uniform distribution.

3. Intersections of sequences. The following proposition is valid in any of the four settings, and its proof is straightforward.

PROPOSITION 3.1. *Let $(M_k : k = 1, 2, \ldots)$ be a jointly regenerative sequence of random sets adapted to a common filtration. Then $\bigcap_{k=1}^\infty M_k$ is a regenerative set adapted to the same filtration.*

Although this proposition is stated for infinite sequences, it also applies to finite sequences, since infinitely many M_k may equal the *identity* ${}_0M$ for the operation of intersection, which, with probability one, equals $\mathbf{Z}^+$, $\mathbf{R}^+$, $\mathbf{Z}$, or $\mathbf{R}$ according to the setting.

The preceding proposition is motivation for studying regenerative sets under the operation of intersection.

We adopt the convention of corresponding subscripts so that if, for instance, M_k is being discussed in the $\mathbf{Z}^+$-setting, then, in accordance with (2.5) and this convention, $u_{k,3} = P(3 \in M_k)$. Another convention is that in statements that are true with probability one, this qualification will sometimes not be mentioned.

3.1. $\mathbf{Z}^+$-setting: intersections of sequences. The following proposition is an immediate consequence of (2.5).

PROPOSITION 3.2. *Let $(M_k : k = 1, 2, \ldots)$ be an independent sequence of regenerative sets in $\mathbf{Z}^+$ with corresponding potential measures U_k, $k = 1, 2, \ldots$. Then the potential derivative of $\bigcap_k^\infty M_k$ is the term-by-term product*

$$\prod_{k=1}^{\infty} U_k' \,.$$

EXAMPLE 3.1. Let M_1, M_2, M_3 be independent regenerative sets in $\mathbf{Z}^+$ all having the gap distribution given in Example 2.1. From the preceding proposition we can immediately obtain the potential derivative of $M = M_1 \cap M_2 \cap M_3$. Stirling's asymptotic formula shows that $\sum_{t=0}^{\infty} u_t < \infty$ and, hence by (2.9) that $f_\infty > 0$ — even though $f_{1,\infty} = f_{2,\infty} = f_{3,\infty} = 0$. The regenerative set M has the distribution of the set of times when a certain random walk is at $(0,0,0)$ in $\mathbf{Z}^3$, this random walk starting at $(0,0,0)$ in $\mathbf{Z}^3$ and having step distribution which assigns probability $\frac{1}{8}$ to each of $(\pm 1, \pm 1, \pm 1)$. Thus, as is well known, this random walk has probability $f_\infty > 0$ of never returning to $(0,0,0)$.

The gap distribution and potential derivative of the identity ${}_0M = \mathbf{Z}^+$ are given by:

$$\begin{aligned} {}_0F &= (1, 0, 0, \ldots, 0)\,; \\ {}_0U' &= (1, 1, 1, \ldots)\,. \end{aligned}$$

At the other extreme is the regenerative set ${}_\infty M$ that equals $\{0\}$ with probability one and plays a role similar to that played by ∞ for sums of nonnegative random variables. Its gap distribution and potential derivative are given by:

$$\begin{aligned} {}_\infty F &= (0, 0, 0, \ldots, 1)\,; \\ {}_\infty U' &= (1, 0, 0, \ldots)\,. \end{aligned}$$

A criterion for the intersection $\bigcap_{k=1}^\infty M_k$ of an independent sequence of regenerative sets in $\mathbf{Z}^+$ to equal ${}_\infty M = \{0\}$ is an immediate consequence of the preceding proposition. It is that $\prod_{k=1}^\infty u_{k,t} = 0$ for $t > 0$. We prefer a criterion in terms of gap distributions, and that is what the next theorem, taken from [Fristedt 1981], gives.

THEOREM 3.1. *Let $(M_k : k = 1, 2, \ldots)$ be an independent sequence of regenerative sets in $\mathbf{Z}^+$ adapted to a common filtration. Let $b \in (0,1)$. Then*

$$\bigcap_{k=1}^{\infty} M_k = {}_{\infty}M$$

if and only if at least one of the following three conditions holds:

(i) $f_{k,\infty} = 1$ *for some* k;

(ii) $\#\{k : \max\{f_{k,s} : s \leq t\} \leq b\} = \infty$ *for every* $t < \infty$;

(iii) $\sum_{k=1}^{\infty}(1 - f_{k,n})I_{\{j : f_{j,n} > b\}}(k) = \infty$ *for some* $n < \infty$.

Proof. For $k = 1, 2, \ldots$, let U'_k denote the potential derivative of M_k and let U' denote the potential derivative of $\bigcap_{k=1}^{\infty} M_k$. As noted in Proposition 3.2,

$$U' = \prod_{k=1}^{\infty} U'_k \, .$$

Thus, the theorem is equivalent to the statement that

$$0 = \prod_{k=1}^{\infty} u_{k,t} \quad \text{for all } t > 0 \tag{3.1}$$

if and only if at least one of (i), (ii), and (iii) holds.

If (i) holds, then $U_k = {}_{\infty}U$ for some k, so, for that k and all $t > 0$, $u_{k,t} = 0$. Therefore, (3.1) follows.

Suppose that (ii) holds. Fix t. We will use induction to prove that $u_{k,s} \leq 1 - (1-b)^s$ for each of the infinitely many k for which $\max\{f_{k,s} : s \leq t\} \leq b$, from which (3.1) will immediately follow, since t is arbitrary. It is true for $s = 1$ since $u_{k,1} = f_{k,1}$. The second induction step, which uses (2.6) and (2.7) is

$$\begin{aligned} u_{k,s} &= f_{k,s} + \sum_{n=1}^{s-1} f_{k,n}u_{k,s-n} \leq f_{k,s} + (1 - f_{k,s})(1 - (1-b)^{(s-1)}) \\ &= 1 - (1 - f_{k,s})(1-b)^{(s-1)} \leq 1 - (1-b)^s \, . \end{aligned}$$

Suppose that (iii) holds, and choose n as in (iii). First consider $t = qn$ for some positive integer q. Then $t \notin M_k$ if the first $q - 1$ gaps in M_k each equal n, the qth gap is different from n, and the $(q+1)^{\text{st}}$ equals n. Since $u_{k,t} = P(t \in M_k)$, we obtain

$$1 - u_{k,t} \geq f_{k,n}^{q-1}(1 - f_{k,n})f_{k,n} \geq b^q(1 - f_{k,n})I_{\{j : f_{j,n} > b\}}(k) \, .$$

Hence,

$$\sum_{k=1}^{\infty}(1 - u_{k,t}) = \infty \tag{3.2}$$

from which it follows that

$$\prod_{k=1}^{\infty} u_{k,t} = 0 \,. \tag{3.3}$$

Now consider other $t > 0$. Thus, $(q-1)n < t < qn$ for some positive integer q. Then $t \notin M_k$ if the first q gaps in M_k each equal n, from which it follows that

$$1 - u_{k,t} \geq f_{k,n}^{q} \geq b^q I_{\{j : f_{j,n} > b\}}(k) \,.$$

Again, (3.2) and, therefore, (3.3) follow.

We now suppose that none of (i), (ii), and (iii) hold. Choose $t < \infty$ so that

$$K = \{k : \max\{f_{k,s} : s \leq t\} \leq b\}$$

is a finite set. Set

$$r = t! \prod_{k \in K} \min\{n : f_{k,n} > 0\} \,,$$

which is finite because (i) fails. Clearly, $u_{k,r} > 0$ for $k \in K$, so

$$\prod_{k \in K} u_{k,r} > 0 \,.$$

We will use the fact that (iii) fails for all $n \leq t$ to prove that

$$\prod_{k \notin K} u_{k,r} > 0$$

thereby finishing the proof. Clearly,

$$u_{k,r} \geq f_{k,q(k)}^{r/q(k)} \geq f_{k,q(k)}^{r} \,,$$

where $q(k) \leq t$ is the smallest $s \leq t$ for which $f_{k,s} > b$, which exists for $k \notin K$. Thus, we only need show

$$\prod_{k \notin K} f_{k,q(k)}^{r} > 0$$

which is equivalent to

$$\sum_{k \notin K} (1 - f_{k,q(k)}^{r}) < \infty \,.$$

But this sum is bounded by

$$\begin{aligned}\sum_{k\notin K} r(1-f_{k,q(k)}) &= \sum_{k\notin K} r(1-f_{k,q(k)})I_{\{j: f_{j,q(j)}>b\}}(k)\\ &\le r\sum_{n=1}^{t}\sum_{k=1}^{\infty}(1-f_{k,n})I_{\{j:f_{j,n}>b\}}(k) < \infty\,. \bullet\end{aligned}$$

In the next four examples, designed to illustrate the preceding theorem, it is to be understood that the regenerative sets M_k are independent.

EXAMPLE 3.2. Let $f_{k,n} = \frac{1}{3}$ for $n = 1,2,3$ and all k. Applying the preceding theorem with $b = \frac{1}{2}$, we see that the second condition holds, so $\bigcap_{k=1}^{\infty} M_k = \{0\}$ almost surely. If, instead, we were to apply the same theorem with $b = \frac{1}{4}$ we would get the same conclusion, but this time because the third condition would hold.

In each of the next three examples each condition that holds for one choice of $b \in (0,1)$ holds for all choices of $b \in (0,1)$.

EXAMPLE 3.3. Let $f_{k,2} = \frac{1}{k+1} = 1 - f_{k,1}$. The third condition in the preceding theorem holds (with $n = 1$) for every choice of b, so $\bigcap_{k=1}^{\infty} M_k = \{0\}$. Intuitive understanding might go along the following lines. Since $f_{k,1}$ is close to one for large k, there is tendency for most of the regenerative sets to include every integer up to some fixed $t-1$. Then the divergence of $\sum_{k=1}^{\infty} f_{k,2}$ means that infinitely many of these regenerative sets will exclude t by having a gap of 2 between $t-1$ and $t+1$.

EXAMPLE 3.4. Let $f_{k,n} = \frac{1}{k}$ for $n \le k$. The second condition in the preceding theorem holds for every choice of b, so, $\bigcap_{k=1}^{\infty} M_k = \{0\}$. For any fixed t, there are infinitely many of the regenerative sets that contain none of the integers $1, 2, \ldots, t$.

EXAMPLE 3.5. Let $f_{k,1} = \frac{1}{(k+1)^2} = 1 - f_{k,2}$. All three conditions of the preceding theorem fail for every choice of b, so, $\bigcap_{k=1}^{\infty} M_k \neq \{0\}$. In fact there is positive probability that 1 does not belong to any M_k, so that necessarily 2 is a member of every M_k.

3.2. $\mathbf{R}^+$-setting: intersections of sequences. The notations λ for Lebesgue measure and δ_x for the probability measure supported by $\{x\}$ will be used.

The describer and potential measure, respectively, of the identity ${}_0M = \mathbf{R}^+$ are given by:

$$\begin{aligned}{}_0(\gamma,\nu) &= (1,0)\,,\\ {}_0U &= \lambda\,.\end{aligned}$$

At the other extreme is the regenerative set ${}_\infty M$ that equals $\{0\}$ with probability one. Its describer and potential measure, respectively, are given by:

$${}_\infty(\gamma,\nu) = (0,\delta_\infty)\,;$$

$$ {}_{\infty}U = \delta_0 . $$

If M_1 and M_2 are independent regenerative sets in $\mathbf{R}^+$, then so is $M_1 \cap M_2$. Thus, a natural problem is that of identifying the potential measure or describer of $M_1 \cap M_2$ in terms of the corresponding quantities associated with M_1 and M_2. As far as I know, this problem, in full generality, is still open, but progress has been made in terms of the potential measures.

Let U_1, U_2 and U be the potential measures of M_1, M_2, and $M_1 \cap M_2$, respectively. Hawkes [1977] has shown that in case U_1 has a continuous monotone derivative, then, for some positive constant c,

$$ \frac{dU}{dU_2} = cU_1' \quad \text{if} \quad \int_0^1 U_1' \, dU_2 < \infty $$

and

$$ U = {}_{\infty}U \quad \text{if} \quad \int_0^1 U_1' \, dU_2 = \infty . $$

Fitzsimmons [1987] has shown that if U is absolutely continuous, then so are U_1 and U_2 and, moreover, $U' = U_1' U_2'$ almost everywhere. I am not aware of an example where $U \neq {}_{\infty}U$ and neither U_1 nor U_2 is absolutely continuous. The discussion in [Hawkes 1975, Section 6] indicates that the issue of identifying, in all cases, the intersection of two independent regenerative sets in $\mathbf{R}^+$ may be quite delicate.

EXAMPLE 3.6. Let M_1 and M_2 be independent regenerative sets in $\mathbf{R}^+$ each having the potential measure described in Example 2.2. Since

$$ \int_0^1 \left(\frac{1}{2\pi t^{1/2}} \right)^2 dt = \infty , $$

Hawkes's [1977] result mentioned above implies that $M_1 \cap M_2 = \{0\}$ with probability one. The coordinate processes of two-dimensional Brownian motion starting at $(0,0)$ are independent one-dimensional Brownian motions. Therefore, the set of times when the two-dimensional Brownian motion is at $(0,0)$ has the same distribution as $M_1 \cap M_2$. We thus obtain the well known fact that two-dimensional Brownian motion never returns to its starting point.

3.3. Z-Setting: intersections of sequences. We need some notation and terminology. If F is a gap distribution of a regenerative set in $\mathbf{Z}$, then its *period*, denoted by β, is defined by

$$ \beta = \gcd\{n : f_n > 0\} . $$

The convention of corresponding subscripts continues in force, so, for instance, in the following paragraph β_i and α_i correspond to M_i and F_i.

Let M_1 and M_2 be independent regenerative sets in $\mathbf{Z}$ and set $M = M_1 \cap M_2$. Then, as is clear, the gap distribution F of M can be obtained from F_1 and F_2, the gap distributions of M_1 and M_2, by treating F_1 and F_2 as gap distributions for regenerative sets in $\mathbf{Z}^+$, calculating the corresponding potential derivatives, and treating the term-by-term product as the potential derivative corresponding to F for a regenerative set in $\mathbf{Z}^+$. The nonempty probability α of M can be determined from F_1, F_2, F, and the nonempty probabilities α_1 and α_2 of M_1 and M_2 as follows. From the equality $P(t \in M) = P(t \in M_1)P(t \in M_2)$ and (2.15) we obtain

$$\alpha = \frac{\alpha_1 \alpha_2 \sum_{n=1}^{\infty} n f_n}{\left(\sum_{m=1}^{\infty} m f_{1,m}\right)\left(\sum_{n=1}^{\infty} n f_{2,n}\right)} \,. \tag{3.4}$$

The renewal theorem [Feller 1968, XIII.3, Theorem 4] yields an alternative formula:

$$\alpha = \frac{\alpha_1 \alpha_2}{\gcd(\beta_1, \beta_2)} \,.$$

From (3.4) we immediately obtain the following criterion for an infinite intersection to equal $\emptyset$ with probability one.

THEOREM 3.2. *Let $(M_k : k = 1, 2, \dots)$ be an independent sequence of regenerative sets in $\mathbf{Z}$ adapted to a common filtration. Then*

$$\bigcap_{k=1}^{\infty} M_k = {}_{\infty}M$$

if and only if the gap distributions and nonempty probabilities satisfy

$$\prod_{k=1}^{\infty} \frac{\alpha_k}{\sum_{n=1}^{\infty} n f_{k,n}} = 0 \,. \tag{3.5}$$

(Since the numerators in (3.5) are no larger than 1 and the denominators are no smaller than 1, (3.5) could be replaced by the union of two conditions each having a simpler appearance than (3.5).)

4. Convergence concepts. In Section 3 it was observed that the intersection of an infinite sequence of jointly regenerative sets is regenerative. These intersections can be regarded as limits of intersections of finite sequences of intersections provided an appropriate concept of convergence is established. Such a concept for each of the four settings will be introduced in this section.

The following three propositions have several things in common:

- They are valid in all four settings.
- They are out of place here because they only become meaningful subsequent to the definitions in the subsections.
- They are rather easily proved on the basis of the subsections, but the proofs will not be given.

PROPOSITION 4.1. *Let* $(M_k : k = 1, 2, \ldots)$ *be a sequence of regenerative sets in* $\mathbf{Z}^+$. *If*

$$\lim_{k\to\infty} M_k = M$$

almost surely, then M *is regenerative.*

If the regenerative sets M_k in the preceding proposition are jointly regenerative with respect to some filtration, it may be that the limit M is not regenerative with respect to the same filtration, even though it is regenerative with respect to some filtration.

PROPOSITION 4.2. *Let* $(M_k : k = 1, 2, \ldots)$ *be a jointly regenerative sequence adapted to some filtration. Then*

$$\lim_{k\to\infty} \bigcap_{j=1}^{k} M_j = \bigcap_{j=1}^{\infty} M_j \ .$$

PROPOSITION 4.3. *The space of distributions of regenerative sets (in any of the four settings) is compact.*

4.1. $\mathbf{Z}^+$**-setting: convergence concepts.** Let $(B_k : k = 1, 2, \ldots)$ be an infinite sequence of subsets of $\mathbf{Z}^+$. We say that this sequence *converges* to $B \subseteq \mathbf{Z}^+$ if, for each $t \in \mathbf{Z}^+$,

$$B_k \cap \{0, 1, 2, \ldots, t\} = B \cap \{0, 1, 2, \ldots, t\}$$

for all sufficiently large k. A consequence of this definition is that if a sequence $(M_k : k = 1, 2, \ldots)$ of regenerative sets in $\mathbf{Z}^+$ converges almost surely to M, then

$$\begin{aligned} &(\forall C \subseteq \mathbf{Z}^+)\,(\forall t \in \mathbf{Z}^+)\ \lim_{k\to\infty} P(M_k \cap \{0, 1, \ldots, t\} = C) \\ &\qquad = P(M \cap \{0, 1, \ldots, t\} = C) \ . \end{aligned} \tag{4.1}$$

We say that a sequence $(M_k : k = 1, 2, \ldots)$ of regenerative sets in $\mathbf{Z}^+$ *converges in distribution* to the regenerative set M if (4.1) holds. It is straightforward to use (2.10) and (2.6)-(2.9) to obtain necessary and sufficient conditions for convergence in distribution.

PROPOSITION 4.4. *Let* M *and* M_k, $k = 1, 2, \ldots$, *be regenerative sets in* $\mathbf{Z}^+$. *The following statements are equivalent:*

(i) $M_k \to M$ *in distribution as* $k \to \infty$;

(ii) $f_n = \lim_{k\to\infty} f_{k,n}$, $1 \le n < \infty$;

(iii) $u_t = \lim_{k\to\infty} u_{k,t}$, $t \in \mathbf{Z}^+$.

Notice that no mention is made of f_∞ in the second condition of the preceding proposition. However, if that condition is satisfied, then f_∞ is determined by the requirement that the sum of the terms of F is equal to 1.

4.2. $\mathbf{R}^+$-setting: convergence concepts. Let $(B_k : k = 1, 2, \ldots)$ be a sequence of closed subsets of $\mathbf{R}^+$. We say that this sequence *converges* to the closed subset B of $\mathbf{R}^+$ if $\rho(B_k, B) \to 0$ as $k \to \infty$, where ρ denotes the modified Hausdorff metric defined at (2.11). A consequence of this definition is that if a sequence $(M_k : k = 1, 2, \ldots)$ of regenerative sets in $\mathbf{R}^+$ converges almost surely to M, then

$$\lim_{k\to\infty} P(\rho(M_k, C) \le r) = P(\rho(M, C) \le r) \tag{4.2}$$

for every closed subset C of $\mathbf{R}^+$ and every $r \ge 0$ at which the function $s \rightsquigarrow P(\rho(M, C) \le s)$ is continuous.

We say that a sequence $(M_k : k = 1, 2, \ldots)$ of regenerative sets in $\mathbf{R}^+$ *converges in distribution* to the regenerative set M if (4.2) holds for each closed subset C of $\mathbf{R}^+$ and every $r \ge 0$ at which the function $s \rightsquigarrow P(\rho(M, C) \le s)$ is continuous. By using standard facts about subordinators, necessary and sufficient conditions for convergence can be obtained.

PROPOSITION 4.5. *Let M and M_k, $k = 1, 2, \ldots$, be regenerative sets in $\mathbf{R}^+$. The following statements are equivalent:*

(i) $M_k \to M$ *in distribution as* $k \to \infty$;

(ii) $\nu(x, \infty] = \lim_{k\to\infty} \nu_k(x, \infty]$ *if* $\nu\{x\} = 0$, $0 < x < \infty$, *and*

$$\begin{aligned}\gamma &= \lim_{\varepsilon\downarrow 0} \limsup_{k\to\infty} \left(\gamma_k + \int_{(0,\varepsilon)} x\, \nu_k(dx)\right) \\ &= \lim_{\varepsilon\downarrow 0} \liminf_{k\to\infty} \left(\gamma_k + \int_{(0,\varepsilon)} x\, \nu_k(dx)\right) ;\end{aligned}$$

(iii) $U_k[0, t] \to U[0, t]$ *as* $k \to \infty$ *for every* t *for which* $U\{t\} = 0$.

4.3. Z-setting: convergence concepts. With the replacements $\mathbf{Z}$ for $\mathbf{Z}^+$ and $\{-t, \ldots, -1, 0, 1, \ldots, t\}$ for $\{0, 1, \ldots, t\}$, the convergence concepts for the $\mathbf{Z}^+$-setting carry over to the current setting.

PROPOSITION 4.6. *A sequence $(M_k : k = 1, 2, \ldots)$ of regenerative sets in $\mathbf{Z}$ converges in distribution to a regenerative set M if and only if either*

$$\alpha = 0 = \lim_{k\to\infty} \frac{\alpha_k}{\sum_{n=1}^{\infty} n f_{k,n}} \tag{4.3}$$

or both

$$f_n = \lim_{k\to\infty} f_{k,n} \; , \quad 1 \le n < \infty \; , \tag{4.4}$$

and

$$\alpha = \left(\sum_{n=1}^{\infty} n f_n\right) \lim_{k\to\infty} \frac{\alpha_k}{\sum_{n=1}^{\infty} n f_{k,n}} \; . \tag{4.5}$$

Proof. For the proof in one direction we suppose that $(M_k : k = 1, 2, \ldots)$ converges to M in distribution. If $\alpha = 0$, we obtain, using (2.15),

$$0 = P(0 \in M) = \lim_{k\to\infty} P(0 \in M_k) = \lim_{k\to\infty} \frac{\alpha_k}{\sum_{n=1}^{\infty} n f_{k,n}} .$$

Suppose that $\alpha \neq 0$. From (2.15) and $P(0 \in M_k) \to P(0 \in M)$ we obtain (4.5). The relation (4.4) follows from Proposition 4.4 applied to the regenerative sets in $\mathbf{Z}^+$ corresponding to M and M_k.

For the converse, suppose that either (4.3) or both (4.4) and (4.5) hold. Then the construction of a regenerative set M corresponding to (α, F) (F arbitrary in case $\alpha = 0$) can be carried out as in Subsection 2.3. It is clear from the construction that $(M_k : k = 1, 2, \ldots)$ converges to M in distribution.

(The fact that it is possible that all three of (4.3), (4.4), and (4.5) hold creates no problem in this proof.) •

EXAMPLE 4.1. Let us apply the preceding proposition in case $\alpha_k = 0$ and $f_{k,k} = 1 - f_{k,1} = \frac{1}{k}$. Clearly, (4.4) and (4.5) hold with $f_1 = 1$, $f_n = 0$ for $n > 1$, and $\alpha = \frac{1}{2}$.

Here is intuition relevant to this fair coin-flip between $\emptyset$ and $\mathbf{Z}$. Fix a large finite interval I of $\mathbf{Z}$. Then when k is very large compared to the length of I, there is probability close to $\frac{1}{2}$ that I lies inside some single gap, that is, that $M_k \cap I = \emptyset$. Also, there is probability close to $\frac{1}{2}$ that the left endpoint of I belongs to M_k and when that happens it is very likely that every member of I belongs to M_k. Thus, the probability is close to $\frac{1}{2}$ that $M_k \cap I = \mathbf{Z} \cap I$.

4.4. R-setting: convergence concepts. The criterion for convergence in this setting is what one would expect in view of Propositions 4.5 and 4.6.

PROPOSITION 4.7. *A sequence $(M_k : k = 1, 2, \ldots)$ of regenerative sets in $\mathbf{R}$ converges in distribution to a regenerative set M if and only if either*

$$\alpha = 0 = \lim_{k\to\infty} \frac{\alpha_k}{\gamma_k + \int_{(0,\infty)} x \nu_k(dx)} \tag{4.6}$$

or all three of the following:

$$\nu(x, \infty) = \lim_{k\to\infty} \nu_k(x, \infty) \ \ \textit{if} \ \ \nu\{x\} = 0 \ , \ \ 0 < x < \infty \ ,$$

and

$$\begin{aligned} \gamma &= \lim_{\varepsilon \downarrow 0} \limsup_{k\to\infty} \left(\gamma_k + \int_{(0,\varepsilon)} x \, \nu_k(dx) \right) \\ &= \lim_{\varepsilon \downarrow 0} \liminf_{k\to\infty} \left(\gamma_k + \int_{(0,\varepsilon)} x \, \nu_k(dx) \right) \end{aligned}$$

and

$$\alpha = \left(\gamma + \int_{(0,\infty)} x\nu(dx)\right) \lim_{k\to\infty} \frac{\alpha_k}{\gamma_k + \int_{(0,\infty)} x\nu_k(dx)}\,. \tag{4.7}$$

The formulas (4.7) and (4.6) for α follow from Theorem 1 of [Taksar 1980]. Since the numerators in (4.6) are no larger than 1 and the denominators are no smaller than 1, (4.6) could be replaced by the union of two conditions each having a simpler appearance than (4.6).

5. Infinite divisibility. A regenerative set M is said to be *infinitely divisible* for the operation of intersection if, for each positive integer k, M has the same distribution as $\bigcap_{j=1}^{k} M_{j,k}$, where $M_{1,k}, M_{2,k}, \ldots, M_{k,k}$ are independent identically distributed regenerative sets.

We will be constructing infinitely divisible regenerative sets as the complements of unions of random, possibly overlapping, intervals. Work of several decades ago concerns such unions. Robbins [1944] studied moments of relevant distributions when the lengths of the intervals are fixed and their locations are random. Domb [1947] focused on the distributions themselves for the same setting. Hammersley [1953] and Takács [1956] permitted the lengths and positions to be random, but did not consider the possibility of infinitely many intervals intersecting a fixed finite interval. More recent work by Kendall [1967, Section 11] describes the connection with infinitely divisible regenerative sets. Also, see [Kingman 1972, Sections 1.5 and 2.5]. Mandelbrot [1972] and Shepp [1972] treated the issue of whether M equal ${}_{\infty}M$ when M is constructed as the complement of the union of random intervals.

5.1. $\mathbf{Z}^+$-setting: infinite divisibility. The following theorem is extracted from [Kendall 1967].

THEOREM 5.1. *Let $u = (u_0 = 1, u_1, u_2, \ldots)$ be a bounded sequence of positive numbers. The following conditions are equivalent:*

(i) u is the potential derivative of an infinitely divisible regenerative set in $\mathbf{Z}^+$;

(ii) $u_t^2 \le u_{t-1}u_{t+1}$ for $t \ge 1$;

(iii) for some sequence $(v_1, v_2, \ldots, v_\infty)$ of nonnegative real numbers which sum to a finite value,

$$u_t = \exp\left(-\sum_{r=1}^{\infty+}(t\wedge r)v_r\right) \tag{5.1}$$

for all t, where '$\infty+$' indicates that the term for $r=\infty$ is included in the sum.

When the above conditions hold, the sequence $(v_1, v_2, \ldots, v_\infty)$ is uniquely determined by (5.1).

Remarks: Sequences satisfying (ii) were treated by Kaluza [1928]. Notice the assumption that $u_t \neq 0$. Cases in which $u_t = 0$ for some t will be discussed at the end of this subsection.

Proof in four parts.

Part 1, (iii) $\Rightarrow$ (i): Suppose the sequence u satisfies (iii) for the sequence $v = (v_1, v_2, \ldots, v_\infty)$. For $t = 1, 2, 3, \ldots$ and $r = 1, 2, \ldots, \infty$ let

$$(5.2) \qquad B_{t,r} = \begin{cases} \{s \geq t : s - t < r\} & \text{with probability } \ 1 - e^{-v_r} \\ \emptyset & \text{with probability } \ e^{-v_r} \end{cases}$$

independently for various t and r. It is easy to see that the complement of $\bigcup_{t,r} B_{t,r}$ is a regenerative set and to show that its potential derivative is given by (5.1). Thus, we have shown that (iii) implies that u is the potential derivative of a regenerative set in $\mathbf{Z}^+$. Taking the kth root of each term of u corresponds to dividing each term of v by k. Such a division creates a new sequence of nonnegative numbers which sums to a finite value. Thus, $(u_0^{1/k}, u_1^{1/k}, \ldots)$ is the potential derivative of a regenerative set. Since k is arbitrary, we conclude that $(u_0, u_1, \ldots)$ is the potential derivative of an infinitely divisible regenerative set, as desired.

Part 2, (i) $\Rightarrow$ (iii): Suppose that u is the potential derivative U' of an infinitely divisible regenerative set in $\mathbf{Z}^+$. Let $U'_k = (u_{k,0}, u_{k,1}, \ldots)$ be the sequence of kth roots of the terms of U'. It is a consequence of infinite divisibility that U'_k is the potential derivative of a regenerative set and, thus, that there is a corresponding gap distribution F_k. From $u_{k,t}^k = u_t$ we deduce

$$k \ln(1 - (1 - u_{k,t})) = \ln u_t$$

and, hence,

$$(5.3) \qquad \lim_{k \to \infty} k(1 - u_{k,t}) = -\ln u_t \ .$$

Setting $t = 1$ and using $u_{k,1} = f_{k,1}$, we obtain

$$(5.4) \qquad \lim_{k \to \infty} k(1 - f_{k,1}) = -\ln u_1 \ .$$

From (5.4), $\sum_{n=2}^{\infty+} f_{k,n} = 1 - f_{k,1}$, and Fatou's Lemma, it follows that there exist nonnegative numbers $v_1, v_2, \ldots, v_\infty$ and a strictly increasing sequence $(k_1, k_2, \ldots)$ of positive integers such that

$$(5.5) \qquad \lim_{i \to \infty} k_i \, f_{k_i,n} = v_{n-1} \ , \ \ 2 \leq n < \infty \ ,$$

and

$$(5.6) \qquad \sum_{r=1}^{\infty+} v_r = -\ln u_1 \ .$$

We use (2.7) to substitute for $u_{k,t}$ in (5.3). The result is

$$-\ln u_t = \lim_{k\to\infty} k\left(1-\sum_{n=1}^{t} f_{k,n}u_{k,t-n}\right)$$
$$= \lim_{i\to\infty} k_i\left(1-\sum_{n=1}^{t} f_{k_i,n}u_{k_i,t-n}\right)$$

$$(5.7)= \lim_{i\to\infty}\left[k_i(1-f_{k_i,1})+f_{k_i,1}k_i(1-u_{k_i,t-1})-\sum_{n=2}^{t} k_i f_{k_i,n}u_{k_i,t-n}\right].$$

By (5.4) and (5.6) the limit of the first term on line (5.7) is

$$-\ln u_1 = \sum_{r=1}^{\infty+} v_r \,.$$

By (5.4) $f_{k,1}\to 1$ as $k\to\infty$. Thus, by (5.3), the limit of second term on line (5.7) is $-\ln u_{t-1}$. By (5.3), $u_{k,t-n}\to 1$ as $k\to\infty$. Thus, by (5.5), the limit of the third term on line (5.7), itself a sum, is $-\sum_{n=2}^{t} v_{n-1}$. Combining these facts we obtain

$$-\ln u_t = -\ln u_{t-1} + \sum_{r=t}^{\infty+} v_r \,.$$

From this equality and (5.6) we obtain (5.1) as desired.

Part 3, (iii) ⇒ (ii): Suppose (iii) holds. Then

$$\ln\frac{u_{t-1}u_{t+1}}{u_t^2} = (\ln u_{t-1}-\ln u_t)-(\ln u_t - \ln u_{t+1})$$
$$= \sum_{r=t}^{\infty+} v_r - \sum_{r=t+1}^{\infty+} v_r = v_t \geq 0\,,$$

from which (ii) follows.

The preceding paragraph also shows that when (iii) holds, it holds for only one sequence $(v_1, v_2, \dots, v_\infty)$.

Part 4, (ii) ⇒ (iii): Suppose that (ii) holds. Then $(u_0/u_1, u_1/u_2, \dots)$ is a decreasing sequence. It is bounded below by 1 since the sequence u is bounded. Thus, we may set

$$v_\infty = \ln\left(\lim_{t\to\infty}\frac{u_{t-1}}{u_t}\right).$$

We also set

$$v_r = \ln\frac{u_{r-1}\,u_{r+1}}{u_r^2}\,,\quad 1\leq r<\infty\,.$$

It is a straightforward calculation to use these relations and $u_0 = 1$ to show that (5.1) holds. •

Motivated by part 1 of the preceding proof, we call the sequence $v = (v_1, v_2, \ldots, v_\infty)$ identified in (iii) of the preceding theorem, the *covering describer* for the corresponding infinitely divisible regenerative set in $\mathbf{Z}^+$. Those $B_{t,r}$, defined at (5.2), that are different from $\emptyset$ are the (random) *covering intervals* corresponding to M.

EXAMPLE 5.1. Let $u = (c^0, c^1, c^2, \ldots)$ for some $c \in (0,1)$. Theorem 5.1 is applicable since the sequence u is bounded and all its terms are positive. Moreover, condition (ii) of that theorem is satisfied. Therefore, u is the potential derivative of some infinitely divisible regenerative set in $\mathbf{Z}^+$. Using part 4 of the proof of that theorem we can calculate the corresponding covering describer: $(0, 0, 0, \ldots, \ln \frac{1}{c})$. The uncovered set M equals $\{0, 1, \ldots, t\}$ if and only if $[t+1, \infty)$ is a covering interval but none of $[s, \infty)$ for $s \leq t$ is a covering interval. Therefore we can immediately obtain the distribution of the regenerative set M:

$$P(M = \{0, 1, \ldots, t\}) = (1-c)c^t \; .$$

EXAMPLE 5.2. Let $c \in (0,1)$ and $u = (1, c, c, c, \ldots)$. As in the preceding example we are in an infinitely divisible situation. The covering describer is $(\ln \frac{1}{c}, 0, 0, \ldots, 0)$. Therefore, membership in the corresponding regenerative set M is determined by an independent sequence of coin flips, one for each positive integer; the probability that any particular positive integer belongs to M is c.

One can obtain other infinitely divisible regenerative sets by multiplying each member of any of the infinitely divisible regenerative sets described in Theorem 5.1 by some positive integer $d > 1$. Then one obtains an infintely divisible regenerative set that, with probability one, contains only multiples of d. There is only one other infinitely divisible regenerative set in $\mathbf{Z}^+$ for which neither the preceding sentences nor Theorem 5.1 has accounted — namely, ${}_\infty M = \{0\}$.

5.2. R$^+$-setting: infinite divisibility. The *covering describer* in this setting is a measure μ on $(0, \infty]$ which is finite away from 0 and does not grow too rapidly at 0. The precise condition given in [Fitzsimmons, Fristedt, and Shepp 1985] is:

$$\int_0^1 \exp\left(\int_t^1 \mu(r, \infty] \, dr\right) dt < \infty \; . \tag{5.8}$$

The potential measure U of the corresponding infinitely divisible regenerative set M is absolutely continuous and satisfies

$$U'(t) = c \exp\left(\int_t^1 \mu(r, \infty] \, dr\right) \tag{5.9}$$

for an appropriate positive constant c [which is determined by the normalizing value 1 that we have chosen to place on the right side of (2.12)]. As in the preceding section, M is the set of points not covered by any *covering intervals*. The random collection of covering intervals corresponds to a Poisson point process in $(0,\infty) \times (0,\infty]$; To the ordered pair (s,r) corresponds the interval $(s, s+r)$. The measure governing the Poisson point process is $\lambda \times \mu$, where λ denotes Lebesgue measure on $(0,\infty)$.

Some infinitely divisible regenerative sets in $\mathbf{R}^+$ not encompassed in the above description are those whose members are, with probability one, random multiples of some fixed positive number. Multiplication by the reciprocal of the fixed positive number turns such an infinitely divisible regenerative set in $\mathbf{R}^+$ into an infinitely divisible regenerative set in $\mathbf{Z}^+$. In particular, ${}_\infty M = \{0\}$ is an infinitely divisible regenerative set in $\mathbf{R}^+$. This regenerative set could be obtained starting from a 'covering describer' provided we drop condition (5.8). From this point of view, the set of uncovered points equals $\{0\}$ almost surely if and only if the left side of (5.8) equals ∞. Of course, we would lose the one-to-one correspondence because many 'covering describers' would correspond to ${}_\infty M$.

I do not know whether we have now accounted for all infinitely divisible regenerative sets in $\mathbf{R}^+$. Fitzsimmons [1987] has shown that if the potential measure U of an infinitely divisible regenerative set in $\mathbf{R}^+$ is absolutely continuous with a decreasing derivative, then it is given by (5.9) for some covering describer μ.

It is straightforward to check that any potential measure U satisfying (5.9) for some covering describer μ has a decreasing derivative U' that is log-convex. Conversely, any right-continuous decreasing log-convex function on $(0,\infty)$ whose integrals over bounded intervals are finite can be represented in the form (5.9) and thus is the derivative of the potential measure of some infinitely divisible regenerative set in $\mathbf{R}^+$.

EXAMPLE 5.3. Consider the covering describer μ defined by

$$\frac{d\mu}{d\lambda}(r) = \frac{b}{r}\,, \quad 0 < r < \infty\,, \text{ and } \mu\{\infty\} = 0\,,$$

where λ denotes Lebesgue measure and $0 \le b < 1$. It is easy to check that this condition on b is exactly what is needed for (5.8) to hold. From (5.9) we see that we obtain ${}_0M$ by taking $b = 0$ and the regenerative sets described in Example 2.3 by taking $b \in (0,1)$. Hence, the regenerative sets corresponding to strictly stable subordinators are infinitely divisible.

5.3. Z-setting: infinite divisibility. In this setting a *covering describer* is a sequence v of nonnegative numbers satisfying

$$\sum_{r=1}^{\infty+} r v_r < \infty \tag{5.10}$$

and, in particular, $v_\infty = 0$. Distributions of infinitely divisible regenerative sets in $\mathbf{Z}$ are characterized by ordered pairs (α, v), where α is a nonempty probability and the sequence v is a *covering describer;* except that the one-to-one correspondence breaks down if $\alpha = 0$ in which case v is either arbitrary or undefined, depending on one's point of view. The infinitely divisible regenerative set in $\mathbf{Z}$ corresponding to (α, v) equals $\emptyset$ with probability $1 - \alpha$ and with probability α can be constructed as described in the first three sentences of part 1 of the proof of Theorem 5.1, except that here t (but not r) ranges over all integers.

5.4. R-setting: infinite divisibility. The story for this setting has roughly the same relationship to Subsection 5.2 as that of Subsection 5.3 has to Subsection 5.1. Shepp [1972] shows that the *covering describers* μ appropriate for this setting are those measures that satisfy the following strengthening of condition (5.8) in Subsection 5.2:

$$\int_0^1 \exp\left(\int_t^\infty \mu(r,\infty]\,dr\right)\,dt < \infty\,.$$

Notice that this condition entails $\mu\{\infty\} = 0$.

I suspect that the distributions identified above are, in combination with arbitrary nonempty probabilities, the only ones that correspond to infinitely divisible regenerative sets in $\mathbf{R}$, but I do not know a proof.

6. Triangular arrays. Consider a *triangular array* of regenerative sets:

$$\begin{array}{llll} M_{1,1} & & & \\ M_{1,2}, & M_{2,2} & & \\ M_{1,3}, & M_{2,3}, & M_{3,3}, & \\ \dots & \dots & \dots & \dots\,. \end{array} \tag{6.1}$$

Suppose, for each k, that $(M_{j,k} : j \leq k)$ is an independent finite sequence and let

$$M_k = \bigcap_{j=1}^{k} M_{j,k}\,.$$

The issue we treat is that of convergence in distribution of the sequence $(M_k : k = 1, 2, \dots)$. (Thus, independence among rows of the triangular array or lack of it is irrelevant.)

One way of creating examples of triangular arrays is to begin with an infinitely divisible regenerative set M and for each k let $M_{j,k}$, $j \leq k$, be independent identically distributed regenerative sets whose intersection has the same distribution as M. Then each M_k has the same distribution as M, so, of course, we have convergence in distribution to M.

6.1. $\mathbf{Z}^+$-setting: triangular arrays. In the $\mathbf{Z}^+$-setting, the triangular array (6.1) is *null* if, for each $t \in \mathbf{Z}^+$,

$$\lim_{k\to\infty} \max_{j\leq k} P(M_{j,k} \cap \{0, 1, \dots, t\} \neq \{0, 1, \dots, t\}) = 0\,.$$

PROPOSITION 6.1. *In the $\mathbf{Z}^+$-setting, the following are equivalent:*
(i) the triangular array (6.1) is null;
(ii) $\lim_{k\to\infty} \min\{u_{j,k,t} : j \le k\} = 1$ *for each* t ;
(iii) $\lim_{k\to\infty} \min\{f_{j,k,1} : j \le k\} = 1$.

THEOREM 6.1. *The sequence of rowwise intersections of a null independent-within-rows triangular array (6.1) of regenerative sets in $\mathbf{Z}^+$ converges in distribution to the infinitely divisible regenerative set ${}_\infty M$ if*

$$\lim_{k\to\infty} \sum_{j=1}^{k} (1 - f_{j,k,1}) = \infty .$$

It converges in distribution to the infinitely divisible regenerative set with covering describer $(v_1, v_2, \ldots, v_\infty)$ if

$$\lim_{k\to\infty} \sum_{j=1}^{k} f_{j,k,n} = v_{n-1} \ , \quad 2 \le n < \infty \ ,$$

and

$$\lim_{k\to\infty} \sum_{j=1}^{k} (1 - f_{j,k,1}) = \sum_{r=1}^{\infty+} v_r \ .$$

In no other case is there convergence in distribution.

We will omit the proof. It is quite similar to the proof of Theorem 5.1. (Recall, in the classical case where real-valued random variables are treated, that it is quite easy to go from characterizing infinitely divisible laws to the triangular array problem in the case where the random variables in each row are identically distributed. When the random variables in each row are not necessarily identically distributed, the classical triangular array problem becomes significantly more difficult because of the possibility that one summand can partially cancel another. This cancellation potential is not present for intersections of regenerative sets. It is for this reason that the second sentence in this paragraph is accurate.)

6.2. $\mathbf{R}^+$-setting: triangular arrays. In the $\mathbf{R}^+$-setting, the triangular array (6.1) is *null* if, for each $\varepsilon > 0$,

$$\lim_{k\to\infty} \max_{j \le k} P(\rho(M_{j,k}, \mathbf{R}^+) > \varepsilon) = 0 \ .$$

PROPOSITION 6.2. *In the $\mathbf{R}^+$-setting, the following are equivalent:*
(i) the triangular array (6.1) is null;
(ii) $\lim_{k\to\infty} \min\{U_{j,k}[0,t] : j \le k\} = t$ *for each* $t < \infty$;
(iii) $\lim_{k\to\infty} \max\{\nu_{j,k}(x,\infty] : j \le k\} = 0$ *for each* $x > 0$.

The convergence problem for null triangular arrays of regenerative sets in $\mathbf{R}^+$ awaits a complete solution.

6.3. Z-setting: triangular arrays. The definition of a *null* triangular array in the $\mathbf{Z}$-setting is the same as that for the $\mathbf{Z}^+$-setting except that $\{0, 1, \ldots, t\}$ should be replaced by

$$\{-t, \ldots, -1, 0, 1, \ldots, t\} .$$

PROPOSITION 6.3. *In the* $\mathbf{Z}$*-setting, the triangular array (6.1) is null if and only if*

$$\lim_{k \to \infty} \min_{j \leq k} \frac{\alpha_{j,k}}{\sum_{n=1}^{\infty} n f_{j,k,n}} = 1 .$$

THEOREM 6.2. *The sequence of rowwise intersections of a null independent-within-rows triangular array (6.1) of regenerative sets in* $\mathbf{Z}$ *converges in distribution to the infinitely divisible regenerative set* ${}_\infty M = \emptyset$ *if*

$$\lim_{k \to \infty} \sum_{j=1}^{k} \left(1 - \frac{\alpha_{j,k}}{\sum_{n=1}^{\infty} f_{j,k,n}}\right) = \infty .$$

It converges in distribution to the infinitely divisible regenerative set with nonempty probability $\alpha > 0$ *and covering describer* $(v_1, v_2, \ldots)$ *if*

$$\lim_{k \to \infty} \sum_{j=1}^{k} f_{j,k,n} = v_{n-1} , \quad 2 \leq n < \infty ,$$

and

$$\lim_{k \to \infty} \sum_{j=1}^{k} \left(1 - \frac{\alpha_{j,k}}{\sum_{n=1}^{\infty} n f_{j,k,n}}\right) = \sum_{r=1}^{\infty} r v_r - \ln \alpha .$$

In no other case is there convergence in distribution.

A proof of this theorem is given in [Fristedt 1981].

6.4. R-setting: triangular arrays. The definition of a *null* triangular array in the $\mathbf{R}$-setting is the same as that for the $\mathbf{R}^+$-setting except that $\mathbf{R}^+$ should be replaced by $\mathbf{R}$.

PROPOSITION 6.4. *In the* $\mathbf{R}$*-setting, the triangular array (6.1) is null if and only if*

$$\lim_{k \to \infty} \min_{j \leq k} \frac{\alpha_{j,k} \gamma_{j,k}}{\gamma_{j,k} + \int_{(0,\infty)} x \, \nu_{j,k}(dx)} = 1 .$$

REFERENCES

[1] Leo Breiman. *Probability*. Addison-Wesley, Reading, Massachusetts, 1968.

[2] C. Domb. The problem of random intervals on a line. *Proc. Cambridge Philos. Soc.*, 43:329–341, 1947.

[3] Richard Durrett. *Probability: Theory and Examples.* Wadsworth, Pacific Grove, California, 1991.

[4] William Feller. *An Introduction to Probability Theory and Its Applications.* Wiley, New York, third edition, 1968.

[5] P. J. Fitzsimmons, 1987. personnal communication.

[6] P. J. Fitzsimmons, Bert Fristedt, and B. Maisonneuve. Intersections and limits of regenerative sets. *Z. Wahrscheinlichkeitstheorie verw. Geb.*, 70:157–173, 1985.

[7] P. J. Fitzsimmons, Bert Fristedt, and L. A. Shepp. The set of real numbers left uncovered by random covering intervals. *Z. Wahrscheinlichkeitstheorie verw. Geb.*, 70:175–189, 1985.

[8] Bert Fristedt. The central limit problem for, infinite products of, and Lévy processes of renewal sequences. *Z. Wahrscheinlichkeitstheorie verw. Geb.*, 58:479–507, 1981.

[9] J. M. Hammersley. On counters with random dead time. I. *Proc. Cambridge Philos. Soc.*, 49:623–637, 1953.

[10] John Hawkes. On the potential theory of subordinators. *Z. Wahrscheinlichkeitstheorie verw. Geb.*, 33:113–132, 1975.

[11] John Hawkes. Intersections of Markov random sets. *Z. Wahrscheinlichkeitstheorie verw. Geb.*, 37:243–251, 1977.

[12] J. Hoffmann-Jørgensen. Markov sets. *Math. Scand.*, 24:145–166, 1969.

[13] Joseph Horowitz. Semilinear Markov processes, subordinators and renewal theory. *Z. Wahrscheinlichkeitstheorie verw. Geb.*, 24:167–193, 1972.

[14] Kiyosi Itô and Henry McKean, Jr. *Diffusion Processes and Their Sample Paths.* Academic Press, New York, 1965.

[15] Olav Kallenberg. Splitting at backward times in regenerative sets. *Ann. Probab.*, 9:781–799, 1981.

[16] D. G. Kendall. Renewal sequences and their arithmetic. In D. A. Kappos, editor, *Symposium on Probability Methods in Analysis*, pages 147–175. Springer-Verlag, Berlin, 1967. Lecture Notes in Mathematics #31. (Reprinted on pp 47–72 of *Stochastic Analysis* edited by D. G. Kendall and E. F. Harding, Wiley, London, 1973.)

[17] D. G. Kendall. Delphic semigroups, infinitely divisible regenerative phenomena, and the arithmetic of p-functions. *Z. Wahrscheinlichkeitstheorie verw. Geb.*, 9:163–195, 1968. (Reprinted on pp 73–114 of *Stochastic Analysis* edited by D. G. Kendall and E. F. Harding, Wiley, London, 1973.)

[18] J. F. C. Kingman. *Regenerative Phenomena.* Wiley, London, 1972.

[19] N. V. Krylov and A. A. Yushkevich. Markov random sets. *Trans. Mosc. Math. Soc.*, 13:127–153, 1965. English translation.

[20] Paul Lévy. *Processus Stochastiques et Mouvement Brownien.* Gauthier-Villars, Paris, 1948.

[21] B. Maisonneuve. Ensembles régénératifs temps locaux et subordinateurs. In P. A. Meyer, editor, *Séminaire de Probabilités V*, pages 151–161. Springer-Verlag, Berlin, 1970. Lecture Notes in Mathematics #191.

[22] B. Maisonneuve. Ensembles régénératifs de la droite. *Z. Wahrscheinlichkeitstheorie verw. Geb.*, 63:501–510, 1983.

[23] B. Maisonneuve and Ph. Morando. Temps locaux pour les ensembles régénératifs. In P. A. Meyer, editor, *Séminaire de Probabilités IV*, pages 151–161. Springer-Verlag, Berlin, 1970. Lecture Notes in Mathematics #124.

[24] Benoit B. Mandelbrot. Renewal sets and random cutouts. *Z. Wahrscheinlichkeitstheorie verw. Geb.*, 22:145–157, 1972.

[25] P. A. Meyer. Ensembles régénératifs, d'après Hoffmann-Jørgensen. In P. A. Meyer, editor, *Séminaire de Probabilités IV*, pages 133–150. Springer-Verlag, Berlin, 1970. Lecture Notes in Mathematics #124.

[26] H. E. Robbins. On the measure of a random set. *Ann. Math. Statist.*, 15:70–74, 1944.

[27] L. A. Shepp. Covering the line with random intervals. *Z. Wahrscheinlichkeitsthe-*

orie verw. Geb., 23:163–170, 1972.

[28] L. Takács. On a probability problem arising in the theory of counters. *Proc. Cambridge Philos. Soc.*, 52:488–498, 1956.

[29] M. I. Taksar. Regenerative sets on real line. In J. Azéma and M. Yor, editors, *Séminaire de Probabilités XIV*, pages 437–474. Springer-Verlag, Berlin, 1980. Lecture Notes in Mathematics #784.

[30] M. I. Taksar. Stationary Markov sets. In J. Azéma, P. A. Meyer, and M. Yor, editors, *Séminaire de Probabilités XXI*, pages 303–340. Springer-Verlag, Berlin, 1987. Lecture Notes in Mathematics #1247.

RANDOM PROCESSES OF THE FORM $X_{n+1} = a_n X_n + b_n \pmod p$ WHERE b_n TAKES ON A SINGLE VALUE

MARTIN HILDEBRAND*

Summary. This paper studies random processes of the form $X_{n+1} = a_n X_n + b_n \pmod p$ where b_n has only one possible value and a_n is a or 1 with probability 1/2 each. For values of p satisfying certain constraints imposed by a_n and b_n, X_n gets close to uniformly distributed on $\mathbf{Z}/p\mathbf{Z}$ for large enough n. This paper explores how large n needs to be as a function of p. Adapting techniques used by Chung, Diaconis, and Graham and techniques previously developed by the author, this paper shows that if $n > c_1 \log p \log\log p$ then X_n gets close to uniformly distributed as p approaches infinity and that if $n > c_2 \log p$ then X_n approaches the uniform distribution for almost all p satisfying the constraints. Furthermore, if $a = 2$, this paper shows that for certain p, if X_n gets close to uniformly distributed, then $n > c_3 \log p \log\log p$.

1. Introduction. Some pseudo-random number generators on computers are of the form

$$X_{n+1} = aX_n + b \pmod p$$

where a and b are constants. The sequence $X_0, X_1, \ldots$ has some of the characteristics of random numbers although such sequences are deterministic. See Knuth [Kn] for more details.

To get truly random numbers, one may replace a with a_n and b with b_n where a_n and b_n are independent random variables with small support. For instance, a_n could be 2 always and b_n could be $-1, 0$ or 1 with probability 1/3 each. Chung, Diaconis, and Graham [CDG] studied this example and asked how large n should be so that X_n is close to uniformly distributed on the integers mod p. Certain parity restrictions are placed on p to ensure that such a value for n exists; in the example, p is restricted to odd integers. Arguments more general than those in [CDG] were presented by the author in [Hi] and [Hi2], but those works left some gaps between the upper and lower bounds of the time it takes to get close to uniform. This paper reduces some of these gaps and extends techniques found in [CDG], [Hi], and [Hi2] to do so.

The principal results of this paper are (in the notation of the next section)

THEOREM 1. If b_n is a constant non-zero integer and a_n is a sequence of i.i.d. random variables with $P(a_n = a) = P(a_n = 1) = 1/2$ where a is a constant integer greater than 1, then there exist constants c_1 and c_2 such that for all integers p with $(a,p) = (a-1,p) = (b_n,p) = 1$ if

* Research supported in part by a Rackham Faculty Fellowship at the University of Michigan. Department of Mathematics, University of Texas at Austin, Austin, TX 78712.

$n > c_1 \log p \log\log\log p$, then $\|P_n - U\| \to 0$ as $p \to \infty$ and for almost all such integers p if $n > c_2 \log p$, then $\|P_n - U\| \to 0$. By "almost all such integers p", we mean that $(a-1,p) = (a,p) = (b_n,p) = 1$ and that we may exclude a fraction of the possible integers p between 1 and m so long as this fraction goes to 0 as $m \to \infty$.

THEOREM 2. If $b_n = 1$ and a_n is a sequence of i.i.d. random variables with $P(a_n = 2) = P(a_n = 1) = 1/2$, then there exists a constant c_3 such that if $n < c_3 \log p \log\log\log p$, then $\|P_n - U\|$ is bounded away from 0 as $t \to \infty$ if $p = 2^t - 1$.

These bounds are similar to bounds presented in [CDG]; the difference is that here b_n instead of a_n takes on a single value.

2. Background. Let P and Q be probability distributions on a finite group G. Define the variation distance of P from Q by

$$\|P - Q\| := \frac{1}{2} \sum_{s \in G} |P(s) - Q(s)|.$$

It follows that

$$\|P - Q\| = \max_{A \subseteq G} |P(A) - Q(A)|.$$

We shall use the following lemma of Diaconis and Shahshahani.

UPPER BOUND LEMMA: Let P be a probability distribution on the integers mod p, and let U be the uniform distribution on the integers mod p. Then

$$\|P - U\|^2 \le \frac{1}{4} \sum_{k \neq 0} |\hat{P}(k)|^2$$

where

$$\hat{P}(k) := \sum_{j=0}^{p-1} P(j) e^{2\pi i j k/p}$$

is the Fourier transform of P on the integers mod p.

Chapter 3 of Diaconis [Di] discusses this lemma generalized for an arbitrary finite group.

Let $P_n(j) = P(X_n = j)$, and let $\hat{P}_n(k)$ be the Fourier transform of $P_n(j)$. As shown by the author [Hi2], if $(a,p) = 1$ when $P(a_n = a) > 0$, then

$$\hat{P}_{n+1}(k) = \Big(\sum_{a \ge 1} P(a_n = a) \hat{P}_n(ak)\Big)\Big(\sum_{b \in \mathbf{Z}} P(b_n = b) q^{bk}\Big) \tag{2.1}$$

where $q := q(p) := e^{2\pi i/p}$. This recurrence gives rise to a "phase cancellation" which is used in [Hi2] to get an upper bound on n which is large enough to make $\|P_n - U\| \to 0$ as $p \to \infty$. In this paper, we shall use this cancellation more carefully to get a better bound.

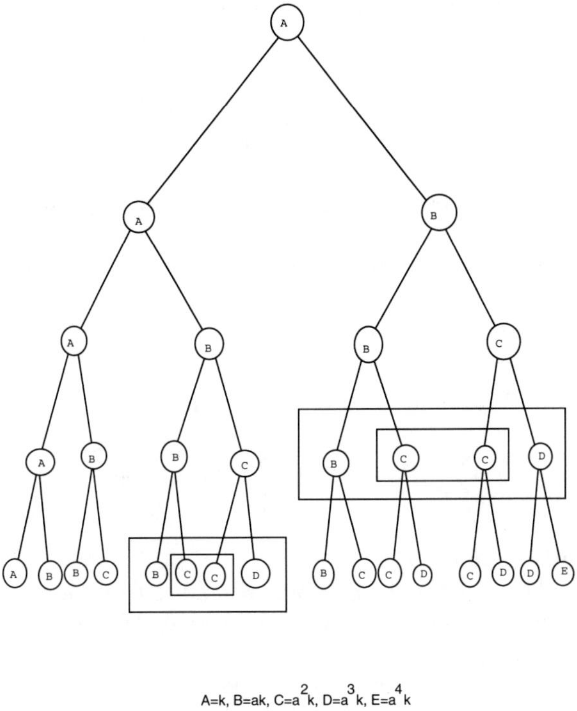

FIG. 1.

3. Upper bound. In this section, we shall show Theorem 1.
Let

$$a_n = \begin{cases} a & \text{with probability } 1/2 \\ 1 & \text{otherwise} \end{cases}$$

Restrict p to positive integers with $(a,p) = (a-1,p) = (b_n,p) = 1$. Using (2.1), we see that

$$\hat{P}_{n+1}(k) = \frac{1}{2} q^{b_n k} \hat{P}_n(k) + \frac{1}{2} q^{b_n k} \hat{P}_n(ak).$$

We shall use this recurrence relation repeatedly.

The tree in Figure 1 provides a guide to the repeated use of the recurrence relation. If n_i is a node in the tree, let $val(n_i)$ be the label on the node. For each node n_0, let $Prod(n_0)$ be the product of $q^{b_n \ val(n_i)}$ over all nodes n_i on a path descending from the root to n_0. The tree has been constructed so that

$$\hat{P}_n(k) = \sum_{n_j \in Row(b)} \frac{1}{2^b} Prod(n_j)(\hat{P}_{n-b}(val(n_j)) + \hat{P}_{n-b}(a\ val(n_j)))$$

where $b \leq n$ and $Row(b)$ is the set of nodes in row b of the tree. (The rows are counted from the top.) Note that since $X_0 = 0$ with probability 1, then $\hat{P}_0(k) = 1$ for all k and hence

$$\hat{P}_n(k) = \sum_{n_j \in Row(n)} \frac{2}{2^n} Prod(n_j).$$

We would like to bound $\sum_{k \neq 0} |\hat{P}_n(k)|^2$ for appropriate values of n. To do so, we use

LEMMA 3.1. If $M(1,n)$, $M(2,n)$, ..., $M(p-1,n)$ are such that

$$|\hat{P}_n(k)| \leq M(k,n) + \frac{\epsilon}{p^{r+1}}$$

for all $k \neq 0$ in $\mathbf{Z}/p\mathbf{Z}$ and for $0 < \epsilon < 1$, then

$$\sum_{k \neq 0} |\hat{P}_n(k)|^2 \leq (\sum_{k \neq 0} M(k,n)^2) + \frac{3\epsilon}{p^r}$$

for any positive number r.

Proof. : The proof is straightforward and left to the reader. Note that we may without loss of generality assume that $M(k,n) \leq 1$ for all k and $n \geq 0$ since $|\hat{P}_n(k)| \leq 1$. □

To find useful values for $M(k,n)$, we use the "phase cancellation" in the recurrence relation. Let $N(i,j)$ be the set of nodes which are in or above row n and which are are a distance i below the first node labeled j on a path descending from the root. The following lemma bounds the "phase cancellation":

LEMMA 3.2.

$$\begin{aligned} | \sum_{n_0 \in N(2, a^{3(\ell-1)} k)} \frac{1}{2^{row(n_0)}} & Prod(n_0)(\hat{P}_{n-row(n_0)}(val(n_0)) \\ & + \hat{P}_{n-row(n_0)}(a\ val(n_0)))| \\ & \leq \prod_{j=0}^{\ell-1} \left(\frac{1}{2} + \frac{1}{2} |\cos(\frac{2\pi b_n k a^{3j}}{p} \frac{(a^2-a)}{2})| \right) \end{aligned}$$

where $row(n_0)$ is the row where n_0 is located.

Proof. Consider the labeled tree in Figure 1. Form a path by starting at the root and at each step going to the left or the right with equal probability. This corresponds to the possibilities for a_n. For a path from the root, note the first nodes labeled ak, a^4k, ..., $a^{3(\ell-1)+1}k$. With probability 1/2, the path will have one left and one right step in the two steps after the node is reached. In Figure 1, this corresponds to going through an inner box. In this case, some phase cancellation occurs.

Form a string S of length ℓ by making the i-th position 1 if the 2 steps after a node labeled $a^{3(i-1)+1}k$ include 1 left step and 1 right step; otherwise make the i-th position a 0.

The probability that S is a given string S_0 of length ℓ and that the first node labeled $a^{3(\ell-1)+1}k$ is reached on or before row $n-2$ is less than or equal to $1/2^\ell$. Let $N(2, a^{3(\ell-1)+1}k, S_0)$ be the nodes in $N(2, a^{3(\ell-1)+1}k)$ which give rise to the string S_0. Due to the phase cancellation, we may conclude

$$\begin{aligned} \Big| \sum_{n_0 \in N(2, a^{3(\ell-1)+1}k, S)} \frac{1}{2^{row(n_0)}} & Prod(n_0)(\hat{P}_{n-row(n_0)}(val(n_0)) \\ & + \hat{P}_{n-row(n_0)}(a\ val(n_0)))\Big| \\ & \leq (\frac{1}{2})^\ell \prod_{i:S(i)=1} \frac{|q^{b_n k a^{3(i-1)+1}} + q^{b_n k a^{3(i-1)+2}}|}{2} \\ & = (\frac{1}{2})^\ell \prod_{i:S(i)=1} |\cos(\frac{2\pi b_n k a^{3(i-1)}}{p} \frac{(a^2-a)}{2})| \end{aligned}$$

Summing over all strings and noting that the possible values for $i-1$ range from 0 to $\ell - 1$, we get the lemma. □

By similar reasoning, we may show

COROLLARY.

$$\Big| \sum_{n_0 \in Row(n)} \frac{2}{2^n} Prod(n_0) \Big| \leq \prod_{j=0}^{\ell-1} (\frac{1}{2} + \frac{1}{2} |\cos(\frac{2\pi b_n k a^{3j}}{p} \frac{(a^2-a)}{2})|) + p_1$$

where p_1 is the proportion of nodes in $Row(n)$ such that the path from the root to the node does not reach the first node labeled $a^{3(\ell-1)+1}k$ on or before row $n-2$.

Let

$$L(k,n) := \prod_{j=0}^{n-1} (\frac{1}{2} + \frac{1}{2} |\cos(\frac{2\pi b_n k a^{3j}}{p} \frac{(a^2-a)}{2})|).$$

The next lemma, which bounds the quantity p_1, shows that we can use Lemma 3.1 to bound the upper bound lemma sum with $M(k, 7n) = L(k, n)$ and $r = 1$.

LEMMA 3.3. There exists a constant c_3 such that if $n > c_3 \log p$ and p is large enough, then after $7n$ steps, the proportion of paths through the tree with less than $3n$ steps to the right is under ϵ/p^2.

The proof of this lemma is left to the reader. □

How large should n be to make $\sum_{k \neq 0} (L(k,n))^2 \to 0$?

To find out, we use techniques developed in [CDG] and expanded in Chapter 3 of [Hi] to prove the following lemma.

LEMMA 3.4. For some constant c_1, if $n \geq c_1 \log p \log\log p$, then $\sum_{k\neq 0}(L(k,n))^2 \to 0$. For some constant c_2, if $n \geq c_2 \log p$, then $\sum_{k\neq 0}(L(k,n))^2 \to 0$ for almost all p.

By almost all p, we mean that the proportion of eligible integers p between 1 and m for which the statement is true approaches 1 as m approaches infinity.

Proof. : Using the restriction $(b_n, p) = (a, p) = (a-1, p) = 1$, we see that multiplication by $b_n(a^2 - a)/2$ is an isomorphism on $\mathbf{Z}/p\mathbf{Z}$. So we may eliminate that multiple from the sum.

Let $g(x) := \frac{1}{2} + \frac{1}{2}|\cos(2\pi x)|$.

Let

$$h_0(x) := \begin{cases} 1 & \text{for } x \in [0, 1/8) \cup [3/8, 5/8) \cup [7/8, 1) \\ b & \text{for other } x \in [0,1) \end{cases}$$

where $b := (1/2) + (\sqrt{2}/4)$. Observe that $g(x) \leq h_0(x)$ for $x \in [0, 1)$.

Let

$$h(x) := \begin{cases} 1 & \text{for } x \in [0, 1/4) \cup [3/4, 1) \\ b & \text{for other } x \in [0,1) \end{cases}$$

Note that $h_0(x) = h(\{2x\})$ where $\{x\}$ denotes the fractional part of x.

So

$$\begin{aligned}\sum_{k=1}^{p-1}(L(k,n))^2 &\leq \textstyle\sum_{k=1}^{p-1}\prod_{i=0}^{n-1}(h_0(\{\frac{k}{p}a^{3i}\}))^2 \\ &= \textstyle\sum_{k=1}^{p-1}\prod_{i=0}^{n-1}(h(\{\frac{2k}{p}a^{3i}\}))^2 \\ &= \textstyle\sum_{k=1}^{p-1}\prod_{i=0}^{n-1}(h(\{(a^{3i})\frac{k}{p}\}))^2\end{aligned}$$

The final equality holds since $(2, p) = 1$ and hence multiplication by 2 is an isomorphism of $\mathbf{Z}/p\mathbf{Z}$.

By an argument similar to one in Chapter 3 of [Hi] (and contained in the appendix of this paper), the lemma follows. □

Together Lemmas 3.1, 3.3, 3.4 and the corollary to Lemma 3.2 prove Theorem 1. □

4. Lower bound for $a_n = 2$ or 1, $b_n = 1$, p odd. In this section, we shall show Theorem 2. We are considering the case where $a_n = 2$ or 1 with probability 1/2 each and $b_n = 1$ always. The values p are such that $p = 2^t - 1$ for some integer t (and so 2 is of low order in the multiplicative group of $\mathbf{Z}/p\mathbf{Z}$.)

The argument presented here is partly modeled on a lower bound argument for a different process which is described in Chung, Diaconis, and Graham [CDG].

Strategy. Consider the "separating" function f which appears in Chung, Diaconis, and Graham [CDG]. They relate the expected value of f and of $f\bar{f}$ under P_n to the Fourier transforms for the random processes they were considering. However, the Fourier transforms in our case differ from those in [CDG]. Nonetheless, we will still conclude that $E_{P_n}(f)$ is close to $t\Pi_1^r$ and $E_{P_n}(f\bar{f})$ is close to $t\sum_{j=0}^{t-1}|\Pi_j|^r$ for appropriate choice of n, r, and Π_j (which will be somewhat different from the values Π_j in [CDG] but will share some of the same properties.) The approximations are good enough for us to conclude that $c_3 \log p \log\log p$ steps are needed to make $||P_n - U|| \to 0$ for the values of p in Theorem 2.

The separating function. Let $p = 2^t - 1$ for some integer t. Let $q = e^{2\pi i/p}$, and let

$$f(k) = \sum_{j=0}^{t-1} q^{k2^j}.$$

The following lemma is proved in Chung, Diaconis, and Graham [CDG].

LEMMA 4.1.

$$\begin{aligned}
E_{P_n}(f) &= \sum_{j=0}^{t-1} \hat{P}_n(2^j) \\
E_{P_n}(f\bar{f}) &= \sum_{j=0}^{t-1}\sum_{j'=0}^{t-1} \hat{P}_n(2^j - 2^{j'}). \\
E_U(f) &= 0 \\
E_U(f\bar{f}) &= t
\end{aligned}$$

Unfortunately, (2.1) does not give a nice expression for $\hat{P}_n$ like what appears in [CDG]. Still it is worthwhile to explore $\hat{P}_n(1)$, $\hat{P}_n(2)$, $\hat{P}_n(4)$, ... on the computer.

Apparent results from the computer. Using the recurrence relation and the equality $\hat{P}_0(k) = 1$ for all k, we can use a computer to find $\hat{P}_n(k)$, $\hat{P}_n(2k)$, $\hat{P}_n(4k)$, ..., provided that n and t are not too large. For even moderate values of t, extreme care must be taken to avoid serious problems with round-off. Let's look at what happens to $\hat{P}_n(1)$, $\hat{P}_n(2)$, $\hat{P}_n(4)$, ... as n grows. Table 1 contains computer generated values for $\hat{P}_n(2^k)$ for the given values of n and k when $p = 2^{60} - 1$.

At first most of the values are close to 1, which is hardly surprising. After awhile, many of the values are close to the value $-.01004 + .08367i$. After yet awhile longer, many of the values are close to the value $(-.01004 + .08367i)^2$. After yet awhile longer, many of the values are close to the value $(-.01004 + .08367i)^3$.

Some pattern seems to be developing. When t is larger than 60, the

TABLE 1
COMPUTER GENERATED VALUES FOR THE FOURIER TRANSFORM

k	n=1	n=131	n=261	n=391
0	1.00000+ 0.00000i	0.02033+ 0.10807i	-0.00817+-0.00097i	0.00020+-0.00061i
1	1.00000+ 0.00000i	0.00950+ 0.10119i	-0.00785+-0.00121i	0.00021+-0.00060i
2	1.00000+ 0.00000i	0.00211+ 0.09584i	-0.00759+-0.00137i	0.00021+-0.00059i
3	1.00000+ 0.00000i	-0.00275+ 0.09184i	-0.00740+-0.00149i	0.00021+-0.00058i
4	1.00000+ 0.00000i	-0.00583+ 0.08897i	-0.00725+-0.00156i	0.00021+-0.00058i
5	1.00000+ 0.00000i	-0.00770+ 0.08699i	-0.00715+-0.00161i	0.00021+-0.00057i
6	1.00000+ 0.00000i	-0.00879+ 0.08568i	-0.00707+-0.00164i	0.00021+-0.00057i
7	1.00000+ 0.00000i	-0.00940+ 0.08485i	-0.00701+-0.00166i	0.00021+-0.00057i
8	1.00000+ 0.00000i	-0.00973+ 0.08433i	-0.00697+-0.00167i	0.00021+-0.00056i
9	1.00000+ 0.00000i	-0.00990+ 0.08403i	-0.00695+-0.00168i	0.00021+-0.00056i
10	1.00000+ 0.00000i	-0.00998+ 0.08386i	-0.00693+-0.00168i	0.00022+-0.00056i
11	1.00000+ 0.00000i	-0.01001+ 0.08377i	-0.00692+-0.00168i	0.00022+-0.00056i
12	1.00000+ 0.00000i	-0.01003+ 0.08372i	-0.00691+-0.00168i	0.00022+-0.00056i
13	1.00000+ 0.00000i	-0.01004+ 0.08369i	-0.00691+-0.00168i	0.00022+-0.00056i
14	1.00000+ 0.00000i	-0.01004+ 0.08368i	-0.00690+-0.00168i	0.00022+-0.00055i
15	1.00000+ 0.00000i	-0.01004+ 0.08367i	-0.00690+-0.00169i	0.00023+-0.00055i
16	1.00000+ 0.00000i	-0.01004+ 0.08367i	-0.00690+-0.00169i	0.00023+-0.00055i
17	1.00000+ 0.00000i	-0.01004+ 0.08367i	-0.00690+-0.00169i	0.00023+-0.00054i
18	1.00000+ 0.00000i	-0.01004+ 0.08367i	-0.00690+-0.00170i	0.00024+-0.00054i
19	1.00000+ 0.00000i	-0.01004+ 0.08367i	-0.00690+-0.00171i	0.00024+-0.00053i
20	1.00000+ 0.00000i	-0.01004+ 0.08367i	-0.00689+-0.00172i	0.00025+-0.00052i
21	1.00000+ 0.00000i	-0.01004+ 0.08367i	-0.00688+-0.00174i	0.00025+-0.00051i
22	1.00000+ 0.00000i	-0.01004+ 0.08367i	-0.00687+-0.00176i	0.00026+-0.00050i
23	1.00000+ 0.00000i	-0.01004+ 0.08367i	-0.00686+-0.00179i	0.00026+-0.00049i
24	1.00000+ 0.00000i	-0.01004+ 0.08367i	-0.00683+-0.00182i	0.00027+-0.00047i
25	1.00000+ 0.00000i	-0.01004+ 0.08367i	-0.00680+-0.00186i	0.00027+-0.00045i
26	1.00000+ 0.00000i	-0.01005+ 0.08367i	-0.00676+-0.00191i	0.00028+-0.00043i
27	1.00000+ 0.00000i	-0.01006+ 0.08368i	-0.00670+-0.00198i	0.00028+-0.00041i
28	1.00000+ 0.00000i	-0.01008+ 0.08368i	-0.00662+-0.00205i	0.00028+-0.00039i
29	1.00000+ 0.00000i	-0.01011+ 0.08368i	-0.00652+-0.00213i	0.00029+-0.00036i
30	1.00000+ 0.00000i	-0.01017+ 0.08368i	-0.00639+-0.00221i	0.00029+-0.00034i
31	1.00000+ 0.00000i	-0.01026+ 0.08367i	-0.00623+-0.00231i	0.00028+-0.00031i
32	1.00000+ 0.00000i	-0.01041+ 0.08365i	-0.00605+-0.00240i	0.00028+-0.00028i
33	1.00000+ 0.00000i	-0.01063+ 0.08359i	-0.00582+-0.00250i	0.00027+-0.00025i
34	1.00000+ 0.00000i	-0.01097+ 0.08348i	-0.00557+-0.00259i	0.00027+-0.00022i
35	1.00000+ 0.00000i	-0.01145+ 0.08328i	-0.00527+-0.00267i	0.00026+-0.00020i
36	1.00000+ 0.00000i	-0.01211+ 0.08293i	-0.00495+-0.00274i	0.00025+-0.00017i
37	1.00000+ 0.00000i	-0.01300+ 0.08236i	-0.00459+-0.00279i	0.00024+-0.00014i
38	1.00000+ 0.00000i	-0.01415+ 0.08148i	-0.00421+-0.00282i	0.00022+-0.00012i
39	1.00000+ 0.00000i	-0.01556+ 0.08018i	-0.00382+-0.00282i	0.00021+-0.00010i
40	1.00000+ 0.00001i	-0.01723+ 0.07834i	-0.00341+-0.00280i	0.00020+-0.00008i
41	1.00000+ 0.00001i	-0.01912+ 0.07585i	-0.00300+-0.00275i	0.00018+-0.00006i
42	1.00000+ 0.00002i	-0.02114+ 0.07261i	-0.00259+-0.00267i	0.00017+-0.00004i
43	1.00000+ 0.00005i	-0.02317+ 0.06856i	-0.00220+-0.00256i	0.00015+-0.00003i
44	1.00000+ 0.00010i	-0.02507+ 0.06370i	-0.00183+-0.00243i	0.00014+-0.00002i
45	1.00000+ 0.00019i	-0.02667+ 0.05808i	-0.00148+-0.00228i	0.00013+-0.00001i
46	1.00000+ 0.00038i	-0.02783+ 0.05183i	-0.00117+-0.00212i	0.00012+ 0.00000i
47	1.00000+ 0.00077i	-0.02841+ 0.04516i	-0.00089+-0.00196i	0.00011+ 0.00001i
48	1.00000+ 0.00153i	-0.02831+ 0.03832i	-0.00066+-0.00178i	0.00010+ 0.00001i
49	1.00000+ 0.00307i	-0.02753+ 0.03159i	-0.00046+-0.00162i	0.00009+ 0.00002i
50	0.99998+ 0.00614i	-0.02609+ 0.02527i	-0.00029+-0.00146i	0.00008+ 0.00002i
51	0.99992+ 0.01227i	-0.02407+ 0.01965i	-0.00017+-0.00131i	0.00007+ 0.00002i
52	0.99970+ 0.02454i	-0.02161+ 0.01498i	-0.00009+-0.00117i	0.00007+ 0.00002i
53	0.99880+ 0.04907i	-0.01884+ 0.01153i	-0.00006+-0.00106i	0.00006+ 0.00002i
54	0.99518+ 0.09802i	-0.01586+ 0.00956i	-0.00009+-0.00095i	0.00006+ 0.00001i
55	0.98079+ 0.19509i	-0.01253+ 0.00936i	-0.00021+-0.00086i	0.00006+ 0.00000i
56	0.92388+ 0.38268i	-0.00792+ 0.01114i	-0.00047+-0.00070i	0.00006+-0.00002i
57	0.70711+ 0.70711i	0.00099+ 0.01343i	-0.00087+-0.00024i	0.00003+-0.00006i
58	0.00000+ 1.00000i	0.01743+ 0.00407i	-0.00068+ 0.00108i	-0.00007+-0.00007i
59	-1.00000+ 0.00000i	-0.00812+-0.03685i	0.00276+ 0.00029i	-0.00007+ 0.00021i

pattern is more conspicuous in the results from the computer. This pattern will give us the technique to approximate $E_{P_n}(f)$.

Approximating the Fourier Transform. To find an approximation to the Fourier transform, we shall develop a function g which in some circumstances has recurrence properties similar to the recurrence in (2.1).

Define $g(1,n,j,k)$ as follows ($n = 0,1,2,...$; $j = 0,1,...,t$; $k \in \mathbf{Z}/p\mathbf{Z}$):

1. $g(1,n,t,k) = 1$ for all n, k.
2. $g(1,0,j,k) = 1$ for all j, k.
3. $g(1,n+1,j,k) = \frac{1}{2}q^{k2^j}g(1,n,j,k) + \frac{1}{2}q^{k2^j}g(1,n,j+1,k)$ for $j = 0,1,...,t-1$.

Note that we will later consider variations where the first argument is not 1.

$g(1,n,j,k)$ is meant to correspond to $\hat{P}_n(k2^j)$; the difference is that the recurrence 3 stops when $j = t$ in g but not in the recurrence involving Fourier transforms. Adding 1 to the 3rd argument of g is meant to correspond with doubling of the argument of the Fourier transform.

The following lemma finds a limit involving g:

LEMMA 4.2.

$$\lim_{n\to\infty} g(1,n,0,k) = \prod_{j=0}^{t-1} \frac{\frac{1}{2}q^{k2^j}}{1-\frac{1}{2}q^{k2^j}}$$

Proof. Let $a = \lim_{n\to\infty} g(1,n,t-1,k)$. (By using techniques such as the one shown in the proof of the next lemma, one can readily show the existence of such a limit.)

Note that

$$a = \frac{1}{2}q^{k(2^{t-1})}1 + \frac{1}{2}q^{k(2^{t-1})}a$$

or

$$a = \frac{\frac{1}{2}q^{k(2^{t-1})}}{1-\frac{1}{2}q^{k(2^{t-1})}}.$$

What is $\lim_{n\to\infty} g(1,n,t-2,k)$? Let $b = \lim_{n\to\infty} g(1,n,t-2,k)$.

$$b = \frac{1}{2}q^{k2^{t-2}}a + \frac{1}{2}q^{k2^{t-2}}b.$$

Thus

$$\begin{aligned} b &= \frac{\frac{1}{2}q^{k2^{t-2}}}{1-\frac{1}{2}q^{k2^{t-2}}}a \\ &= \frac{\frac{1}{2}q^{k2^{t-2}}}{(1-\frac{1}{2}q^{k2^{t-2}})}\frac{\frac{1}{2}q^{k2^{t-1}}}{(1-\frac{1}{2}q^{k2^{t-1}})}. \end{aligned}$$

Eventually we conclude

$$\lim_{n\to\infty} g(1,n,0,k) = \prod_{j=0}^{t-1} \frac{\frac{1}{2}q^{k2^j}}{1-\frac{1}{2}q^{k2^j}}.$$

□

Let

$$\Pi_1 := \prod_{j=0}^{t-1} \frac{\frac{1}{2}q^{2^j}}{1-\frac{1}{2}q^{2^j}}.$$

Note that Π_1 varies with t, but $|\Pi_1|$ is bounded away from 0 and 1 as $t \to \infty$.

The following lemma tells us how fast $g(1,n,0,k)$ converges to Π_1 for $k = 1, 2, 4, 8, \ldots$:

LEMMA 4.3. When $n \geq 2t+t^{3/4}$ and t is sufficiently large, $|g(1,n,0,k)-\Pi_1| \leq 1/t^m$ where m is an arbitrary fixed positive number.

Proof. Consider the random walk which starts at 0 and adds 0 or 1 with probability $\frac{1}{2}$ at each step. After $2t+t^{3/4}$ additions, the random walk is at t or larger with probability at least $1-\frac{1}{t^m}$ for sufficiently large t and an arbitrary fixed m. Let P be the probability the random walk has NOT reached t.

To see that $P \leq \frac{1}{t^m}$, use Stirling's formula as follows:

$$\begin{aligned}
P &\leq \sum_{i=0}^{t-1} \binom{2t+t^{3/4}}{i} \frac{1}{2^{2t+t^{3/4}}} \\
&\leq \frac{t(2t+t^{3/4})!}{2^{2t+t^{3/4}}(t+t^{3/4})!t!} \\
&\approx \frac{t(2t+t^{3/4})^{2t+t^{3/4}+\frac{1}{2}}e^{-(2t+t^{3/4})}\sqrt{2\pi}}{2^{2t+t^{3/4}}(t+t^{3/4})^{t+t^{3/4}+\frac{1}{2}}e^{-(t+t^{3/4})}\sqrt{2\pi}t^{t+\frac{1}{2}}e^{-t}\sqrt{2\pi}} \\
&= \frac{t2^{2t+t^{3/4}+\frac{1}{2}}(1+\frac{1}{2}t^{-1/4})^{2t+t^{3/4}+\frac{1}{2}}t^{2t+t^{3/4}+\frac{1}{2}}}{2^{2t+t^{3/4}}t^{2t+t^{3/4}+1}(1+t^{-1/4})^{t+t^{3/4}+\frac{1}{2}}\sqrt{2\pi}} \\
&= \frac{\sqrt{2t}}{\sqrt{2\pi}} \frac{(1+\frac{1}{2}t^{-1/4})^{2t+t^{3/4}+\frac{1}{2}}}{(1+t^{-1/4})^{t+t^{3/4}+\frac{1}{2}}} \\
&= \sqrt{\frac{t}{\pi}} \left(\frac{1+t^{-1/4}+\frac{1}{4}t^{-1/2}}{1+t^{-1/4}}\right)^{t+t^{3/4}+\frac{1}{2}} \frac{1}{(1+\frac{1}{2}t^{-1/4})^{t^{3/4}+\frac{1}{2}}} \\
&\leq \frac{e^{at^{1/2}}}{e^{bt^{1/2}}}
\end{aligned}$$

$$\leq \frac{1}{e^{\cdot 2t^{1/2}}}$$

$$\leq \frac{1}{t^m}$$

for sufficiently large t where a is a constant near $\frac{1}{4}$ and b is a constant near $\frac{1}{2}$. Note that by $t^{3/4}$, we really mean the nearest integer to it, and $t^{-1/4}$ really should be $1/t$ times the nearest integer to $t^{3/4}$. (This abuse of notation does not change the result.)

Expand $g(1, n, 0, k)$ by using the recurrence equation for g repeatedly until the 3rd argument is t or the 2nd argument is 0. When $n \geq 2t + t^{3/4}$, the expansion will be identical except for terms whose coefficients sum to less than $\frac{1}{t^m}$. So

$$|g(1, n, 0, k) - \Pi_1| \leq \frac{1}{t^m}$$

for these n and $k = 1, 2, 4, \ldots$ □

The following lemma relates the Fourier transform to Π_1 for some values of k:

LEMMA 4.4. For $k = 1, 2, 4, 8, \ldots, 2^{t-2t^{3/4}}$ and sufficiently large values of t,

$$|\hat{P}_{2t+t^{3/4}}(k) - \Pi_1| \leq \frac{3}{t^m}$$

where m is an arbitrary fixed positive number.

Proof. First observe that

$$|\hat{P}_{2t+t^{3/4}}(k) - \Pi_1| \leq |\hat{P}_{2t+t^{3/4}}(k) - g(1, 2t + t^{3/4}, 0, k)| + |g(1, 2t + t^{3/4}, 0, k) - \Pi_1|.$$

The second term is less than or equal to $\frac{1}{t^m}$ by the previous lemma.

To find $|\hat{P}_{2t+t^{3/4}}(k) - g(1, 2t + t^{3/4}, 0, k)|$, we shall repeatedly use the recursion $\hat{P}_{n+1}(k) = \frac{1}{2}q^k \hat{P}_n(k) + \frac{1}{2}q^k P_n(2k)$. At each use of the recursion, the argument of the Fourier transform either stays the same or doubles mod p. After $2t + t^{3/4}$ uses of the Fourier transform, most of the arguments will have doubled (mod p) at least t times but no more than $t + t^{3/4}$ times; the exceptions correspond to terms which sum to less than $1/t^m$. The terms corresponding to most of the arguments will be close to the corresponding terms in the expansion for $g(1, n, 0, k)$; the difference will sum to no more than $\ell t / 2^{t^{3/4}}$ for some constant ℓ; this fact results from the assumption that $k \leq 2^{t-2t^{3/4}}$ and hence the result that the terms $q^{2^t k}$, $q^{2^{t+1} k}, \ldots, q^{2^{t+t^{3/4}} k}$ will all be close to 1.

In conclusion,

$$|\hat{P}_{2t+t^{3/4}}(k) - g(1, 2t + t^{3/4}, 0, k)| \leq \frac{2}{t^m}$$

for fixed m and sufficiently large t, and

$$|\hat{P}_{2t+t^{3/4}}(k) - \Pi_1| \leq \frac{3}{t^m}.$$

□

For large values of t, the value Π_1 will be close to $-.01004 + .08367i$; the previous lemma explains why this number appears in computer results.

The following lemma explains the appearance of the square of the number, the cube of the number, etc.

LEMMA 4.5. For $k = 1, 2, 4, 8, ..., 2^{t-(r+1)t^{3/4}}$ and m being an arbitrary fixed integer,

$$|\hat{P}_{r(2t+t^{3/4})}(k) - \Pi_1^r| \leq \frac{3r}{t^m}$$

provided that t is large enough and r is an integer which may vary with p but is small enough so that $t - rt^{3/4} \geq 2t^{3/4}$.

Proof. Define $g(r, n, j, k)$ like $g(1, n, j, k)$ except that the 1's on the right side of the equations in conditions 1 and 2 should be replaced by Π_1^{r-1} if $k = 1, 2, 4, ...$ (We need not currently concern ourselves about other values of k.)

Observe that

$$|\hat{P}_{r(2t+t^{3/4})}(k) - \Pi_1^r| \leq |\hat{P}_{r(2t+t^{3/4})}(k) - g(r, 2t + t^{3/4}, 0, k)| + |g(r, 2t + t^{3/4}, 0, k) - \Pi_1^r|.$$

Also observe that

$$|g(r, 2t + t^{3/4}, 0, k) - \Pi_1^r| \leq |\Pi_1|^{r-1} \frac{1}{t^m}$$

by arguments identical to the case where $r = 1$.

Furthermore, note

$$|\hat{P}_{r(2t+t^{3/4})}(k) - g(r, 2t + t^{3/4}, 0, k)| \leq \frac{2}{t^m} + \frac{\ell t}{2^{t^{3/4}}} + \frac{3(r-1)}{t^m}.$$

The first term comes from terms in the recurrence expression which have arguments which double more than $t + t^{3/4}$ times or less than t times on the last $2t + t^{3/4}$ steps. The middle term comes from the terms which have arguments which double on the last $2t + t^{3/4}$ steps slightly more than t times (as in the argument for when $r = 1$). The final term is a result of the error in comparing $\hat{P}_{(r-1)(2t+t^{3/4})}(\tilde{k})$ with Π_1^{r-1} for the values $\tilde{k}$ in $2^t k$, $2^{t+1}k$, ..., $2^{t+t^{3/4}}k$.

Note that t may be chosen to be sufficiently large uniformly for all values of r less than or equal to a multiple of $\log t$ and for all values of k being considered. □

The relation between $E_{P_n}(f)$ and $|\Pi_1|^r$ is given by

LEMMA 4.6. If $r = r_0 \log t$ where r_0 is a constant such that $t|\Pi_1|^r > t^{4/5}$,

$$\left|\frac{\sum_{j=0}^{t-1} \hat{P}_{r(2t+t^{3/4})}(2^j)}{t|\Pi_1|^r}\right| \to 1.$$

Proof. From Lemma 4.5, observe

$$\begin{aligned}\left|\left(\sum_{j=0}^{t-1} \hat{P}_{r(2t+t^{3/4})}(2^j)\right) - t\Pi_1^r\right| &\leq \frac{3r}{t^{m-1}} + (1+|\Pi_1|^r)(r+1)t^{3/4} \\ &\leq \frac{3r}{t^{m-1}} + 2(r+1)t^{3/4}.\end{aligned}$$

The result follows by dividing both sides by $t|\Pi_1|^r$ and letting m be sufficiently large. ($m = 1$ suffices.) □

Looking at the Variance and Conclusion. Now that we have a good idea what's going with the expected value of f under P_n, we wish to explore the expected value of $f\bar{f}$ to see that the variance is small enough for us to use the complex Chebyshev's inequality as in [CDG].

Let

$$\Pi_j := \prod_{j'=0}^{t-1} \frac{\frac{1}{2}q^{(2^j-1)2^{j'}}}{1-\frac{1}{2}q^{(2^j-1)2^{j'}}}.$$

Note $\Pi_j = \overline{\Pi_{t-j}}$; so we need only consider $0 \leq j \leq \frac{t}{2}$. For the values of r given previously, we can use a similar argument to show that

$$\left|\frac{\sum_{\alpha=0}^{t-1} \hat{P}_{r(2t+t^{3/4})}(2^\alpha(2^j-1))}{t|\Pi_j|^r}\right| \to 1.$$

We do need to be little careful of the α we choose to exclude in the argument (analogous to making $2^{t-(r+1)t^{3/4}}$ the largest value for k in the argument for $E_{P_n}(f)$. Here we need to exclude α where $\alpha + j$ is near t or α is near t.)

Considering that the convergence in the previous paragraph can be made to be uniform in j, we thus note

$$\left|\frac{\sum_{j=1}^{\lfloor t/2\rfloor} |\sum_{\alpha=0}^{t-1} \hat{P}_{r(2t+t^{3/4})}(2^\alpha(2^j-1))|}{t\sum_{j=1}^{\lfloor t/2\rfloor} |\Pi_j|^r}\right| \to 1.$$

So

$$E_{P_n}(f\bar{f}) \quad \leq \quad t + 2\left|\sum_{j=1}^{\lfloor t/2\rfloor}\sum_{\alpha=0}^{t-1} \hat{P}_{r(2t+t^{3/4})}(2^\alpha(2^j-1))\right|$$

$$\leq \quad t + 2\sum_{j=1}^{\lfloor t/2 \rfloor} |\sum_{\alpha=0}^{t-1} \hat{P}_{r(2t+t^{3/4})}(2^\alpha(2^j-1))|$$

$$\leq \quad t + 2\beta t \sum_{j=1}^{\lfloor t/2 \rfloor} |\Pi_j|^r$$

where $\beta \to 1$ as $t \to \infty$.

Note that for fixed m (in $\frac{1}{t^m}$), since we get uniform bounds on $|\Pi_1|$, ..., $|\Pi_{\lfloor t/2 \rfloor}|$, we can get uniform lower bounds on t covering all places where we say "for sufficiently large t".

By arguments akin to those mentioned in [CDG], we can show

LEMMA 4.7. $|\Pi_j| \leq |\Pi_1|$ for $1 \leq j \leq t-1$ and

$$\frac{|\Pi_j|}{|\Pi_1|^2} \leq 1 + \frac{c_0}{2^j}$$

for $t^{1/3} \leq j \leq \frac{t}{2}$ at least for sufficiently large t.

Proof. See appendix. □

So we can show via arguments essentially the same as those in [CDG] (and summarized at the end of the appendix) that

$$(Var_{P_n}(f))^{1/2} = o(E_{P_n}(f)),$$

and we then can use the complex Chebyshev's inequality to get the disjoint sets analogous to those in [CDG].

From this fact, we may conclude Theorem 2. □

5. Problems for further study. One may wonder if the argument for Theorem 2 can be generalized to other probability distributions for a_n; a good example is $a_n = a$ with probability 1/2 and $a_n = 1$ otherwise. Although the techniques in the proof of Theorem 2 remain valid if $p = a^t - 1$, the problem is that these values of p are subject to "parity" restrictions. If $a - 1 > 1$ and $(a-1)|p$, then it can be readily shown that the probability distribution for X_n can not approach the uniform distribution. There may be other values for p where parity does not pose a problem for the convergence to the uniform distribution but where a constant times $\log p \log\log p$ steps are necessary for the convergence of the uniform distribution.

Another question left open here is the actual upper bound corresponding to a process where b_n has a single value and a_n has a different probability distribution on $\mathbf{Z}^+$ than was considered here. It seems reasonable that in a constant times $\log p \log\log p$ steps, the random process will be close to uniform if "parity problems" do not intervene. Perhaps the techniques described in this paper can be extended further to cover this case.

Acknowledgement. The author would like to thank the referee for a few suggestions concerning style.

Appendix. Techniques used to complete the proof of Lemmas 3.4 and 4.7. The techniques used in completing these proofs involve expressions which are closely related to those found in [CDG]; the techniques used in completing the proof of Lemma 3.4 are also described in Chapter 3 of [Hi].

Interchanging. The technique developed by [CDG] and called "interchanging" shall be used in the proofs in this appendix. If $a \leq c$ and $b \leq d$ are real numbers, then $bc + ad \leq cd + ab$. By applying this repeatedly, one can show the

INTERCHANGING LEMMA

$$\sum_{j=1}^{s} \prod_{i=1}^{r} a_{\pi_i(j)} \leq \sum_{j=1}^{s} \prod_{i=1}^{r} a_j = \sum_{j=1}^{s} a_j^r$$

where π_i is a permutation of $1, ..., s$, and $a_j \geq 0$.

The proof is shown in [Hi2].

A bound for all p. LEMMA A.1. With h as in the proof of Lemma 3.4, if $n \geq c \log p \log\log p$ for some constant c, then

$$\sum_{k=1}^{p-1} \prod_{i=0}^{n-1} (h(\{(a^{3i})k/p\}))^2 \to 0$$

as $p \to \infty$ provided that $(a, p) = 1$.

Proof. Let $a' = a^3$. Let $F(n,p) = \frac{1}{4} \sum_{k=1}^{p-1} \prod_{i=0}^{n-1} (h(\{(a^{3i})k/p\}))^2$.

Write $x \in [0, 1)$ in its a'-ary expansion $x = \alpha_1\alpha_2\alpha_3...$ with $\alpha_i = 0, 1, ..., a' - 1$ where $\alpha_i \neq a' - 1$ infinitely often.

Note $h(x) = b$ if $\alpha_1 \neq \alpha_2$ or $\alpha_1 \notin \{0, a' - 1\}$.

Define the integer t to satisfy

$$a'^{t-1} < p < a'^{t}.$$

Choose n to be of the form $n = rt$ where r is an integer depending on t (as in the case in [CDG]).

Define blocks B_{ki}, each of length t, on the a'-ary expansion in the same way as for the case in [CDG]. Partition the string for k/p into r disjoint blocks B_{ki}, $1 \leq i \leq r$, each of length t:

$$B_{ki} := \alpha_{(i-1)t+1}\alpha_{(i-1)t+2}...\alpha_{it}.$$

Observe that

$$\prod_{\alpha=0}^{n-1} h(\{\frac{a'^{\alpha}k}{p}\}) \leq \prod_{i=1}^{r} (\frac{1}{b})^{-A(B_{ki})}$$

where A is defined slightly differently than in the case in [CDG]. Define it to be the number of "generalized alternations" (i.e. where $\alpha_i \neq \alpha_{i+1}$)

unless that number is 0. If the block is all 0's or all $a'-1$'s, define A to be 0. (Note that all $a'-1$'s is impossible and all 0's occurs only at 0.) Otherwise define $A=1$.

Since $(a,p)=1=(a'^t,p)$, the set of blocks $\{B_{ki}\}$ is identical with the set of blocks $\{B_{k1}\}$. Also note all blocks B_{k1} are distinct.

So

$$\begin{aligned} F(n,p) &\le \frac{1}{4}\sum_{k\neq 0}\prod_{i=1}^{r}(\frac{1}{b})^{-A(B_{ki})} \\ &\le \frac{1}{4}\sum_{k\neq 0}(\frac{1}{b})^{-rA(B_{k1})} \end{aligned}$$

by interchanging.

$$F(n,p)\le\frac{1}{4}\sum_{\substack{\text{length}B=t\\ A(B)>0}}(\frac{1}{b})^{-rA(B)}.$$

Let $M(j)$ denote the number of blocks B of length t with $A(B)=j$.

$$\begin{aligned} M(j) &\le a'\binom{t-1}{j}(a'-1)^j+\delta_{j1}(a'-2) \\ &\le 2a'\binom{t-1}{j}(a'-1)^j \\ &\le 2a'\binom{t}{j}(a'-1)^j \end{aligned}$$

Thus

$$\begin{aligned} F(n,p) &\le \frac{1}{4}\sum_{j=1}^{t}M(j)(\frac{1}{b})^{-rj} \\ &\le \frac{2a'}{4}\sum_{j=1}^{t}\binom{t}{j}((\frac{1}{b})^{-r}(a'-1))^j \\ &= \frac{a'}{2}((1+[(a'-1)(\frac{1}{b})^{-r}])^t-1) \\ &\le \frac{a'}{2}(e^{t(a'-1)(1/b)^{-r}}-1). \end{aligned}$$

For $r\ge\frac{\log(t(a'-1))}{\log\frac{1}{b}}+d$, $F(n,p)\le\frac{a'}{2}(e^{(1/b)^{-d}}-1)$. As $d\to\infty$, $F(n,p)\to 0$. Thus $c\log p\log\log p$ steps suffice to make $F(n,p)\to 0$ for some constant c. □

A bound for almost all p. One can extend the argument in [CDG] to show

LEMMA A.2. With h as in the proof of Lemma 3.4, if $n > c \log p$ steps for some constant c, then for almost all p such that $(a,p) = 1$,

$$\sum_{k=1}^{p-1} \prod_{i=0}^{n-1} (h(\{(a^{3i})k/p\}))^2 \to 0$$

as $p \to \infty$.

Proof. Let a' and F be as in the previous proof.

This argument shall be slightly looser on the bounds than [CDG]; thus one does not have to consider a term like S_3 in [CDG]. Suppose $n \geq d \log p$ where $d > 2$ will be a constant determined later.

Let

$$f_k(p) := \prod_{\alpha=0}^{n-1} h(\{a'^{\alpha} k/p\})$$

$$f(p) := \sum_{k=1}^{p-1} f_k(p).$$

Note that $F(n,p) = \frac{1}{4} f(p)$.

$f_k(p) \leq (\frac{1}{b})^{-A_{k/p}(n)}$ where $A_{k/p}(n)$ is the number of "generalized alternations" in the first n digits of the a'-ary expansion of k/p. A generalized alternation has one of the following types:

1. A place where the digit differs from the following digit

or

2. A place where the digits in a block (of size to be defined later in this proof) are identical and not all 0 or $a' - 1$.

Choose a large fixed integer t and consider the set of all $f_k(p)$ where $a'^{t-1} < p < a'^t, 1 \leq k \leq p-1$.

Consider only p such that $(a', p) = 1$.

Define

$$S := \sum_{a'^{t-1} < p < a'^t} \sum_{k=1}^{p-1} f_k(p).$$

Note that as in the version in [CDG], if $k/p = k'/p'$, then $f_k(p) = f_{k'}(p')$.

Let $M(r/s)$ denote the number of pairs (k,p) with $a'^{t-1} < p < a'^t, 1 \leq r \leq p-1$ which satisfy $k/p = r/s$ with $(r,s) = 1$. $M(r/s) \leq a'^t/s$.

$$S = \sum_{\substack{1 \leq s < a'^t \\ 1 \leq r \leq s-1 \\ (r,s)=1 \\ (s,a)=1}} M(\frac{r}{s}) f_r(s)$$

$$= \sum_{\omega=1}^{t} {\sum}' M(\frac{r}{s}) f_r(s)$$

$$\leq a'^t \sum_{\omega=1}^{t} {\sum}' f_r(s)/s$$

$$\leq a'^t \sum_{\omega=1}^{t} {\sum}' \frac{1}{s} (\frac{1}{b})^{-A_{r/s}(n)}$$

with $\sum'$ denoting sum over all r and s such that $a'^{\omega-1} < s < a'^{\omega}$, $1 \leq r \leq s-1$, $(r,s) = 1$, and $(s,a) = 1$.

As in the case for [CDG], partition the sum $S^* := \sum_{\omega=1}^{t} \sum' \frac{1}{s} (\frac{1}{b})^{-A_{r/s}(n)}$, but this time into

$$S_1 := \sum_{1 \leq \omega \leq t/\log t} {\sum}' \frac{1}{s} (\frac{1}{b})^{-A_{r/s}(n)}$$

$$S_2 := \sum_{t/\log t < \omega \leq t} {\sum}' \frac{1}{s} (\frac{1}{b})^{-A_{r/s}(n)}$$

Note that by choice of d, $t \leq n/2$, and so the third sum in the argument used for the case in [CDG] is vacuous here

Consider both S_1 and S_2. Let $n' = \lfloor n/2\omega \rfloor$ and for each $r/s = \alpha_1 \alpha_2 ... \alpha_n ...$, partition the first $2\omega n'$ a'-ary digits of r/s into n' blocks $B_i(r/s)$, each of length 2ω. Note that since $(a,s) = 1$, then for $R(s) := \{1 \leq x < s : (x,s) = 1\}$, $\{a'r \pmod s : r \in R(s)\} = R(s)$.

Thus all the sets $\{B_i(r/s) : r \in R(s)\}, 1 \leq i \leq n'$ are equal.

For $a'^{\omega-1} < s, s' < a'^{\omega}, (r,s) = (r',s') = 1$, if $r/s \neq r'/s'$, then

$$|r/s - r'/s'| \geq 1/ss' > a'^{-2\omega}.$$

Hence all the blocks $B_1(r/s), a'^{\omega-1} < s < a'^{\omega}, 1 \leq r \leq s-1, (r,s) = 1$ are distinct. Also note that $B_1(r/s)$ has at least 1 generalized alternation.

$$\sum_{r \in R(s)} (\frac{1}{b})^{-\sum_{i=1}^{n'} A(B_i(r/s))} \leq \sum_{r \in R(s)} (\frac{1}{b})^{-n' A(B_1(r/s))}$$

by interchanging.

$$\sum{}' \frac{1}{s} (\frac{1}{b})^{-A_{r/s}(n)} \leq a' a'^{-\omega} \sum_{\substack{\text{length}(B)=2\omega \\ A(B)\geq 1}} (\frac{1}{b})^{-n' A(B)}$$

$$<< a'^{-\omega} \sum_{j=1}^{2\omega} 2a' \binom{2\omega - 1}{j} (a'-1)^j (\frac{1}{b})^{-n' j}.$$

(The notation $a(p) << b(p)$ means that $\limsup a(p)/b(p) < \infty$.) Note that the multiple a' comes from the choice of the first digit in the a'-ary expansion. The $\binom{2\omega-1}{j}$ tells where the j alternations of type 1. are. The $(a'-1)^j$ tells the choice of changed digits for each of the j alternations of type 1. The factor of 2 allows one to consider, for $j=1$, the $a'-2$ blocks of type 2. and for $j>1$ does no harm to the inequality.

$$\sum{}' \frac{1}{s}(\frac{1}{b})^{-A_{r/s}(n)} << a'^{-\omega} 2a' \sum_{j=1}^{2\omega} \binom{2\omega}{j} ((\frac{1}{b})^{-n'}(a'-1))^j$$

$$<< a'^{-\omega}((1+((\frac{1}{b})^{-n'}(a'-1)))^{2\omega}$$

Now, for S_1,

$$\sum{}' \frac{1}{s}(\frac{1}{b})^{-A_{r/s}(n)} << a'^{-\omega}(e^{((1/b)^{-n'}(a'-1))2\omega} - 1)$$

$$<< a'^{-\omega}((\frac{1}{b})^{-n'}(a'-1))2\omega$$

provided that d is large enough. To find such a value for d, examine

$$(\frac{1}{b})^{-n'} \leq (\frac{1}{b})^{-d\log t \log a'/2}$$

$$= e^{(\log b)d \log t \log a'/2}$$

$$= t^{(\log b)d \log a'/2}$$

and choose d so that $(\log b)d\log a'/2 < -1$. Note that $\log b < 0$. Thus for this value of d, on S_1,

$$(\frac{1}{b})^{-n'}(a'-1)2\omega$$

will be in a bounded range.

$$S_1 << \sum_{1\leq\omega\leq t/\log t} \omega a'^{-\omega}(\frac{1}{b})^{-n'}$$

$$<< (\frac{1}{b})^{-(n\log t)/2t}$$

since $\sum_{\omega=1}^{\infty} \omega a'^{-\omega}$ is finite.

With $n = dt$, $S_1 << (\frac{1}{b})^{-d\log t/2} \to 0$ as $t \to \infty$.

For S_2, note

$$S_2 << \sum_{t/\log t<\omega\le t} a^{-\omega}((1+(\frac{1}{b})^{-n'}(a'-1))^{2\omega}-1).$$

Choose d so that

$$1+(\frac{1}{b})^{-\lfloor d/2\rfloor}(a'-1)<\sqrt{a'}$$

and it satisfies earlier conditions $d>2$ and $(\log b)d\log a'/2<-1$.

Then

$$S_2 << t \max_{t/\log t<\omega\le t} a'^{-\omega}((1+(\frac{1}{b})^{-n'}(a'-1))^{2\omega}-1).$$

$n'\ge\lfloor d/2\rfloor$ if $a'\ge 3$.
So

$$S_2 << t \max_{t/\log t<\omega\le t}(\tilde{c})^{\omega}$$

where $\tilde{c}=\frac{1}{a'}(1+(1/b)^{-\lfloor d/2\rfloor}(a'-1))^2$. $S_2 << t(\tilde{c})^{t/\log t}\to 0$ as $t\to\infty$.

So for $n=dt$, $S^*\to 0$ as $t\to\infty$. So, $S={a'}^t o(1)$. Since there are at least ${a'}^{t-2}$ integers p in $({a'}^{t-1},{a'}^t)$ such that $(a,p)=1$, then for almost all such p, the process converges in $c\log p$ steps for some constant c. "Almost all" means that for a given $\epsilon>0$, $F(n,p)<\epsilon$ for an arbitrarily large percentage of such p in $[1,M]$ by picking M large enough. □

Conclusions about Π_j. In this section, we develop techniques to show some relationship between Π_1 and Π_j for $2\le j\le t/2$. These techniques, in fact, are nearly identical to some in [CDG].

Let

$$G(x)=|\frac{\frac{1}{2}e^{2\pi ix}}{1-\frac{1}{2}e^{2\pi ix}}|.$$

Note that G is a continuous and differentiable function. Also note

$$|\Pi_j|=\prod_{i=0}^{t-j-1}G(\frac{2^{i+j}-2^i}{p})\prod_{i=0}^{j-1}G(\frac{2^{i+t-j}-2^i}{p}).$$

LEMMA A.3. For large enough t, $|\Pi_j|\le|\Pi_1|$ for all j such that $2\le j\le t/2$. The value of t does not necessarily depend on j.

Proof. Set up the same correspondence as in [CDG]:

$$x \longleftrightarrow \pi(x)$$

$$G(\frac{2^{i+j}-2^i}{p}) \longleftrightarrow G(\frac{2^{i+j-1}}{p}) \text{ for } 0 \le i \le t-j-2$$

$$G(\frac{2^{t-1}-2^{t-j-1}}{p}) \longleftrightarrow G(\frac{2^{t-1}}{p})$$

$$G(\frac{2^{i+t-j}-2^i}{p}) \longleftrightarrow G(\frac{2^i}{p}) \text{ for } 0 \le i \le j-3$$

$$G(\frac{2^{t-1}-2^{j-1}}{p})G(\frac{2^{t-2}-2^{j-2}}{p}) \longleftrightarrow G(\frac{2^{j-2}}{p})G(\frac{2^{t-2}}{p})$$

$x \le \pi(x)$ for the first and third cases. If you multiply the other cases, the left side is less than the right side if $2 \le j \le t/2$ and if t is larger than some value which does not depend on j. For these values of j and t, $|\Pi_j| \le |\Pi_1|$.

Next we shall show

LEMMA A.4. For sufficiently large t and $t^{1/3} \le j \le t/2$.

$$\frac{|\Pi_j|}{|\Pi_1|^2} \le 1 + \frac{c_0}{2^j}$$

where c_0 is some constant.

Proof. Again we set up a correspondence:

$$G(\frac{2^{t-i}-2^{t-i-j}}{p})G(\frac{2^{t-i}-2^{j-i}}{p}) \longleftrightarrow G(\frac{2^{t-i}}{p})$$

for $1 \le i \le j$. Note that while each term on the left is not equal to the square of the term on the right, they are close to each other. Also note that the product of each column is not $|\Pi_j|$ or $|\Pi_1|$ due to some terms which are not included.

Consider $G(x)$ for x in $[0, 1]$. This is a differentiable function everywhere on the interval (including the one-sided derivatives at the endpoints.) In fact, at the endpoints, the one-sided derivative is 0. Also note that $|G(x)|$ is bounded away from 0.

Now consider $G(\frac{2^{t-i}-2^{t-i-j}}{p})G(\frac{2^{t-i}-2^{j-i}}{p})$ and $(G(\frac{2^{t-i}}{p}))^2$. They are close to each other, but the first one is slightly larger than the second one for these values of i and j. However,

$$\frac{G(\frac{2^{t-i}-2^{t-i-j}}{p})}{G(\frac{2^{t-i}}{p})} \le 1 + \frac{d}{2^{i+j}}$$

where d is a constant while

$$\frac{G(\frac{2^{t-i}-2^{j-i}}{p})}{G(\frac{2^{t-i}}{p})} < 1 + \frac{d}{2^{t-j+i}} < 1 + \frac{d}{2^{t/2+i}}.$$

Thus the product of the left column is less than the product of squares of the right column times $1 + \frac{c_1}{2^j}$ where c_1 is some constant.

What happens to the other terms? Both $|\Pi_1|$ and $|\Pi_j|$ will contain terms $G(x)$ not included in the columns. Note that $|x| \leq 2/2^j$ and that $G(x) = 1 + O(x^2)$. So the product of the remaining terms $G(x)$ is greater than

$$\begin{aligned}(1 - \frac{c_2}{2^{2j}})^t &\geq e^{-c_3 t/2^j 2^j} \\ &\geq e^{-c_4/2^j} \text{ for large enough } t \text{ since } j \geq t^{1/3} \\ &\geq 1 - \frac{c_4}{2^j}\end{aligned}$$

where c_2, c_3, and c_4 are all constants. Thus Lemma A.4 follows.

One last step in the lower bound argument is to show that the variance of f is small compared with its expected value. We just use the argument from [CDG]:

$$\sum_{t^{1/3} \leq j \leq t/2} |(\frac{|\Pi_j|}{|\Pi_1|^2})^r - 1| \leq \frac{c_5 tr}{2^{t^{1/3}}} < \frac{c_6}{2^{t^{1/4}}}$$

So

$$\begin{aligned}\frac{1}{t}\sum_{j=0}^{t-1}(\frac{|\Pi_j|}{|\Pi_1|^2})^r &\leq \frac{2}{t}\Big(\sum_{0 \leq j \leq t^{1/3}} (\frac{|\Pi_j|}{|\Pi_1|^2})^r + \sum_{t^{1/3} \leq j \leq t/2} (\frac{|\Pi_j|}{|\Pi_1|^2})^r\Big) \\ &= 1 + o(1).\end{aligned}$$

From earlier results, we can make our conclusions about the variance of f under P_n, and the lower bound we desire follows.

REFERENCES

[CDG] Chung, F., Diaconis, P., and Graham, R.L. "A random walk problem arising in random number generation." *Annals of Probability* **15**, 1148-1165. 1987.

[Di] Diaconis, P. *Group Representations in Probability and Statistics.* Hayward, Calif.: Institute of Mathematical Statistics, 1988.

[Hi] Hildebrand, M. "Rates of Convergence of Some Random Processes on Finite Groups." Ph.D. thesis, Harvard University Department of Mathematics, 1990.

[Hi2] Hildebrand, M. "Random Processes of the Form $X_{n+1} = a_n X_n + b_n \pmod p$)." *Annals of Probability* **21**, 710-720. 1993.

[Kn] Knuth, D. *The Art of Computer Programming.* Vol. II, 2nd ed. Menlo Park, Calif.: Addison-Wesley, 1981.

THE SECOND MOMENT METHOD, CONDITIONING AND APPROXIMATION*

SVANTE JANSON†

Abstract. The second moment method has been fruitfully combined with conditioning by several authors, for example Robinson and Wormald [7], [8] and Frieze and Janson [2]. We try to give a general exposition of this method, and of its relation to approximations of random variables by simpler variables. We discuss several examples where the method has been used in different forms.

1. The second moment method. One of the classical methods to show the existence (with large probability) of certain substructures in a random graph is the second moment method. We can describe a general version of the method as follows.

Consider some random combinatorial structure G. In order to study the existence of a certain type of substructure (for example, a Hamilton cycle, a perfect matching, an induced $C_4, \ldots$), we let $X = X(G)$ be the number of such substructures in G. Thus X is a random variable taking non-negative integer values, and the problem is to estimate $\mathrm{P}(X > 0)$. We assume that $\mathrm{P}(X > 0) > 0$ (otherwise the substructures never appear) and that $\mathrm{E}X^2 < \infty$ (in fact, we usually study finite random structures, where X is bounded).

Typically, we are interested in asymptotic results. We consider a sequence of such random structures G (a more precise notation would be G_n, but we will usually ignore the subscript n) and want to show $\mathrm{P}(X > 0) \to 1$. The second moment method asserts that this follows if we can show

$$\operatorname{Var} X/(\mathrm{E}X)^2 \to 0 \tag{1.1}$$

or, equivalently,

$$\mathrm{E}X^2/(\mathrm{E}X)^2 \to 1. \tag{1.1'}$$

The proof is immediate by Chebyshev's inequality:

$$\mathrm{P}(X = 0) \le \frac{\operatorname{Var} X}{(\mathrm{E}X)^2} = \frac{\mathrm{E}X^2}{(\mathrm{E}X)^2} - 1. \tag{1.2}$$

Alternatively, the Cauchy–Schwarz inequality gives

$$(\mathrm{E}X)^2 = \big(\mathrm{E}(XI(X > 0))\big)^2 \le \mathrm{E}X^2\mathrm{P}(X > 0)$$

* Supported by the Göran Gustafsson Foundation for Research in Natural Sciences and Medicine. Research carried out during a visit to the IMA, Minneapolis, MN.

† Department of Mathematics, Uppsala University, P.O. Box 480, S-751 06 Uppsala, Sweden. Email address: svante.janson@math.uu.se .

and thus

$$P(X=0) = 1 - P(X>0) \le 1 - \frac{(EX)^2}{EX^2}. \tag{1.3}$$

The bound in (1.3) is better than the bound in (1.2), but for our purposes both bounds are equally powerful.

Note that (1.1) is much stronger than the conclusion $P(X>0) \to 1$. In fact, (1.1) can be written $\mathrm{Var}(X/EX) \to 0$ or $E(X/EX-1)^2 \to 0$, i.e.

$$X/EX \xrightarrow{L^2} 1. \tag{1.4}$$

In particular, by Chebyshev's inequality again,

$$X/EX \xrightarrow{p} 1; \tag{1.5}$$

i.e., for every $\varepsilon > 0$,

$$P\big((1-\varepsilon)EX < X < (1+\varepsilon)EX\big) \to 1. \tag{1.6}$$

Hence, the second moment method is applicable only when X can be well approximated (in the sense of (1.4)–(1.6)) by the non-random value EX; in other words, the distribution of X has to be (asymptotically) concentrated at a single value.

2. Conditioning. There are many cases where the second moment method fails because (1.4)–(1.6) do not hold; for example, X/EX may converge in distribution to a non-degenerate random variable. In such cases, the method may be combined with conditioning as follows.

Let us choose another random variable $Y = Y(G)$; Y does not have to be real-valued, for example, Y may be a sequence $(Y^{(1)}, Y^{(2)}, \dots)$ of real random variables. For simplicity we assume that Y is discrete, i.e. Y assumes only a countable number of values. (For the approach to be useful, we also want Y to be simpler than X in some sense.) Two examples, considered again later, are Y = the number of edges in a random graph $G(n,p)$, and $Y = (d_1, \dots, d_n)$, where d_i is the degree of vertex i in a random graph or hypergraph.

Recall that the conditional expectation $E(X \mid Y)$ is a random variable which is a function of Y; more precisely, it equals $g(Y)$, where the function g is defined by $g(y) = E(X \mid Y=y)$ whenever $P(Y=y) > 0$. Similarly, the conditional variance is the random variable

$$\mathrm{Var}(X \mid Y) = E\big((X - E(X \mid Y))^2 \mid Y\big) = E(X^2 \mid Y) - E(X \mid Y)^2.$$

We can now apply (1.2) after conditioning on Y (i.e., conditioning on $Y = y$ for each y) and obtain

$$P(X=0 \mid Y) \le \frac{\mathrm{Var}(X \mid Y)}{E(X \mid Y)^2} \tag{2.1}$$

whenever $\mathrm{E}(X \mid Y) > 0$. (Recall that both sides of (2.1) are random variables.)

For convenience, we make the right hand side of (2.1) everywhere defined by assigning an arbitrary value in $[0, \infty]$ to it when $\mathrm{E}(X \mid Y) = 0$ (in which case it is of the form $0/0$).

THEOREM 2.1. *Suppose that* $\mathrm{P}\big(\mathrm{E}(X \mid Y) = 0\big) \to 0$ *and*

$$\mathrm{Var}(X \mid Y)/\mathrm{E}(X \mid Y)^2 \xrightarrow{p} 0. \tag{2.2}$$

Then $\mathrm{P}(X = 0) \to 0$.

Proof. Since conditional probabilities are bounded by 1, (2.1) can be rewritten

$$\mathrm{P}(X = 0 \mid Y) \le \min\big(1, \mathrm{Var}(X \mid Y)/\mathrm{E}(X \mid Y)^2\big) + I\big(\mathrm{E}(X \mid Y) = 0\big) \tag{2.3}$$

which by taking expectations yields

$$\mathrm{P}(X = 0) \le \mathrm{E}\min\big(1, \mathrm{Var}(X \mid Y)/\mathrm{E}(X \mid Y)^2\big) + \mathrm{P}\big(\mathrm{E}(X \mid Y) = 0\big). \tag{2.4}$$

The first term on the right hand side tends to 0 by (2.2) and dominated convergence, while the second tends to 0 by assumption. □

The condition (2.2) may be rewritten

$$\mathrm{E}(X^2 \mid Y)/\mathrm{E}(X \mid Y)^2 \xrightarrow{p} 1 \tag{2.5}$$

or

$$\mathrm{P}\big(\mathrm{Var}(X \mid Y) > \varepsilon \mathrm{E}(X \mid Y)^2\big) \to 0, \qquad \text{for every } \varepsilon > 0. \tag{2.6}$$

We may also write it as

$$\mathrm{Var}\Big(\frac{X}{\mathrm{E}(X \mid Y)} \mid Y\Big) \xrightarrow{p} 0 \tag{2.7}$$

and it follows easily that the conditions of Theorem 2.1 imply

$$\frac{X}{\mathrm{E}(X \mid Y)} \xrightarrow{p} 1 \tag{2.8}$$

or, for every $\varepsilon > 0$,

$$\mathrm{P}\big((1-\varepsilon)\mathrm{E}(X \mid Y) < X < (1+\varepsilon)\mathrm{E}(X \mid Y)\big) \to 1. \tag{2.9}$$

Hence the conditional second moment method is applicable only when X can be well approximated (with a relative error that is small in probability) by a function of Y. The converse is not necessarily true, since variables may converge in probability without converging in L^2, but we do not expect this to happen in practical applications, and thus we expect Theorem 2.1 to be applicable whenever X is well approximated by some function $g(Y)$. (We may then take $g(Y) = \mathrm{E}(X \mid Y)$, but some approximation of $\mathrm{E}(X \mid Y)$ may be simpler.) This should be compared to the unconditional second moment method described above, which applies only when the approximation can be chosen as a "constant". (We put "constant" in inverted commas, since we really consider a sequence $G = G_n$ of random structures and everything, including the function g and this "constant", may depend on the parameter n.)

We also remark that when the conditional second moment method is applicable, it typically also yields the asymptotic distribution of X exactly when the unconditional second moment method is *not* applicable. In fact, we have just seen that then $X/g(Y) \xrightarrow{p} 1$ for some variable $g(Y)$ which is not (asymptotically) degenerate. Hence, if for example $g(Y)/\mathrm{E}g(Y)$ or $a_n \log g(Y) - b_n$ (for some constants a_n, b_n) converges to some (non-degenerate) distribution, we obtain the same result for X. (This is closely related to the projection methods for obtaining asymptotic distributions, where, however, one often considers approximations of a variable X by a *linear* function $g(Y)$ only.)

3. Applications. We have so far discussed the theoretical limits of the method. For practical applications, we also have to be able to verify the hypotheses of the theorem. There are at least three different methods to do this. We describe these methods below, and illustrate them by examples where they have been used by various authors (at least implicitly; the descriptions below are reformulations of the arguments in the original references). The methods often lead to long and non-trivial combinatorial calculations, which we omit here; see the references given in the examples below.

I. The first possibility is a straightforward computation of $\mathrm{E}(X \mid Y)$ and $\mathrm{E}(X^2 \mid Y)$.

Example 1. Janson [3] considered the number of spanning subtrees of the random graph $G(n,p)$, with p constant and $n \to \infty$; let us denote this number by X. It is well-known that $\mathrm{P}(X > 0) \to 1$, but the objective in [3] was to find the asymptotic distribution. This was achieved using the argument above, with Y being the number of edges in $G(n,p)$. Note that conditioning on $Y = m$ is the same as considering the random graph $G(n,m)$; hence $\mathrm{E}(X \mid Y = m)$ and $\mathrm{E}(X^2 \mid Y = m)$ can be evaluated as the first two moments of the number of spanning trees in $G(n,m)$. These can be estimated rather easily, see Moon [6], which leads to an asymptotic log-normal distribution of X.

Example 2. Consider the random hypergraph $G(ns, m; s)$ with ns vertices and m random hyperedges consisting of s vertices each. Frieze and Janson [2] showed that if $s \geq 2$ is fixed, and $n \to \infty$ with $m/n^{4/3} \to \infty$, then the random hypergraph contains a perfect matching (i.e., a covering of the vertex set by n disjoint hyperedges) with probability tending to 1. The proof is based on Theorem 2.1, letting X be the number of perfect matchings and $Y = (d_1, \dots, d_{ns})$, where d_i is the degree of vertex i, i.e., the number of hyperedges containing i. Actually, the computations are performed for a slightly different model, namely a random multihypergraph where the hyperedges are obtained by sampling ms vertices with replacement and then grouping them s by s; this allows both multiple hyperedges and defect hyperedges with less than s distinct vertices, but the distribution of Y (which becomes multinomial) and the calculation of $\mathrm{E}(X \mid Y)$ and $\mathrm{E}(X^2 \mid Y)$ are much simplified.

It turns out that if $m \gg n^{3/2}$, then $\mathrm{E}(X \mid Y)$ is concentrated at $\mathrm{E}X$, and the unconditional second moment method applies, see Schmidt and Shamir [9]. When m is smaller, this is no longer true and we need the conditional version. Note also that the proof requires $m \gg n^{4/3}$, although there is no reason to believe that this is necessary for the result (presumably $m \gg n \log n$ is enough, as it is for $s = 2$). The discussion above gives (or at least suggests) the explanation that the number of perfect matchings is well approximated by a function of the vertex degrees only when $m \gg n^{4/3}$ (and by a constant when $m \gg n^{3/2}$); for smaller m the number depends significantly also on other properties of the hypergraph, but it may of course still be non-zero.

In principle it seems possible to improve the result by conditioning on more parameters beside the vertex degrees, cf. Example 4 below, but it seems hopeless to perform the necessary computations.

II. It is often complicated to calculate the conditional variance $\mathrm{Var}(X \mid Y)$ or $\mathrm{E}(X^2 \mid Y)$. A possible alternative is to use an analysis of variance argument as follows, which only requires calculation of the conditional expectation $\mathrm{E}(X \mid Y)$ and the unconditional variance $\mathrm{Var}\, X$. This is essentially the method invented by Robinson and Wormald [7], [8], see also Cooper, Frieze, Molloy and Reed [1] and Janson [5].

By definition, the terms in the decomposition

$$X = \big(X - \mathrm{E}(X \mid Y)\big) + \mathrm{E}(X \mid Y) \tag{3.1}$$

are orthogonal, and

$$\begin{aligned} \mathrm{Var}\, X &= \mathrm{Var}\big(X - \mathrm{E}(X \mid Y)\big) + \mathrm{Var}\big(\mathrm{E}(X \mid Y)\big) \\ &= \mathrm{E}\big(X - \mathrm{E}(X \mid Y)\big)^2 + \mathrm{Var}\big(\mathrm{E}(X \mid Y)\big) \\ &= \mathrm{E}\big(\mathrm{Var}(X \mid Y)\big) + \mathrm{Var}\big(\mathrm{E}(X \mid Y)\big). \end{aligned} \tag{3.2}$$

Recall that the model G, and thus also the variables X and Y, depends on a parameter n. The o below should be interpreted as $n \to \infty$.

THEOREM 3.1. *If*

$$\mathrm{Var}\, X = \mathrm{Var}\big(\mathrm{E}(X \mid Y)\big) + o\big((\mathrm{E}X)^2\big), \tag{3.3}$$

and

$$\lim_{\varepsilon\to 0}\limsup_{n\to\infty} \mathrm{P}\big(\mathrm{E}(X \mid Y) < \varepsilon \mathrm{E}X\big) = 0, \tag{3.4}$$

then $\mathrm{P}(X=0)\to 0$.

Proof. For every $\varepsilon > 0$ and $\delta > 0$,

$$\begin{aligned} &\mathrm{P}\big(\mathrm{Var}(X \mid Y)/\mathrm{E}(X \mid Y)^2 > \varepsilon\big) \\ &\quad \le \mathrm{P}\big(\mathrm{Var}(X \mid Y)/(\mathrm{E}X)^2 > \varepsilon\delta^2\big) + \mathrm{P}\big(\mathrm{E}(X \mid Y) < \delta \mathrm{E}X\big). \end{aligned} \tag{3.5}$$

Moreover, by (3.3) and (3.2), $\mathrm{E}\big(\mathrm{Var}(X \mid Y)\big) = o(\mathrm{E}X^2)$, and thus

$$\mathrm{P}\big(\mathrm{Var}(X \mid Y)/(\mathrm{E}X)^2 > \varepsilon\delta^2\big) \to 0.$$

Hence,

$$\limsup_{n\to\infty} \mathrm{P}\big(\mathrm{Var}(X \mid Y)/\mathrm{E}(X \mid Y)^2 > \varepsilon\big) \le \limsup_{n\to\infty} \mathrm{P}\big(\mathrm{E}(X \mid Y) < \delta \mathrm{E}X\big).$$

The right hand side tends to 0 as $\delta \to 0$, and thus the left hand side vanishes for every $\varepsilon > 0$. This shows $\mathrm{Var}(X \mid Y)/\mathrm{E}(X \mid Y)^2 \xrightarrow{p} 0$. Furthermore, for every $\varepsilon > 0$,

$$\limsup_{n\to\infty} \mathrm{P}\big(\mathrm{E}(X \mid Y) = 0\big) \le \limsup_{n\to\infty} \mathrm{P}\big(\mathrm{E}(X \mid Y) < \varepsilon \mathrm{E}X\big),$$

and thus (3.4) implies $\mathrm{P}\big(\mathrm{E}(X \mid Y) = 0\big) \to 0$. The result now follows by Theorem 2.1. □

Note that (3.3) can be regarded as a condition on the variance of the absolute error $X - \mathrm{E}(X \mid Y)$ in the approximation of X by $\mathrm{E}(X \mid Y)$, while condition (2.2) in Theorem 2.1 is a condition on the (conditional) variance of the relative error $X/\mathrm{E}(X \mid Y) - 1$. The absolute error gives less information when X varies wildly so the theorem is most useful when the variation is not too large. In fact, in all applications so far of this version of the conditional second moment method, we have $\mathrm{Var}\, X = O\big((\mathrm{E}X)^2\big)$.

Example 3. Robinson and Wormald [7] considered X = the number of Hamilton cycles in a random cubic graph. In principle, they used Theorem 3.1 with $Y = (\xi_3, \xi_4, \ldots)$, where ξ_j is the number of cycles of length j, but they also used a truncation argument. In fact, since $X = \xi_n$ (where n is the order of the graph) is one of the components of Y, $\mathrm{E}(X \mid Y) = X$ so nothing at all is gained by conditioning on Y. Instead, they condition on $Y_M = (\xi_3, \ldots, \xi_M)$ and let first $n \to \infty$ for fixed M, and then $M \to \infty$. We can regard their argument as a special case of the following extension of Theorem 3.1. We assume that Y_M is a variable, defined for some sequence $M \to \infty$ and each G_n.

THEOREM 3.2. *Suppose that*

$$\lim_{M\to\infty}\limsup_{n\to\infty}\big[\mathrm{Var}\, X - \mathrm{Var}\big(\mathrm{E}(X \mid Y_M)\big)\big]/(\mathrm{E}X)^2 = 0 \tag{3.6}$$

and

$$\lim_{\varepsilon\to 0}\limsup_{M\to\infty}\limsup_{n\to\infty} \mathrm{P}\big(\mathrm{E}(X \mid Y_M) < \varepsilon \mathrm{E}X\big) = 0. \tag{3.7}$$

Then $\mathrm{P}(X=0)\to 0$.

This follows from a corresponding extension of Theorem 2.1.

THEOREM 3.3. *Suppose that*

$$\lim_{M\to\infty}\limsup_{n\to\infty} \mathrm{P}\big(\mathrm{E}(X \mid Y_M) = 0\big) = 0 \tag{3.8}$$

and, for each $\varepsilon > 0$,

$$\lim_{M\to\infty}\limsup_{n\to\infty} \mathrm{P}\big(\mathrm{Var}(X \mid Y_M) > \varepsilon \mathrm{E}(X \mid Y_M)^2\big) = 0. \tag{3.9}$$

Then $\mathrm{P}(X=0)\to 0$.

Proof. By (2.4),

$$\mathrm{P}(X=0) \le \mathrm{P}\big(\mathrm{Var}(X \mid Y_M) > \varepsilon \mathrm{E}(X \mid Y_M)^2\big) + \varepsilon + \mathrm{P}\big(\mathrm{E}(X \mid Y_M) = 0\big).$$

The result follows from (3.9) and (3.8) by taking the lim sup as $n\to\infty$, $M\to\infty$ and $\varepsilon\to 0$ (in this order). □

Proof of Theorem 3.2. Taking the lim sup as $n\to\infty$, $M\to\infty$ and $\delta\to 0$ (in this order) in (3.5) (with Y_M instead of Y), we obtain (3.9). Similarly, (3.8) follows from (3.7). Hence Theorem 3.3 applies. □

III. Another way to avoid calculating the conditional variance $\mathrm{Var}(X \mid Y)$ is to estimate the unconditional variance of the variable $X/\mathrm{E}(X \mid Y)$. Note that this variable has mean 1 (if defined to be 1 when $\mathrm{E}(X \mid Y) = 0$). If the variance tends to 0, (2.7) and (2.2) follow immediately and we can use Theorem 2.1 provided $\mathrm{P}\big(\mathrm{E}(X \mid Y) = 0\big) \to 0$.

The drawback of this version is that we have to calculate (or at least estimate) the variance of the more complicated variable $X/\mathrm{E}(X \mid Y)$. We may sometimes simplify the calculations by using an approximation $g(Y)$ of $\mathrm{E}(X \mid Y)$. (The function g may also depend on the hidden parameter n.) We assume for simplicity that $g(Y) > 0$.

THEOREM 3.4. *Suppose that* $g(Y)$ *is a positive function of* Y *such that*

$$\mathrm{E}\frac{X}{g(Y)} \to 1, \tag{3.10}$$

$$\mathrm{Var}\,\frac{X}{g(Y)} \to 0. \tag{3.11}$$

Then $\mathrm{P}(X=0)\to 0$.

Proof. This is immediate by Chebyshev's inequality (1.2), applied to $X/g(Y)$.

It is also easy to see that the assumptions imply the assumptions of Theorem 2.1. In fact, let $Z = X/g(Y)$. Then $Z \to 1$ in L^2, and thus, cf. (3.2), $\mathrm{E}(\mathrm{Var}(Z \mid Y)) \to 0$ and $\mathrm{E}(Z \mid Y) \to 1$ in L^2 (and thus in probability). Hence $\mathrm{Var}(X \mid Y)/\mathrm{E}(X \mid Y)^2 = \mathrm{Var}(Z \mid Y)/\mathrm{E}(Z \mid Y)^2 \xrightarrow{p} 0$, and $\mathrm{P}(\mathrm{E}(X \mid Y) = 0) = \mathrm{P}(\mathrm{E}(Z \mid Y) = 0) \to 0$. □

Example 4. Janson [4] studied for example the number of spanning subtrees in $G(n,p)$, where p may depend on n. Denote this number by X and let Y be the number of edges. Then $\mathrm{E}X = n^{n-2}p^{n-1}$, which suggest an approximation of the form

$$\begin{aligned} c_1\left(\frac{Y}{\binom{n}{2}}\right)^{n-1} &= c_2\left(\frac{Y}{\mathrm{E}Y}\right)^{n-1} = c_2 e^{(n-1)\log(Y/\mathrm{E}Y)} \\ &\approx c_2 e^{(n-1)(Y/EY-1)} = c_3 e^{aY}, \end{aligned}$$

with $a = 2p/n$. It is convenient to change this a little, and we choose

$$g(Y) = c^{-1}(1-a)^{-Y}$$

with $c^{-1} = \mathrm{E}\big(X(1-a)^Y\big)$. Then $X/g(Y) = cX(1-a)^Y$ and $\mathrm{E}(X/g(Y)) = 1$. It turns out, cf. [4, (4.15) and Remark 4.1], that if $\limsup p < 1$ and $p \gg n^{-1/2}$, then $\mathrm{Var}(X/g(Y)) \to 0$. Hence Theorem 3.4 applies; moreover, $\log X - \log g(Y) \xrightarrow{p} 0$, which implies that X has the same asymptotic log-normal distribution as $g(Y)$.

Consider now the case $n^{1/2}p \to c > 0$. Then $\mathrm{Var}\, X/g(Y) \to \alpha$ for some $\alpha > 0$. However, it follows by the argument in [4, section 4] that if Y' denotes the number of paths of length 2 in the graph, then $\mathrm{Var}\big(X/g(Y) - h(Y')\big) \to 0$ for a suitable function h. Hence

$$\mathrm{Var}\big(X/g(Y)\big) = \mathrm{Var}\big(\mathrm{E}(X/g(Y) \mid Y')\big) + o(1),$$

and Theorem 3.1 applies to $X/g(Y)$ and Y', which again gives an asymptotic log-normal distribution of X. It follows that, in this case, X can be approximated (with small relative error) by a function of (Y, Y'). We conjecture that this is the case also when $n^{1/2}p \to 0$ as long as $n^{2/3}p \to \infty$; for smaller p we conjecture that X can be well approximated by a function of (Y, Y', Y''), where Y'' is the number of subtrees of order 4 (paths of length 3 and 3-stars), as long as $n^{3/4}p \to \infty$, and so on with more and more terms for smaller p.

The discussion above applies also to some similar variables treated in [4], for example the number of Hamilton cycles and the number of perfect matchings. Note that the result in the latter case corresponds to the hypergraph result in Example 2, and that the condition $m \gg n^{4/3}$ corresponds to $p \gg n^{-2/3}$, which we conjectured to be the right condition for approximation by a function of Y and Y'. Note also that both Y and Y'

are functions of the sequence of vertex degrees, so the conditioning used in Example 2 is stronger than conditioning on (Y, Y').

REFERENCES

[1] C. Cooper, A. Frieze, M. Molloy and B. Reed, *Perfect matchings in random r-regular, s-uniform hypergraphs*, Combinatorics, Probability and Computing, (to appear).

[2] A. Frieze and S. Janson, *Perfect matchings in random s-uniform hypergraphs*, preprint, Uppsala, 1994.

[3] S. Janson, *Random trees in a graph and trees in a random graph*, Math. Proc. Camb. Phil. Soc. **100** (1986), pp. 319–330.

[4] S. Janson, *The numbers of spanning trees, Hamilton cycles and perfect matchings in a random graph*, Combinatorics, Probability and Computing, **3** (1994), pp. 97–126.

[5] S. Janson, *Random regular graphs: asymptotic distributions and contiguity*, preprint, Minneapolis, 1993.

[6] J. W. Moon, *The second moment of the complexity of a graph*, Mathematika **11** (1964), pp. 95–98.

[7] R.W. Robinson and N.C. Wormald, *Almost all cubic graphs are hamiltonian*, Random Struct. Alg. **3** (1992), pp. 117–125.

[8] R.W. Robinson and N.C. Wormald, *Almost all regular graphs are hamiltonian*, Random Struct. Alg. **5** (1994), pp. 363–374.

[9] J. Schmidt and E. Shamir, *A threshold for perfect matchings in random d-pure hypergraphs*, Discrete Math. **45** (1983), pp. 287–295.

HOW FAST AND WHERE DOES A RANDOM WALKER MOVE ON A RANDOM TREE?

RUSSELL LYONS*

Abstract. The oldest model of random trees is that of Galton and Watson. In joint work with Robin Pemantle and Yuval Peres, described informally here, we consider simple random walk on the family tree T of a supercritical Galton-Watson branching process. First we calculate the speed (i.e., rate of escape) of the walk; the answer is simple, yet surprising. Then we show that the resulting harmonic measure has a.s. strictly smaller dimension than that of the whole boundary of T. Concretely, this means that after T is picked but before the random walk takes place, a subtree of T of a smaller exponential growth rate can be specified such that with overwhelming probability, the random walk particle will never exit this subtree (!).

Keywords and phrases. Trees, harmonic measure, boundary, Galton-Watson, branching processes, random walk, Hausdorff dimension, rate of escape.

AMS(MOS) subject classifications. Primary 60J80. Secondary 60J15.

We present an exposition of some recent joint work with Robin Pemantle and Yuval Peres (see Lyons, Pemantle and Peres ([5])). Consider the tree of Figure 1. This was generated pseudorandomly by beginning with one vertex at the top, called the **root**, and adding 1 or 2 children with equal probability (in this case, it was 2 children). Then each of these children also has 1 or 2 children with equal probability and independently, and so on forever (only 9 generations are represented in the figure). We shall refer to this tree as the **1-2 tree**.

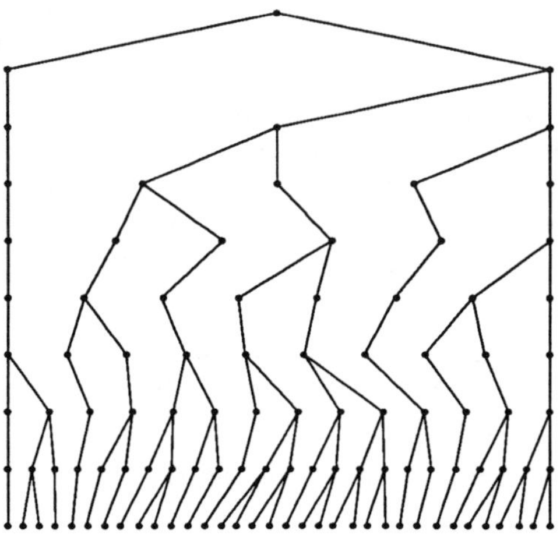

FIG. 1. *A typical random tree.*

* Department of Mathematics, Indiana University, Bloomington, IN 47405-5701.

A random walker starts at the root of the tree and chooses among the edges incident to the root at random, walking to the adjacent vertex. Then the walker again chooses among the incident edges *including the one leading back to the root* at random and moves to the adjacent vertex. This goes on forever. Such a random walk is called **simple random walk**. Since there are, on average, 1.5 edges leading away from the root and only 1 leading toward the root, the walker eventually stops returning to the root and, in fact, escapes to infinity a.s. The questions we address are:

- How fast does the walker escape? In other words, after a long time, about how far is the walker from the root?
- Where does the walker go? In other words, what are the chances of escaping along the various paths to infinity?

The 1-2 tree is actually only an example of the general set-up we treat: In general, each particle (or vertex) has k children with probability p_k, where $\sum_{k=1}^{\infty} p_k = 1$. (We exclude the trivial case of $p_1 = 1$.) This gives the family tree of a **Galton-Watson** branching process. The sequence $\{p_k\}$ is is called the **progeny distribution**. The 1-2 tree is the special case where $p_1 = p_2 = 1/2$. Again, simple random walk is transient by the intuition that on average, there are $m := \sum_{k=1}^{\infty} p_k k > 1$ edges out of each vertex leading away from the root. Note that when we say that simple random walk is transient, we actually mean that for *almost every* Galton-Watson tree for a fixed progeny distribution, the random walk is transient. This intuition was first proved correct in an unpublished supplement to Grimmett and Kesten ([3]).

It is well known that an equivalent assertion comes from electrical networks: Assume that each edge of the tree is a unit conductor. Then transience is equivalent to the effective conductance from the root to infinity being positive (see, e.g., Doyle and Snell ([2])). We illustrate this with a couple of simple examples before turning back to Galton-Watson trees.

First, simple random walk on the nonnegative integers is the same as on the 1-ary tree of Figure 2. If each edge has unit conductance, then each edge also has unit resistance. As these edges are in series and resistances in series add, we see that there is infinite resistance to infinity, whence 0 effective conductance. This proves the well-known fact that such a random walk is recurrent.

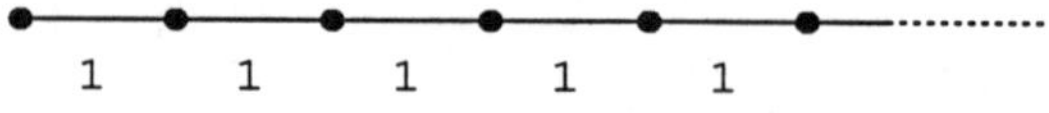

FIG. 2. *The nonnegative integers.*

Second, consider simple random walk on the binary tree (Figure 3). By symmetry, the vertices of a given generation are at the same voltage when a battery is connected between the root and infinity. Therefore, they may be shorted together with no change in the effective conductance.

This produces the network of Figure 4. Since conductances in parallel can be added, we have an equivalent network of Figure 5. Finally, adding the resultant resistances in series gives an effective conductance of 1.

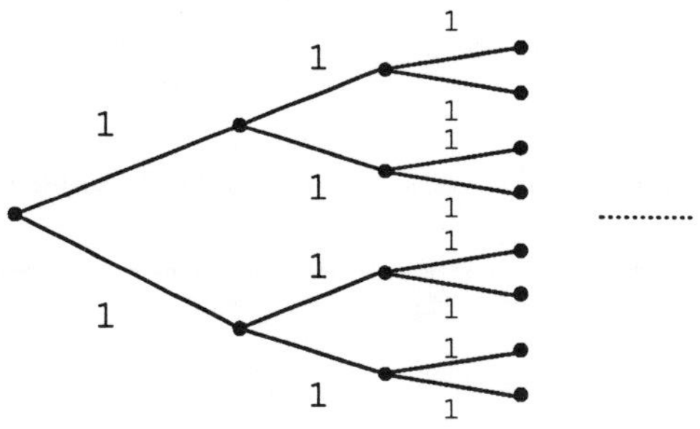

FIG. 3. *The binary tree.*

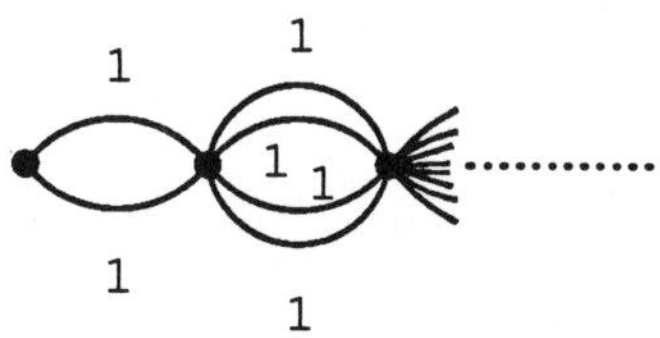

FIG. 4. *Shorting the generations.*

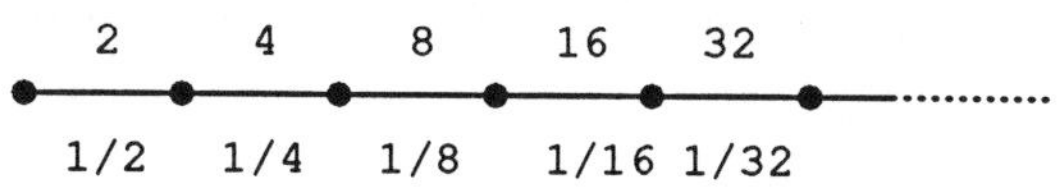

FIG. 5. *Adding the conductances (on top) from Figure 4 and then the resistances (on bottom).*

For the 1-2 tree, it follows by comparison with the binary tree that the conductance is a random variable lying in (0, 1). Its density is shown in Figure 6. (In fact, it is not known that the conductance has an absolutely continuous distribution, but it certainly looks that way from the graph: the derivative of the c.d.f. is never more than 2.5.) From this information, we can calculate that the probability that the random walker never returns to the root is about 29.7%.

Figure 6 is calculated by an iterative solution to a functional equation. If $\mathcal{C}$ is the conductance of a Galton-Watson tree, then the c.d.f. F of $\mathcal{C}/(\mathcal{C}+1)$ satisfies the equation

$$F(s) = \begin{cases} \sum_k p_k F^{*k}\left(\frac{s}{1-s}\right), & \text{if } s \in (0,1); \\ 0, & \text{if } s \leq 0; \\ 1, & \text{if } s \geq 1. \end{cases}$$

The only other solution to this equation is the Heaviside function.

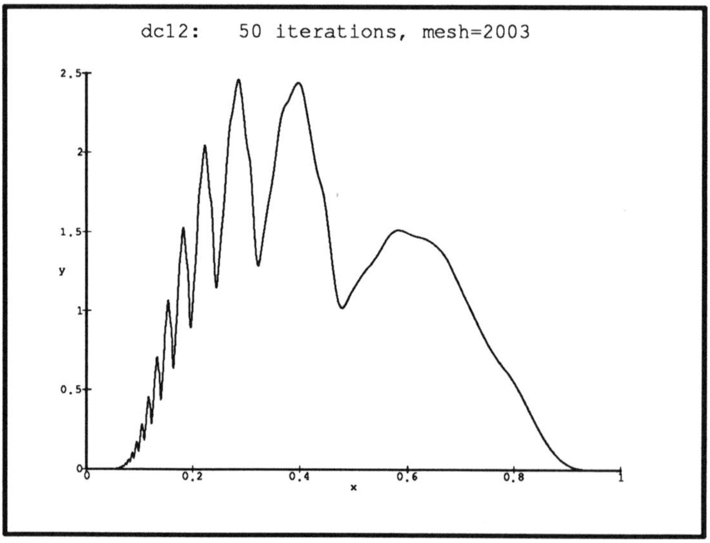

FIG. 6. *The apparent density of the conductance of the 1-2 tree.*

Figure 6 reflects the stochastic self-similarity of the 1-2 tree. For example, suppose that the tree begins without branching, i.e., the root has only one child. Then what follows this child is no bigger than a binary tree and so this descendant subtree has conductance at most 1. Therefore, the whole tree has conductance at most 1/2 (see Figure 7). The wide peak at the right of Figure 6 is thus due entirely to those trees which begin with two children of the root. The peak immediately to its left is due, at first approximation, to an unspecified number of generations without branching, while the nth peak to the left of it is due to n generations without branching (and an unspecified continuation). Of course, the next level of approximation deals with further resolution of the peaks.

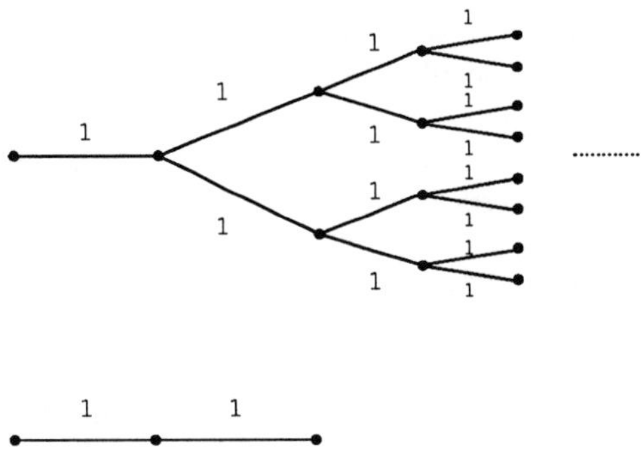

FIG. 7. *The upper bound of a 1-2 tree that starts with one child of the root.*

Let X_n be the position of the random walker on a Galton-Watson tree. Denote by $|X_n|$ its (graph) distance to the root. The question of speed, or rate of escape, is due to David Aldous and asks for the limit

$$\lim_{n\to\infty} \frac{|X_n|}{n} = ?$$

Consider first the m-ary tree ($p_m = 1$). The stochastic process $|X_n|$ is the same as a random walk on $\mathbf{N}$ but with transition probabilities $m/(m+1)$ to the right and $1/(m+1)$ to the left at each positive integer. Since the walk always steps to the right when it is at 0, this is not a stationary process. Nevertheless, it is intuitive that the speed should be given by the strong law of large numbers as

$$\frac{m}{m+1}(+1) + \frac{1}{m+1}(-1) = \frac{m-1}{m+1}\,.$$

Now consider the 1-2 tree. Half of its vertices have 1 child and half have 2, so perhaps the speed is

$$\frac{1}{2}\cdot\frac{1-1}{1+1} + \frac{1}{2}\cdot\frac{2-1}{2+1} = \frac{1}{6}\,.$$

Actually, this seems naive since it doesn't take into account how much time the walker spends at different vertices. Indeed, recall that on a finite graph, the stationary distribution for simple random walk is proportional to the degree of a vertex. From this, we might guess that the speed on the 1-2 tree is

$$\frac{2}{5}\cdot\frac{1-1}{1+1} + \frac{3}{5}\cdot\frac{2-1}{2+1} = \frac{1}{5}\,.$$

Or perhaps neither of these guesses is correct? We shall see in formula (1) below.

To find correctly the speed on Galton-Watson trees, it is useful first to see how to rigorize the calculation of speed for m-ary trees. One way of doing this is to extend the random walk on $\mathbf{N}$ given by the distance process $|X_n|$ to a stationary random walk on $\mathbf{Z}$. We then have true i.i.d. pushes to the right and their averages converge, by the strong law of large numbers, to their expectation.

The analogue we need for Galton-Watson trees is a Markov chain whose state space consists of rooted unlabelled trees. The walk is regarded as being at the root, so that when it takes a step to a neighboring vertex, the new state is the same tree except for the fact that the root has moved. Thus, though the state space is uncountable, from any state, only a finite number of states can be visited next.

Suppose that, somehow, we have a stationary ergodic probability measure μ_{SRW} for this Markov chain. We'd like the speed to be an ergodic average of changes in distance to the initial position X_0. But this suffers from the same problem as did our intuitive calculation of the speed on an m-ary tree: the summands won't then be stationary. One way to fix this, analogous to the rigorization for the m-ary tree, is by a standard trick: we use "horodistance" from a boundary point.

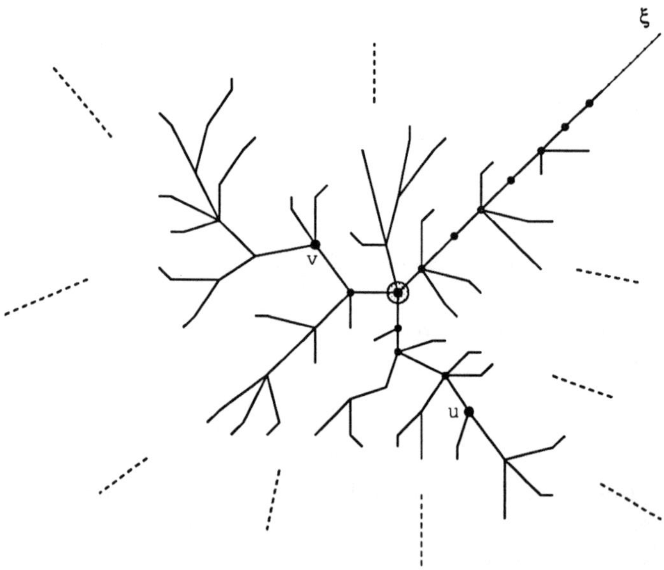

FIG. 8. *The horodistance from v to u as seen from the boundary point ξ is 2.*

A boundary point ξ is an equivalence class of rays, where two rays are **equivalent** if they have an infinite number of vertices in common. In Figure 8, the two rays from v and u going out to the upper right are equivalent. They meet at the shaded vertex, $u \wedge_\xi v$. The **signed distance** from v to u as seen from ξ is defined to be

$$[u - v]_\xi := |u - u \wedge_\xi v| - |v - u \wedge_\xi v|,$$

where $|x-y|$ denotes the distance between x and y in the tree. The intuitive notion of vertices at a given horodistance from a boundary point is given by the concept of a **horosphere**. This is a set of vertices such that any pair of them have signed distance 0 from one to the other as seen from the boundary point. The horospheres are drawn in Figure 9 as horizontal levels. The picture makes clear the additivity of signed distance:

$$[u - w]_\xi = [u - v]_\xi + [v - w]_\xi\,.$$

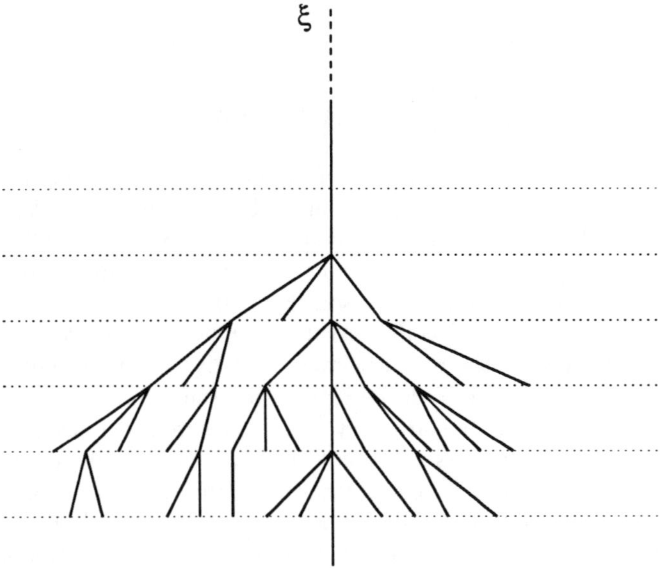

FIG. 9. *Horospheres centered at the boundary point* ξ.

Now the random walk tends to a boundary point in the following sense: The shortest paths $\overline{X_0X_n}$, by virtue of transience of the random walk, have more and more edges eventually constant, defining (a.s.) a limiting ray in an equivalence class which we denote X_∞. Provided the boundary point $\xi \neq X_\infty$, there is a (random) constant c such that for all but finitely many n,

$$|X_n - X_0| = [X_n - X_0]_\xi + c\,.$$

Therefore,

$$\lim_{n\to\infty} \frac{1}{n}|X_n - X_0| = \lim_{n\to\infty} \frac{1}{n}[X_n - X_0]_\xi = \lim_{n\to\infty} \frac{1}{n}\sum_{k=0}^{n-1}[X_{k+1} - X_k]_\xi\,.$$

We can now make the summands stationary as follows. Extend the walk backwards in time and let $X_{-\infty} := \lim_{n\to\infty} \overline{X_0 X_{-n}}$. Since $X_{-\infty} \neq X_\infty$ a.s., we may take $\xi := X_{-\infty}$. The ergodic theorem then gives that the speed is the expectation of, say, the 0th term in the sum above:

$$\iint [X_1 - X_0]_{X_{-\infty}} \, d\overleftrightarrow{X} \, d\mu_{\mathsf{SRW}}(T) \, .$$

To evaluate this, note that X_{-n} is gotten by running an independent simple random walk from X_0. Thus X_1 is independent of $X_{-\infty}$ and is uniform on the neighbors of X_0. That is,

$$[X_1 - X_0]_{X_{-\infty}} = \begin{cases} 1 & \text{with probability } (d-1)/d \\ -1 & \text{with probability } 1/d, \end{cases}$$

where $d := \deg(X_0) = \deg(\mathrm{Root}(T))$. Therefore, the speed is

$$\int \frac{\deg(\mathrm{Root}(T)) - 2}{\deg(\mathrm{Root}(T))} \, d\mu_{\mathsf{SRW}}(T) \, .$$

It remains now to find μ_{SRW}. Of course, there are many stationary ergodic measures for the Markov chain under consideration. As a trivial example, use the measure concentrated on the regular tree, all of whose vertices have degree 3. What we want is a stationary measure that will give information about Galton-Watson trees. But we cannot use the measure given by the standard Galton-Watson branching process: this would not be stationary since the root has stochastically one fewer neighbor than all the other vertices. Instead, we claim that the following "augmented" Galton-Watson measure is stationary. This measure is defined just like ordinary Galton-Watson measure except that the number of children of the root (only) is one more; i.e., the root has $k+1$ children with probability p_k and these children all have independent standard Galton-Watson descendant trees. This is the same as generating a Galton-Watson tree and then adding one more edge to the root, to which we append another independent Galton-Watson tree: see Figure 10. To see that augmented Galton-Watson measure is stationary, consider a random step from the root. The particle is then at a vertex to which is attached

- a Galton-Watson tree labelled **GW** in the figure
- and another edge, at the end of which is another Galton-Watson tree.

But this is precisely the same stochastic figure as the particle sees from the original root.

Now augmented Galton-Watson measure has the property that all vertices, including the root, have degree $k+1$ with probability p_k. We simply plug this measure into the last equation above to get that the speed is

$$\sum_k p_k \frac{k-1}{k+1} \, . \qquad (1)$$

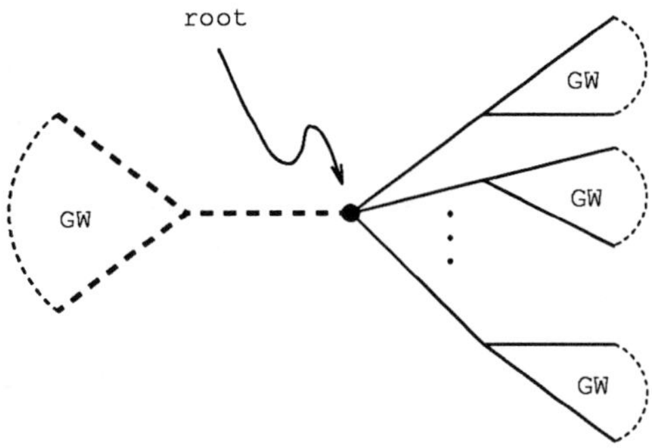

FIG. 10. *An augmented Galton-Watson tree.*

Although this is the speed for simple random walk on augmented Galton-Watson trees, it is intuitively clear that it is also the speed on standard Galton-Watson trees. In particular, the speed on the 1-2 tree is 1/6, as the naive calculation had it. Of course, this is because the stationary measure is *not* proportional to the degree of the root; the intuition from finite graphs turns out to be misleading. By way of comparison, note that the argument from finite graphs would give a speed of $(m-1)/(m+1)$. Concavity of the function $x \mapsto (x-1)/(x+1)$ shows that this is always strictly larger than the true speed (1) except in the trivial case where $p_m = 1$. If m is an integer, then this shows that the speed on a Galton-Watson tree with mean m is smaller than on the corresponding deterministic tree, even though both have the same growth rate.

An "explanation" for the smaller speed on Galton-Watson trees arises from the fact that simple random walk essentially remains on a subtree of strictly smaller exponential growth rate. We shall discuss the proof of this in a moment. This fact raises yet another question: The growth rate of a Galton-Watson tree is a.s. $\log m$; if d denotes the growth rate of this smaller subtree, so that $d < \log m$, then is the speed (1) less than $(e^d-1)/(e^d+1)$?

To prove that there are such smaller subtrees on which the random walk essentially remains, consider the following finitistic problem. Grow a Galton-Watson tree for n generations. Start a simple random walk at the root and kill it when it reaches level n. This gives a probability distribution on the vertices at level n. Given $\alpha \in (0,1)$, let $K_T^n(\alpha)$ be the minimum number of such vertices needed to form a set of probability at least α. How does $K_T^n(\alpha)$ grow with n? It turns out that

$$\text{(2)} \qquad \lim_{n\to\infty} \frac{1}{n} \log K_T^n(\alpha) = d \quad \text{a.s.}$$

For the 1-2 tree, for example, we have that $d \approx 0.38 \approx \log 1.47$.

For the infinite Galton-Watson tree, we said that there is a subtree containing, with probability α arbitrarily close to 1, the path of simple random walk. This subtree is formed, intuitively, from the $K_T^n(\alpha)$ vertices needed to achieve a set of probability α and their "neighbors". Since the walk has positive speed, it can't go very far from the points first visited at level n.

This hitting measure on level n is a finitistic harmonic measure. Its asymptotics (2) are analyzed by considering, instead, harmonic measure on the boundary of the infinite tree. For this purpose, instead of thinking of boundary points as equivalence classes, we choose a representative ray emanating from the root. The distribution of X_∞ is **harmonic measure**, denoted HARM_T. We shall need a metric on ∂T, the boundary of T. If ξ and η are boundary points with n edges in common, then we define the distance to be e^{-n}. (See Figure 11.)

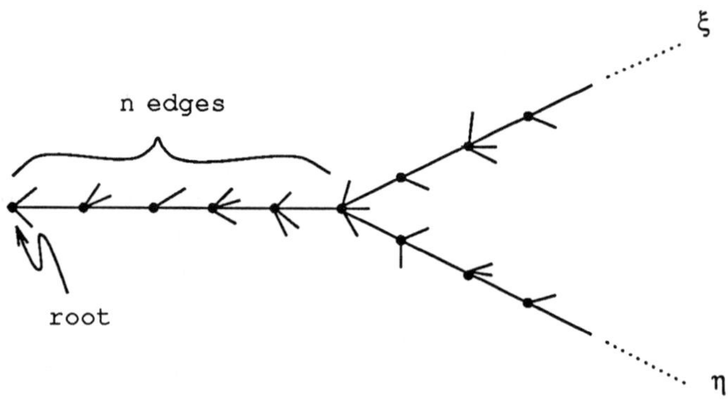

FIG. 11. *The metric on the boundary of a tree.*

Although we shall not use it directly, we briefly discuss Hausdorff dimension. In the above metric, the boundary becomes compact and has Hausdorff dimension $\log m$ a.s. (Hawkes ([4])). Define the dimension of a probability measure θ on ∂T as

$$\dim\theta := \min\{\dim E\,;\ \theta(E) = 1\}\,.$$

This measures how much θ spreads out; in (2), spread is indeed the quantity of interest. The smaller $\dim\theta$ is, the more θ is concentrated on a small set (even though its support may be the whole boundary). Of course, $\dim\theta \le \dim\partial T = \log m$ a.s. for any θ. In particular, the dimension of harmonic measure is the number d of (2) that we want to show is strictly less than $\log m$ (unless $p_m = 1$).

Now a nonnegative function on the vertices of a tree taking the value 1 at the root and having the property that the value at any vertex equals the sum of the values at its children is called a **unit flow**. Borel probability measures θ on ∂T are in 1-1 correspondence with unit flows on T via the assignment to a vertex v the value $\theta(\{\xi\,;\ v \in \xi\})$; i.e., the measure of the set of all rays passing through the vertex v. For harmonic measure, as an example, this is the probability that the walk will have the vertex v as the last point visited in its generation.

We are interested in the dimension of harmonic measure, but before its consideration, it is useful to examine first another measure, called **limit uniform** and denoted UNIF_T. We define it as follows through its associated unit flow. Given k, consider $n > k$ and all the vertices in generation k. Assign temporarily to the vertices in generation k the proportion of their descendants in generation n among all vertices in generation n. This is a probability measure on the vertices of level k. Now let n tend to infinity and take the limit of this probability measure. It turns out that this converges a.s. and so is well defined a.s. Considering now all k, this defines the unit flow of limit uniform measure. By construction, this measure spreads out the most possible; at least, this is so if nothing odd happens in taking the limit. Indeed, one can show that if $\sum_k p_k k \log k < \infty$, then the dimension of limit uniform measure is $\log m$ a.s. (This extends Hawkes ([4]), who assumed that $\sum_k p_k k(\log k)^2 < \infty$.)

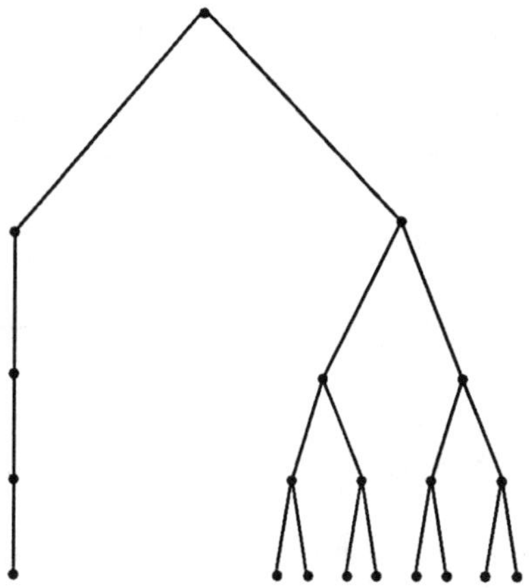

FIG. 12. *The difference between limit uniform and harmonic measures.*

Now an easy argument shows that the dimension of harmonic measure is an a.s. constant, d. What is this constant? Consider the finite tree in Figure 12. Let the height of this tree be n. Imagine that Galton-Watson descendant trees continue from the leaves at the bottom. Then limit uniform measure will put about 2^n as much mass on the right child of the root as on the left. On the other hand, harmonic measure will not be nearly so unbalanced since the random walk will exit the "pipe" on the left with probability of the order of magnitude $1/n$, not $1/2^n$. This is an extreme illustration of the idea that harmonic measure is more concentrated than limit uniform measure. From this intuition, one expects that $d = \dim \mathsf{HARM}_T < \log m$ a.s. This was first conjectured by Robert Kaufman.

To prove that this is indeed the case, we use another notion, that of Hölder exponent. Given a Borel probability measure θ on a metric space, define the **Hölder exponent** of θ at a point x as

$$\lim_{r\downarrow 0} \frac{\log 1/\theta(B_r(x))}{\log 1/r},$$

where $B_r(x)$ denotes the ball of radius r centered at x — if this limit exists. For example, if θ is Lebesgue measure on n-dimensional euclidean space, then it is clear that its Hölder exponent is n at every point. Billingsley ([1]) showed that if the Hölder exponent exists and is constant θ-a.s., then it equals $\dim \theta$. He showed this for euclidean space, but the same proof works even more easily on the boundary of trees.

If $\xi \in \partial T$ is the ray $\langle \xi_0, \xi_1, \xi_2, \ldots \rangle$, where $\xi_0 = \mathrm{Root}(T)$, then the Hölder exponent of θ at ξ is

$$\lim_{n\to\infty} \frac{1}{n} \log \frac{1}{\theta(\xi_n)}. \tag{3}$$

Before using this in the case of primary interest, i.e., that of harmonic measure for simple random walk, let us see how to treat an easier case, that of **simple forward random walk**. For this random walk, the walker chooses each time among the children (only) of the present vertex. It cannot go back towards the root. The corresponding unit flow is the **equally-splitting flow** which divides the mass at every vertex equally among its children. We call the associated measure on the boundary of the tree **visibility measure**, denoted VIS_T. For its Hölder exponent, we consider

$$\frac{1}{n} \log \frac{1}{\mathsf{VIS}_T(\xi_n)} = \frac{1}{n} \sum_{k=0}^{n-1} \log \frac{\mathsf{VIS}_T(\xi_k)}{\mathsf{VIS}_T(\xi_{k+1})}. \tag{4}$$

Now the quotient on the right-hand side is, by definition, the number of children of the vertex ξ_k. Its distribution is, therefore, the progeny distribution $\{p_k\}$. These quotients are also independent along a ray and of the ray chosen, whence the strong law of large numbers gives that the right-hand side converges to $\sum p_k \log k$ a.s. (Note that by concavity of the log function, this is strictly less than $\log \sum p_k k = \log m$ unless $p_m = 1$. That is, the Hölder exponent, and thus also the Hausdorff dimension, of visibility measure is less than that of the full boundary. The intuition behind this from Figure 12 is immediate. It also suggests that the dimension of VIS_T, i.e., $\sum p_k \log k$, should be less than d, but this is unknown.)

In order to treat harmonic measure of simple random walk, we again need to use the ergodic theorem for a Markov chain. This Markov chain has the same state space as before, namely, the space of rooted unlabelled trees, but has the following transition probabilities. Starting from a rooted tree, choose among the children of the root according to their respective harmonic measures (more precisely, the harmonic measures of the sets of rays passing through the children). The new state is then the descendant tree of the child picked, rooted at this child. (See Figure 13.)

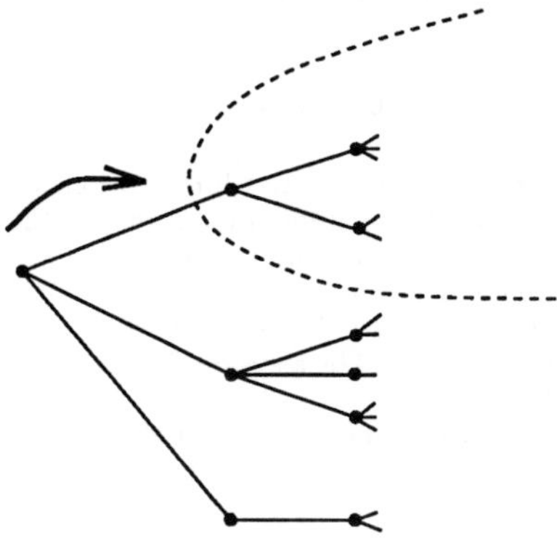

FIG. 13. *A transition of the Markov chain for harmonic measure. Here, the top child is being chosen.*

One can show that this Markov chain is ergodic for Galton-Watson measure, but it is not stationary. To get around this same problem that we encountered before, we induce from augmented Galton-Watson measure, the stationary measure for simple random walk, to a certain set of states called "exit points". The resulting measure is denoted μ_{HARM}. Again, the expression (3) whose limit gives the Hölder exponent is written as a telescoping sum. In (4), this was an average of i.i.d. r.v.'s. Now, it is an ergodic average, so the ergodic theorem gives its limit as

$$\iint \log \frac{1}{\mathsf{HARM}_T(\xi_1)}\, d\mathsf{HARM}_T(\xi)\, d\mu_{\mathsf{HARM}}(T)$$
$$= \int \sum_{|x-\mathrm{Root}(T)|=1} \mathsf{HARM}_T(x) \log \frac{1}{\mathsf{HARM}_T(x)}\, d\mu_{\mathsf{HARM}}(T). \tag{5}$$

Recall Shannon's inequality (the concavity of log): if $\sum_{i=1}^n a_i = \sum_{i=1}^n b_i = 1$, where $a_i,\, b_i \in (0,\, 1)$ and some $a_i \neq b_i$, then

$$\sum_{i=1}^n a_i \log \frac{1}{a_i} < \sum_{i=1}^n a_i \log \frac{1}{b_i}.$$

Since harmonic measure and limit uniform measure are a.s. unequal (except on a deterministic Galton-Watson tree), this gives that the expression in (5) is

$$< \int \sum_{|x-\mathrm{Root}(T)|=1} \mathsf{HARM}_T(x) \log \frac{1}{\mathsf{UNIF}_T(x)}\, d\mu_{\mathsf{HARM}}(T).$$

(Recall that UNIF_T denotes limit uniform measure on ∂T.) To evaluate this, we use a trick with the ergodic theorem and a certain coboundary. It turns out that this integral is equal to $\log m$. Therefore d, the dimension of harmonic measure, is less than $\log m$.

REFERENCES

[1] Billingsley, P. (1965) Ergodic Theory and Information. Wiley, New York.

[2] Doyle, P. and Snell, J. L. (1984) Random Walks and Electric Networks. Mathematical Assoc. of America, Washington, D. C.

[3] Grimmett, G. and Kesten, H. (1984) Random electrical networks on complete graphs, *J. London Math. Soc.* **30**, 171–192.

[4] Hawkes, J. (1981) Trees generated by a simple branching process, *J. London Math. Soc.* **24**, 373–384.

[5] Lyons, R., Pemantle, R. and Peres, Y. (1995) Ergodic theory on Galton-Watson trees: speed of random walk and dimension of harmonic measure, *Erg. Theory Dynam. Systems*, to appear.

A NOTE ON RECURRENCE, AMENABILITY, AND THE UNIVERSAL COVER OF GRAPHS

SAM NORTHSHIELD*

Abstract. Given a regular graph G, and its unversal covering tree T, we give conditions which relate recurrence and amenability of G to the size of a certain subset R of the Martin boundary of T. In particular, G is recurrent if and only if R has full harmonic measure; G is amenable if and only if R has full Hausdorff demension.

Potential theory on graphs begins with the study of trees (see [C]) . The theory is significantly simpler and more highly developed in the acyclic setting and, for this reason, it is desirable to translate questions about arbitrary graphs to questions about trees. In particular, since every regular graph is covered by a regular tree, we shall interpret certain potential theoretic properties of a regular graph as properties of certain subsets of the covering tree's boundary.

The properties in which we are interested are in some ways measures of "smallness" (i.e. if a graph has one of these properties, then so does every "smaller" graph). As we shall see, the properties that will be considered are of interest mainly in infinite graphs which are not trees; finite graphs will possess all these properties whereas infinite regular trees (with degree ≥ 3) will possess none.

Geometry and group theory are sources for such graphs and, ultimately, the problems we shall address have their origins there. Given a finitely generated group G with a minimal (i.e. contains no inverses) generating set G_o, the Cayley graph of G with respect to G_o is the graph with vertex set G and an undirected edge between two vertices x and y if $xy^{-1} \in G_o$. We may also consider the modified Cayley graph where we allow double edges between vertices x and y when xy^{-1} has order 2. In this case, the canonical homeomorphism from the free group F generated by G onto G induces a local homemorphism from the Cayley graph of F (which is a regular tree) onto the modified Cayley graph of G. Let G be an infinite connected graph with constant vertex degree $d \geq 3$. We shall also let G denote the set of vertices of the graph. Since G is connected we may define a metric: given vertices x and y, let $\delta(x, y)$ be the legnth of the shortest path between x and y. We write xEy if x and y share an edge and we define the simple random walk X to be the Markov chain with transition probabilities:

$$p(x, y) = \begin{cases} 1/d \text{ if } xEy \\ 0 \text{ otherwise.} \end{cases}$$

* Department of Mathematics, SUNY College at Plattsburgh, Plattsburgh, NY 12901.

The transition "matrix" p defines a linear transformation from $\mathbb{R}^G$ to itself:

$$pf(x) = \sum_{y \in G} p(x,y)f(y)$$

We say that f is harmonic if $pf = f$ and superharmonic if $pf \leq f$. The Martin boundary, ∂G , is the boundary of a compactification of G such that every positive harmonic function h is of the form

$$h(x) = \int_{\partial G} K(x,\xi)\nu(d\xi)$$

where ν is a Borel measure. This measure is uniquely determined by h if the support of ν is restricted to a certain Borel subset $\triangle G$ of ∂G. $\triangle G$ denotes the minimal Martin boundary and, in the case of regular trees, $\triangle G = \partial G$. Let $p^{(n)}$ denote the n-fold convolution of p. The following definitions are based on standard results (see [KSK], [DK], [G]).

DEFINITION. G is RECURRENT if any of the following equivalent conditions hold:

a1) The simple ramdom walk X almost surely hits every vertex infinitely often,

a2) Green's function $G(x,y) \equiv \sum_{n \geq 0} p^{(n)}(x,y)$ is identically infinite,

a3) all positive superharmonic functions are constant.

G has a TRIVIAL MARTIN BOUNDARY if

b)all positive harmonic functions are constant.

G is LIOUVILLE if

c)all bounded harmonic functions are constant.

G is AMENABLE if any on the following eqivalent conditions hold:

d1) the L^2 operator norm $||p||$ of p is 1,

d2) $\lambda = 0$ where $\lambda \equiv \inf\{\varepsilon : \exists f > 0 : pf \leq (1-\varepsilon)f\}$,

d3) $\propto \equiv \inf_K [\#\text{edges between } K \text{ and } K^C]/[\#\text{vertices in } K] = 0$ where the infimum is taken over all finite subgraphs K of G.

Obviously, recurrence implies trivial Martin boundary implies Liouville. As we shall see, in the case of Cayley graphs, Liouville implies amenable and all these implications are strict.

Dodziuk [D] showed that recurrence implies amenability. A mean on $L_\infty(G)$ is a linear functional on $L_\infty(G)$ such that $m(1) = 1$. A mean m is said to be invarient if $m(xf) = m(f)$ for all $x \in G$, $f \in L_\infty(G)$ where $xf(y) := f(x^{-1}y)$. It is usually taken as the definition of amenability that a group is amenable if the exists an invariant mean on $L_\infty(G)$ (see [P]). For finitely generated groups, the equivalence of the definition and ($d2$) above was shown first by Kesten [K] (see also [G]).

PROPOSITION 1. *For a Cayley graph of a finitely generated group, the Liouville property implies amenability.*

Proof. Let G be the Cayley graph of a finitely generated group with the Liouville property. We shall use an idea of Lyons and Sullivan [LS] to construct an invariant mean on $L_\infty(G)$. Since $\mathbb{N}$ is an amenable semigroup, we may choose an invariant mean M on $L_\infty(\mathbb{N})$ (see [P]). Given a bounded f and $x \in G$, the sequence $p^{(n)}f(x)$ is a bounded function of n and so $M(p^{(\cdot)}f(x))$ is a bounded function of x. Clearly, $m(f) = M(p^{(\cdot)}f(x))$ defines a mean on $L_\infty(G)$. Since $p^{(n)}(zf) = zp^{(n)}f$ for all f and z, the mean m is invariant if,for $f \in L\infty(G)$, $M(p^{(\cdot)}f(x))$ is a constant function of x. Given $f \in L\infty(G)$, let $h(x) = M(p^{(\cdot)}f(x))$. By the $\mathbb{N}-$ invariance of M, and since $p(x,y) \neq 0$ for only finitely many y,

$$ph(x) = \sum_y p(x,y)h(y) = M(\sum_y p(x,y)p^{(\cdot)}f(y)) = M(p^{(\cdot+1)}f(x)) = h(x)$$

and h is harmonic. By the Liouville property, h is constant. □

In the case of planar graphs with bounded vertex degree, Northshield [N1] showed that the Liouville property implied amenability. Aldous [A] asked if, for groups, the Liouville property implies trivial Martin boundary. Northshield [N3] shows that amenability implies the existence of a fixed point in the Martin boundary (under the natural action of the group on its Martin boundary). It is still an open question whether amenability is equivalent to the existence of a fixed point in the Martin boundary. Finally, Aldous's question was answered negatively by Bougerol [B].

A covering of G by a graph G' is a function θ: $G' \longrightarrow G$ which preserves adjacency and vertex degree. A maximal cover of a graph G is a tree and is unique up to isomorphism. Let T denote this universal covering tree.

We now define the boundary for T. Fix $o \in T$ and let Γ be the set of infinite paths in T which start at o. For $z \in T$, let $\Gamma(z) = \{\gamma \in \Gamma : \gamma_{\delta(o,z)} = z\}$ be the set of paths in G which contain z, and let

$$T_z = T(z) = \{w \in T : \exists \gamma \in \Gamma(z) : \gamma_{\delta(o,w)} = w\} \cup \Gamma(z).$$

We define a topology on $T \cup \Gamma$ by defining a base of open sets:

$$T \cup \{T(z)/F : z \in T, F \subset T, F \text{ finite.}\}$$

Thus topologized,$T \cup \Gamma$ is a compactification of T and, in fact,∂T is the Martin boundary of T (see Cartier [C]). The Marin kernel is

$$K(x,\gamma) = \lim_{n\to\infty} G(x,\gamma_n)/G(o,\gamma_n) = (d-1)^{H(\gamma,x)}$$

where $H(\gamma, x)$ is the discrete Busemann function $\lim_{n\to\infty}[\delta(o,\gamma_n)-\delta(x,\gamma_n)]$. Each $K(\cdot,\gamma)$ is minimal harmonic and, as was mentioned above,∂T is the minimal Martin boundary.

The random walk X_n on T starting at o is transient and converges to a point X_∞ on the Martin boundary $P-$almost surely. Define, for a Borel subset A of ∂T,

$$\mu(A) = P(X_\infty \in A),$$

the probability that the random walk starting at o converges to a point of A. Note that μ is characterized by

$$\mu(T_z) = [d(-1)/d](d-1)^{\gamma(o,z)} \text{ for all } z \neq o.$$

Equivalently, μ is the measure in the Martin representation of 1 (i.e. $\int_{\partial T} K(\cdot, \gamma) d\mu(\gamma) \equiv 1$).

We say that two vertices x and y of T are equivalent if $\theta(x) = \theta(y)$ and we let $[x]$ denote the equivalence class containing x.

Let $R = \{\gamma \in G : \gamma_n \in [o] \text{ infinitely often}\}$.

It will be seen that the recurrence of G is equivalent to $\mu(R) = 1$. We show that the amenability of G is necessary and sufficient for R to have Hausdorff dimension 1(where Hausdorff dimension is defined with respect to μ) and therefore a recurrent graph is amenable since $\mu(R) = 1$ implies $\dim(R) = 1$. Furthermore, since every bounded harmonic function on G lifts to a bounded harmonic function on T with boundary values constant on R, if $\mu(R) = 1$, then by the Martin representation theorem,every bounded harmonic function is constant (hence recurrent implies Liouville).

Our original motivation was to investigate discrete analogues of Sullivan and Patterson's work on the "critical exponent" and Hausdorff dimension of the "conical limit set" of a discrete group of isometries in hyperbolic space (see, for example, Nicholls [Ni]). The set R is a discrete analogue of the conical limit set.

If $\mu(R) = 1$, then the simple random walk on T hits [o] infinitely often and thus the simple random walk on G is recurrent. Conversely, if the simple random walk on G is recurrent, the simple random walk on T hits [o] infinitely often and it follows, although not immediately, that $X_\infty \in R$ almost surely. The main difficulty here is this: although X_n visits [o] infinitely often if $X_\infty \in R$, the converse is false–it is possible to trace out a walk which visits [o] infinitely often but whose limiting geodesic ray visits [o] only once (take any geodesic in T which never hits [o] after the first step—for example,the lift of a walk that circles around the origin in $\mathbb{Z}^2$—and follow this geodesic out making infinitely many excursions to [o]). The following theorem is a discrete analogue of theorem 2.4.6 of Nicholls [Ni].

THEOREM 2. *$\mu(R) = 1$ or 0 according to whether the simple random walk on G is recurrent or transient respetively.*

Proof. Let Z_n denote the forward random walk (FRW) on T (Z_n chooses first step uniformly from all d neighbors of starting point but on n^{th} step ($n \geq 2$)chooses uniformly from the $d-1$ neighbors not visited at the $(n-1)^{st}$ step).

By the symmetry of T, Z_∞ and X_∞ have the same distribution and therefore

$$\mu(R) = P(X_\infty \in R) = P(Z_\infty \in R) = P(Z_n \in [o] i.o.) = P(\theta(Z_n) = \theta(o) i.o.).$$

Note that $\theta(Z_n)$ is a FRW on G. Although $\theta(Z_n)$ is not a Markov chain, it is the image of one (namely a Markov chain with state space $\{(x,y,)\in G\times G : xEy \text{ or } x=y\}$ under a "finite-to-one" map. Therefore, $P(\theta(Z_n)=\theta(o) i.o.)=1$ or 0 according to whether $\sum_{n\geq 0} P(Z_n \in [o])=\infty$ or not.

Let G_T and G denote Green's functions for the SRW on T and G respectively. Since θ preserves SRW's, $G(\theta(o),\theta(o)) = \sum_{z\in[o]} G_T(o,z)$. Letting $f\sim g$ mean $C^{-1}<f/g<C$ for some $C\geq 1$, note that

$$\sum_{n\geq 0} P(Z_n=z)=\mu(T_z)\sim (d-1)^{1-\delta(o,z)}/(d-2)=G_T(o,z),$$

and therefore $\sum_{n\geq 0} P(Z_n\in[o])\sim G(\theta(o),\theta(o))$. Since the recurrence of G is equivalent to $G(\theta(o),\theta(o))=\infty$, the proof is complete. □

We define the dimension of a subset of ∂T as in Falconer [F] (see also [L]). For $\propto,\epsilon>0$ and A a Borel subset of ∂T, let

$$\wedge_{\propto,\epsilon}(A)=\inf\{\sum_{z\in S}\mu(T_z)^{\propto} : A\subset \bigcup_{z\in S} T_z \text{ and, for all } z\in S, \mu(T_z)<\epsilon\},$$

$$\wedge_{\propto}(A)=\lim_{\epsilon\searrow 0}\wedge_{\propto,\epsilon}(A),$$

and

$$\dim(A)=\sup\{\propto: \wedge_{\propto}(A)=\inf\{\propto: \wedge_{\propto}(A)=0\}.$$

This is equivalent to the definition of dimension in [L] and, as was pointed out there, it is a natural analouge of Hausdorff dimension. Note that $\wedge_{\propto}(A)=0$ if and only if for all $\epsilon>0$, there exsists a set $S\subset T$ such that $A\subset\bigcup_{z\in S}T_z$ and $\sum_{z\in S}\mu(T_z)^{\propto}<\epsilon$.

The critical exponent of G (in T) is

$$\beta=\inf\{\propto: \sum_{z\in[y]}(d-1)^{-\propto\delta(x,z)}<\infty\}.$$

Note that β is independent of x and y. The Hausdorff dimension of R and the critical exponent are related (as in Nicholls [Ni]).

LEMMA 3. $\dim(R)\leq\beta$

Proof. Since

$$\beta=\inf\{\propto: \sum_{z\in[o]}(d-1)^{-\propto\delta(o,z)}<\infty\}.$$

and

$$\dim(R)=\inf\{\propto: \wedge_{\propto}(R)=0\}.$$

it is enough to show that if $\sum_{z\in[o]}(d-1)^{-\propto\delta(o,z)} < \infty$, then $\wedge_\propto(R) = 0$.

Suppose $\sum_{z\in[o]}(d-1)^{-\propto(o,z)} < \infty$ and let $\epsilon > 0$. Then $[o] = F \cup \{z(1), z(2), \ldots\}$ where F is finite and

$$\sum_{n\geq 0}(d-1)^{\propto\delta(o,z(n))} < \epsilon.$$

Note that $R \subset \bigcup_{n\geq 0} T_{z(n)}$ (since every γ in R contains infinitely many elements of [o]). Since $\mu(T_z) = c(d-1)^{-\delta(o,z)}$ for all $z \neq 0$,

$$\sum_{n\geq 0}[\mu(T_{z(n)})]^\propto < c^\propto \epsilon$$

and, since ϵ is arbitrary, $\wedge_\propto(R) = 0$. □

THEOREM 4. *G is amenable if and only if* $\dim(R) = 1$.

Proof. By lemma 3, if $\dim(R) = 1$, then $\beta = 1$.

Let $S_n(x)$ be the metric sphere $\{y \in T : \delta(x, y,) = n\}$. We define the cogrowth constant of G (see [N2]) by

$$\rho = \lim\sup_{n\to\infty} \mid [y] \cap S_n(x) \mid^{1/n} .$$

Note that

$$\rho = \inf\{\propto: \sum_{z\in T} \propto^{-\delta(x,z)} < \infty\}$$

and therefore is independent of x and y. Since $\beta = \log_{(d-1)}\rho, \rho = d-1$. By corollary of [N2], G is amenable.

The proof of the remaining implication depends heavily on Lyons [L]. Suppose G is amenable. Fix $a \in (0,1)$.Since $p^{(n)}(\theta(o),\theta(o))^{1/n} \to 1$(see, for example, Gerl [G]), we may choose k such that $p^{(k)}(\theta(o),\theta(o)) > a^k$. The paths in T that correspond to returns after k steps form a finite tree T_o of height k and the kth level contains at least $a^k(d-1)^k$ vertices. Let $R' = \{\gamma \in \partial T : \gamma_{nk} \in [o] \text{ for all } n\}$. The union of all paths in R' from a tree T' which is periodic (it is generated by T_o). The growth number ($\equiv \liminf_{n\to\infty} \mid S_n \mid^{1/n}$) equals $\liminf_{n\to\infty} \mid S_{nk} \mid^{1/nk}$ which is at least $a(d-1)$ and therefore, since T' is periodic, $\dim(R' \geq \log_{d-1}(a(d-1))$. Letting $a \longrightarrow 1$, we see that $\dim(R) \geq \dim(R') = 1$ □

COROLLARY 5. *If G is recurrent, then G is amenable.*

Proof. This follows from the fact that, for any Borel set A, if $\mu(A) > 0$, then $\dim(A) = 1$. □

Finally, we relate the Liouville property of a non-recurrent graph G to the size of certain subsets of ∂T. It is unknown at present how this relates to R or to theorems 2 and 5.

Let $\triangle G$ denote the minimal Martin boundary of G and extend θ to a map from ∂T to $\triangle G \cup \theta(o)$:

$$\theta(\gamma) = \begin{cases} \lim_{n\to\infty} \theta(\gamma_n) & \text{if this limit exists and is in } \triangle G \\ \theta(o) & \text{otherwise} \end{cases}$$

THEOREM 6. *G is Liouville if and only if there exsists* $\xi \in \partial G$ *such that* $\mu(\theta^{-1}(\xi)) = 1$.

Proof. Given a measure ν on ∂T, define a measure ν^* on $\triangle G$ by $\nu^* = \nu \circ \theta^{-1}$. We shall show that $\int_{\triangle G} K(\cdot,\xi)\mu^*(d\xi) \equiv 1$ and therefore, since a representing measure of a bounded harmonic function on G is then absolutely continuous with respect to μ^* (see [KSK] 10-42), G is Liouville if and only μ^* is supported on a singleton (i.e. $\mu(\theta^{-1}(\xi)) = 1$ for some ξ).

For any $\xi \in \triangle G$, the function $h(x) = K(\theta(x),\xi)$ is positive harmonic on T and, by the Martin representation theorem [KSK] 10-41, for example), there exists a unique Borel measure $\mu^{(\xi)}$ on ∂T such that

$$h(x) = \int_{\partial T} K_T(x,\gamma)\mu^{(\xi)}(d\gamma),$$

Note that $\mu^{(\xi)}$ is $h-$harmonic measure on ∂T([KSK] 10-40) and so

$$\mu^{(\xi)}(\theta^{-1}(\xi)) = P_h(X_\infty \in \theta^{-1}(\xi)) \geq P_h(\theta(X_n) \longrightarrow \xi).$$

Since θ preserves random walks, $P_h(\theta(X_n) \longrightarrow \xi)$ equals the probability that the random walk on G conditions by $K(\cdot,\xi)$ converges to ξ which, since ξ is in the minimal boundary, is 1 ([KSK] 10-35). Hence, for all $\xi \in \triangle G$,

$$\mu^{(\xi)*}(A) = 1_A(\xi).$$

Let ν be the unique measure on $\triangle G$ such that $\int_{\triangle G} K(\cdot,\xi)\nu(d\xi) \equiv 1$. Since $\int_{\partial T} K_T(\cdot,\gamma)\mu(d\gamma) \equiv 1$, it is not difficult to see (by Fubini's theorem and the fact that, for all x, $K_T(x,\cdot)$ takes on only finitely many values), for Borel subsets E of ∂T,

$$\mu(E) = \int_{\triangle G} \mu^{(\xi)}(E)\nu(d\xi).$$

Then, for Borel subsets A of $\triangle G$,

$$\mu^*(A) = \int_{\triangle G} \mu^{(\xi)*}(A)\nu(d\xi) = \nu(A),$$

and therefore $\int_{\triangle G} K(\cdot,\xi)\mu^*(d\xi) = \int_{\triangle G} K(\cdot,\xi)\nu(d\xi) = 1$. □

Acknowledgments. I wish to thank Russ Lyons and Yuval Peres for supplying the last half of the proof of theorem 4. I thank the IMA for their hospitality and for allowing me to contribute the talk which led to this paper and I thank the CMS (Madison) and their NSF grant, DMS-9122242, for financial support. Lastly, I thank Robin Pemantle for editorial comments.

REFERENCES

[A] Aldous, D. Personal Communication, 1990.

[B] Bougerol, P. and Elie, L., *Existence of positive harmonic functions on groups and on covering manifolds,* prépublication # 198 du Laboratoire de Probabilités de l'Université Paris VI, 1993.

[C] Cartier, P. *Fonctions Harmoniques sur un Arbre,* Symposia Mathematica **9**(1972), 203-270.

[D] Dodziuk, J. *Difference Equations, Isoperimetric Inequality, and Transience of Certain Random Walks.* Trans. Amer. Math. Soc. **284** (1984), 787-794.

[DK] Dodziuk, J. and Karp, L. *Spectral and function theory of combinational Laplacians,* Contemp. Math. vol. 73, Amer. Math. Soc., Providence, RI (1986), 25-40.

[F] Falconer, K.J., The Geometry of Fractal Sets, Cambridge University Press, Cambridge, 1985.

[G] Gerl, P. *Amenable Groups and Amenable Graphs.* Lecture Notes in Mathematics **1359** (1988), 181-190.

[K] Kesten, H., *Full Banach mean values on countable groups.* Math. Scand. **7** (1959), 146-156.

[KSK] Kemeny, J.G., Snell, J.L., and Knapp, A.W., Denumerable Markov Chains, van Nostrand, Princeton, 1966.

[L] Lyons, R. *Random Walks and Percolation on Trees.* Ann. Probab. **18** (1990), 931-968.

[LS] Lyons, T. and Sullivan D., *Function theory, random paths and covering spaces,* J. Differential Geom. **19** (1984), 299-323.

[Ni] Nicholls, P.J., The Ergodic Theory of Discrete Groups, Cambridge University Press, Cambridge, 1989.

[N1] Northshield, S., *Geodesics and bounded harmonic functions on infinite planar graphs,* Proc. Amer. Math. Soc., **113** #1 (1991), 229-233.

[N2] Northshield, S., *Cogrowth of Regular Graphs,* Proc. Amer. Math. Soc., **116** #1 (1992), 203-205.

[N3] Northshield, S., *Amenability and superharmonic functions,* Proc. Amer. Math. Soc., **119** #2 (1993), 561-566.

[P] Paterson, A., Amenability, AMS, Providence, RI, 1988.

ON WHICH GRAPHS ARE ALL RANDOM WALKS IN RANDOM ENVIRONMENTS TRANSIENT?

ROBIN PEMANTLE* AND YUVAL PERES†

Summary. An infinite graph G has the property that a random walk in random environment on G defined by i.i.d. resistances with any common distribution is almost surely transient, if and only if for some $p < 1$, simple random walk is transient on a percolation cluster of G under bond percolation with parameter p.

Key words. RWRE, graph, random electrical networks, percolation

AMS(MOS) subject classifications. Primary: 60K35 ; Secondary 60J15.

Simple random walks on infinite graphs have been widely studied with particular attention to the dichotomy between recurrence and transience. In the last two decades there has been interest in random walks in random environments (RWRE's), which are mixtures of Markov chains in which the transition probabilities are picked according to some prior distribution. For reversible Markov chains, the transition probabilities may be described by associating resistances $R(vw)$ to each edge (vw), the transition probabilities out of any vertex being proportional to the inverses of the resistances on the edges incident to that vertex:

$$p(v,w) = \frac{R(vw)^{-1}}{\sum_z R(vz)^{-1}} \tag{1}$$

where the sum is over all vertices z adjacent to v. One natural model[1], considered in Grimmett and Kesten (1984), involves taking the resistances as i.i.d. positive random variables. A result of Adams and Lyons (1991) implies that if a tree Γ has positive Hausdorff dimension, then for an arbitrary distribution of the resistances the resulting network is almost surely transient. (See Lyons (1990) where the term *branching number* is used for the exponential of the Hausdorff dimension.) The surprising aspect of this is that no conditions are imposed on the tail of the distribution. This note proves a converse of this implication, but on a general graph. Specifically, the implication $(a) \Rightarrow (b)$ in part (ii) of the theorem below shows that almost sure transience for an arbitrary resistance distribution on a tree Γ implies $p_c < 1$ for IID bond percolation. By results in Lyons (1990), this implies that Γ has positive dimension. Part (i) is immediate from the well known connections between random walks and electrical networks (c.f.

* Research supported in part by National Science Foundation grant # DMS 9300191, by a Sloan Foundation Fellowship, and by a Presidential Faculty Fellowship.
Department of Mathematics, University of Wisconsin-Madison, Van Vleck Hall, 480 Lincoln Drive, Madison, WI 53706.

† Department of Statistics, University of California, Berkeley, CA 94720.

[1] see Remark 1 for another RWRE model relevant to the present discussion.

Doyle and Snell (1984)) and is included for completeness.

Theorem: *Let G be a locally finite connected graph. Suppose that a positive random variable $R(e)$ is attached to every edge e of G so that the collection $\{R(e)\}$ is i.i.d.. Let μ denote their common distribution. Consider the RWRE given by (1).*

(i) If $\int x\,d\mu < \infty$ and simple random walk on G is transient, then the resistor network $\{R(e)\}$ has finite net resistance almost surely, thus defining almost surely a transient random walk.

(ii) The following conditions on G are equivalent: (a) for any distribution μ on $(0,\infty)$ the resistor network $\{R(e)\}$ is almost surely transient; (b) there exists some $p < 1$ such that independent bond percolation on G with parameter p contains with positive probability a cluster on which simple random walk is transient.

PROOF: We rely on two facts from the theory relating reversible random walks to the corresponding resistor networks. Firstly, the random walk is transient if and only if the network has finite resistance from some (hence every) vertex to infinity, i.e., there is some nonzero flow $\{F(e)\}$, satisfying Kirchoff's laws except at one source vertex, such that the energy

$$\sum_e F(e)^2 R(e)$$

is finite. Secondly, Rayleigh's principle asserts that the resistance of a network is monotone in each of the individual resistances. Both facts may be found in Doyle and Snell (1984). The proof of (i) follows immediately: G is transient for simple random walk, so some flow $\{F(e)\}$ on G has finite energy

$$\sum_e F(e)^2$$

with respect to unit resistances; the expected energy of this flow in the network of random resistances is

$$\mathbf{E}\sum_e R(e)F(e)^2 = \int x\,d\mu \sum_e F(e)^2 < \infty$$

so the resistance is almost surely finite and hence the network is almost surely transient.

(ii) The direction $(b) \Rightarrow (a)$ is easy: assume (b) for some parameter p and consider the random subgraph of G consisting of those edges e for which $R(e) \le Q_p$, where $1 > \mu(0, Q_p) \ge p$. Resistances have the following linearity: multiplying every resistor by C multiplies the resistance of the network by C. Thus our assumption (b) implies there is a positive probability that some component of this subgraph will have finite resistances when all edges have resistance Q_p. By Rayleigh's law, decreasing the resistances on edges in this component from Q_p to $R(e)$ and decreasing all other resistances from ∞ to $R(e)$ gives a network with finite resistance, and therefore random walk on G with resistances $\{R(e)\}$ is transient with positive probability. Since this is a tail event for the countable collection $\{R(e)\}$, the probability is one.

Now we verify that $(a) \Rightarrow (b)$. Assume that for every $p < 1$, simple random walk on p-percolation clusters is almost surely recurrent. Let $p_k \uparrow 1$ and construct a probability measure μ on $(0, \infty)$ as follows. Fix a vertex $v \in G$ and set $\gamma_1 = 1$. Define a measure μ_1 on $\mathbf{R} \cup \{\infty\}$ by $\mu_1(\{\gamma_1\}) = p_1$ and $\mu_1(\{\infty\}) = 1 - p_1$. The RWRE on G with resistance distribution μ_1 is just simple random walk on the p_1-percolation clusters and is thus recurrent. Therefore there exists N_1 such that when choosing a random environment according to μ_1, the probability is at least $1/2$ that either all edges adjacent to v have infinite resistance (i.e. v is isolated) or else the resulting RWRE started from v returns to v in the first N_1 steps. Let D_1 be the maximal degree of any vertex in G within distance N_1 of v.

For the induction step, let $k \geq 2$ and assume that γ_j, μ_j, N_j and D_j have been defined for $1 \leq j < k$ and $\mu_j(\infty) = 1 - p_j$. Pick $\gamma_k > 2N_{k-1}D_{k-1}\gamma_{k-1}$. Define μ_k by moving mass $p_k - p_{k-1}$ from infinity to γ_k, i.e.

$$\mu_k = \mu_{k-1} + (p_k - p_{k-1})(\delta_{\gamma_k} - \delta_\infty).$$

The RWRE with resistance distribution μ_k is recurrent since it yields an environment whose resistances on the p_k-percolation cluster are bounded between 1 and γ_k [use the aforementioned linearity together with Rayleigh's law]. Thus we may choose an integer N_k, such that if a random environment is chosen according to μ_k, the probability is at most 2^{-k} that v is not isolated and the resulting RWRE started from v fails to return to v within the first N_k steps. Define D_k to be the maximal degree of any vertex of G within distance N_k of v.

Finally, set $\mu(\{\gamma_k\}) = p_k - p_{k-1}$ for all $k \geq 2$, with $\mu(\{\gamma_1\}) = p_1$, so that μ is the weak limit of the μ_k. We construct a μ-RWRE $\{S_n\}$ on the same probability space as a sequence $\left(\{S_n^{(k)}\} : k = 1, 2, 3, \ldots\right)$ of μ_k-RWRE's as follows. Let the resistances $\{R(e)\}$ be i.i.d. with common distribution μ and define

$$R^{(k)}(e) = \begin{cases} R(e) & \text{if } R(e) \leq \gamma_k \\ \infty & \text{if } R(e) > \gamma_k \end{cases}.$$

Observe that this makes $\{R^{(k)}(e)\}$ i.i.d. with distribution μ_k. Let $\{S_n\}$ be a random walk starting from v with transition probabilities determined by the resistances $\{R(e)\}$. Note that if $R(wz) \leq \gamma_k$ then the transition probability from w to z with resistances $\{R^{(k)}(e)\}$ is greater than or equal to the the transition probability from w to z with resistances $\{R(e)\}$. Thus for each k, we may define a random sequence $\{S_n^{(k)}\}$ so as to have transition probabilities determined by $\{R^{(k)}\}$ and so that $S_n^{(k)} = S_n$ for all $n \leq T_k$, where T_k is the least time t for which $R(S_t S_{t+1}) > \gamma_k$.

Define events

$$A_k = \{R(vw) > \gamma_k \text{ for all } w \text{ neighboring } v\} \subseteq \{T_k = 0\}$$

$$B_k = \{R(S_j, S_{j+1}) \geq \gamma_{k+1} \text{ for some } 0 \leq j < N_k\} \setminus A_k = \bigcup_j \{0 \leq T_j < N_k\} \setminus A_k$$

$$C_k = \{S_n^{(k)} \neq v \text{ for all } 1 \leq n \leq N_k\}.$$

The event $G_k = \{S_n \neq v \text{ for all } 1 \leq n \leq N_k\}$ is contained in $A_k \cup B_k \cup C_k$, since on $G_k \setminus C_k$ the time T_k is less than N_k. For $1 \leq j < N_k$, the probability of the event $\{T_k = j\}$ is at most $D_k \gamma_k / \gamma_{k+1}$, since the sum of $R(e)^{-1}$ over edges incident to S_j with resistance greater than γ_k is at most D_k / γ_{k+1} and there is at least one edge incident to S_j with $R(e)^{-1} \geq \gamma_k^{-1}$. Similarly, $\mathbf{P}(\{T_k = 0\} \setminus A_k) \leq D_k \gamma_k / \gamma_{k+1}$. Thus $\mathbf{P}(B_k) \leq N_k D_k \gamma_k / \gamma_{k+1} \leq 2^{-k}$ by construction of γ_{k+1}. But N_k is defined so that $\mathbf{P}(C_k \setminus A_k) \leq 2^{-k}$ and clearly $\mathbf{P}(A_k) \leq 1 - p_k$. Therefore

$$\mathbf{P}(G_k) \leq (1 - p_k) + 2^{1-k} \to 0$$

as $k \to \infty$ and the μ-RWRE is recurrent. □

Remarks:

1. The recent paper of Grimmett, Kesten and Zhang (1991) shows that for Euclidean lattices $\mathbf{Z}^d$ $(d \geq 3)$, random walk on supercritical percolation clusters is almost surely transient, so the theorem is applicable, and all RWRE's with i.i.d. positive resistances are transient. Note that the RWRE models discussed in Durrett (1986) and Sunyach (1987) may be recurrent even when $d \geq 3$. Resistances in Durrett's model are stationary under $\mathbf{Z}^d$-shifts, so his model is close to the present one.
2. By Lyons (1990), the critical probability for bond percolation on a tree Γ is $p_c(\Gamma) = e^{-dim\Gamma}$ and if $dim(\Gamma) > 0$ and $p > p_c$, then simple random walk on any infinite cluster is transient (since the cluster must have positive dimension). For a tree, property (b) is thus equivalent to $dim(\Gamma) > 0$. The Max Flow Min Cut Theorem (as applied in Lyons 1990) allows us to formulate this purely in terms of flows:

> On a tree, property (a) is equivalent to the existence of a flow $\{F(e)\}$ satisfying Kirchoff's laws (except at the root) for unit resistors, and having the exponential decay property:
>
> $$F(e) \leq C\rho^{\text{dist}(e)}$$

where dist(e) is the distance from e to the root of Γ.

Acknowledgement: Thanks to Harry Kesten for showing us a surprising example which led to this note.

REFERENCES

[1] Adams, S. and Lyons, R. (1991). Amenability, Khazdan's property and percolation for trees, groups and equivalence relations. *Israel J. Math.* **75** 341 - 370.

[2] Doyle, P. and Snell, J. L. (1984). Random walks and electrical networks. Mathematical Association of America: Washington.

[3] Durrett, R. (1986). Multidimensional random walks in random environments with subclassical limiting behavior. *Comm. Math. Phys.* **104** 87 - 102.

[4] Grimmett, G. and Kesten, H. (1984). Random electrical networks on complete graphs. *J. London Math. Soc.* **2** 171 - 192.

[5] Grimmett, G., Kesten, H. and Zhang, Y. (1991). Random walk on the infinite cluster of the percolation model. *Preprint.*

[6] Lyons, R. (1990). Random walks and percolation on trees. *Ann. Probab.* **18** 931 - 958.

[7] Sunyach, C. (1987). Sur la transience et la récurrence des marches aléatoires en milieu aléatoire. *Ann. Inst. H. Poin.* **23** 613 - 626.

ENERGY, AND INTERSECTIONS OF MARKOV CHAINS*

THOMAS S. SALISBURY†

Abstract. In an earlier paper, the author and Fitzsimmons resolved several problems about the intersections of Brownian motions and Lévy processes, using projection techniques to construct certain measures of finite energy. The present paper aims to make these techniques more accessible, by treating an analogous but simpler problem. Upper bounds will be obtained on intersection probabilities for discrete time Markov chains. This will be applied to obtain a Wiener type test for intersections of random walks within given sets.

Key words. Capacity, energy, path intersections

AMS(MOS) subject classifications. 60J10, 60J15, 60J45

1. Introduction. In [FS], a relationship was established between energy and capacity on the one hand, and hitting probabilities for multiparameter Markov processes on the other. The argument there was cast in the language of Kuznetsov measures (see [G] for an exposition), which indeed allows an elegant argument that works in great generality. Applications were given to two problems concerning the multiple points of Lévy processes, and multiple points of Brownian motion within given sets.

These applications could have been obtained using less sophisticated machinery, for the idea underlying the technique is in fact quite simple. This note is an attempt to meet requests from several people, for an exposition of this idea in elementary terms.

In continuous time the problem is to characterize the sets K that get hit with positive probability by some multiparameter stochastic process. For example, to see if two independent copies X_s and Y_t of an E-valued stochastic process will cross, one can ask if the process $Z_{s,t} = (X_s, Y_t)$ will hit the diagonal $K = \{(x,x)\,;\, x \in E\}$. The characterization is often that, for K to be hit, it is necessary and sufficient that K supports a measure η of finite energy $e(\eta) = \int G(z,w)\eta(dz)\eta(dw)$, with respect to some known kernel $G(z,w)$. For the above applications, there had been several techniques for establishing sufficiency, yet [FS] gave the first proof of necessity. It should be mentioned that Yuval Peres has recently developed a new and elegant technique for establishing necessity (see [P]). This technique has other applications, apparently not obtainable by the arguments of [FS]. On the other hand, it does not seem to apply to the Brownian sheet, for which the arguments of [FS] do work. Thus both techniques are of interest.

If Z were a one parameter Markov process, the usual proof of necessity takes η be the law of Z_T, where T is the first time Z meets K. With

---

* Research supported in part by a grant from NSERC.

† York University, Department of Mathematics and Statistics, 4700 Keele Street, North York, Ontario M3J IP3. Email address: salt@nexus.yorku.ca.

this argument, the fact that η has finite energy is closely linked to the strong Markov property of Z at T. The difficulty in the multiparameter setting is that there may be no "first hitting time" at which any kind of Markov property applies. It turns out that if T is *any* random time at which $Z_T \in K$, then the law of Z_T has finite energy, and that this can usually be proved using projection techniques. The idea of constructing additive functionals using projections goes back to Azéma [A].

In order to avoid technical difficulties, we will work with discrete time Markov chains. In this context, hitting a set with positive probability is no longer an issue. The problem instead becomes one of estimating this probability using energy. We will focus on the estimate that corresponds to necessity, namely an upper bound on the hitting probability. For clarity, we'll first carry out the argument for a one parameter Markov chain X_n. We'll then modify the argument to apply to the two parameter process $Z_{m,n} = (X_m, Y_n)$, where X and Y are independent Markov chains.

Even in the context of [FS], the results involve explicit estimates for hitting probabilities. A natural assumption in this context is that each Markov process X^i involved be symmetric with respect to some excessive measure $m^i(dy)$. Thus hitting probabilities are estimated for processes in a kind of equilibrium. This made the relation with earlier results clearer. However, the results, as stated in [FS], were not restricted to this setting. To get quantitative inequalities for Markov processes started with arbitrary initial distributions $\mu^i(dy)$, one would work with respect to the excessive measures $m^i(dy) = \int g^i(x, dy)\mu^i(dx)$, together with the kernels this choice forces upon one.

As far as I know, the first real applications of such quantitative inequalities are due to Benjamini, Pemantle and Peres, as developed in [BPP]. They observed that, for processes started at fixed points, the appropriate kernels to use in energy computations are Martin kernels. The emphasis on such estimates in the present paper is at the suggestion of Yuval Peres.

I'd like to thank Jay Rosen for suggesting the idea of this paper, and Russ Lyons for his comments. I'd also like to thank Robin Pemantle and Yuval Peres for their comments and encouragement, at great personal risk to their ability to crack jokes about the complexity of [FS].

2. Energy for one chain. Consider a Markov chain with countable state space E and transition function $p(x, x')$. For convenience, we assume that we are given a process X_n and a family of probabilities P^μ on its underlying measurable space, such that for every probability μ on E, X_n is a Markov chain under P^μ, with transition function p and initial measure μ. We write $\mu(x)$ for $\mu(\{x\})$. If μ is a point mass at x, we write P^x for P^μ. A property holds *almost surely* if it holds P^x-a.s. for every x. We write ζ for the lifetime of X. That is, there is a cemetery state $\Delta \notin E$, and $\zeta \leq \infty$ is the last value of n such that $X_n \neq \Delta$. The Green function of X is

$$g(x,x') = E^x\left[\sum_{n=0}^{\infty} 1_{\{x'\}}(X_n)\right].$$

THEOREM 1. *Let X_n be a Markov chain with Green function g and values in a countable state space E. Let μ be its initial measure. Assume that g is finite everywhere, and that ζ is finite almost surely. Set $\gamma(x') = \sum_x g(x,x')\mu(x)$. Assume that $\gamma(x') > 0$ for every x', and set $g^\mu(x,x') = g(x,x')/\gamma(x')$. Let K be a finite subset of E and let T be any random element of $\{\, n \geq 0\,;\, X_n \in K\,\}$, if this set is nonempty, and $T = \infty$ otherwise. Set*

$$u(x) = P^\mu(X_T = x,\, T < \infty).$$

Then

$$P^\mu(X_n \in K \textit{ for some } n \geq 0) \leq 16\frac{\left(\sum_x u(x)\right)^2}{\sum_{x,x'} g^\mu(x,x')u(x)u(x')}$$

Remark 1. Any Markov chain can be modified to make g finite and ζ finite almost surely, simply by killing it at a geometric time. Instead of assuming that $\gamma > 0$, we could have taken $g^\mu(x,x') = \infty$ if $\gamma(x') = 0$.

Remark 2. A classical result of this form is that if T is the last exit time from K, and if 16 is replaced by 1, then the above inequality becomes an equality. (To see this, first prove that $P^x(T=0) = u(x)/\gamma(x)$. From this it follows that $\sum_{x'} g^\mu(x,x')u(x') = P^x(X \text{ hits } K)$, which equals 1 for $x \in K$. Equality now follows. For T the first hit of K, one uses the same argument together with time reversal.) The generalization to multiparameter time will be possible because no such restriction on T will be made.

Remark 3. If K is finite, and $v : K \to [0,\infty)$, we may define a probability measure η on K by requiring that its density with respect to counting measure be a multiple of v. The kind of fraction appearing in the upper bound may be thought of as the reciprocal of the *energy* $e(\eta)$ of η,

$$e(\eta) = \sum_{x,x'} g^\mu(x,x')\eta(x)\eta(x') = \frac{\sum_{x,x'} g^\mu(x,x')v(x)v(x')}{\left(\sum_x v(x)\right)^2}$$

The *capacity* $\Gamma(K)$ of K may be defined as the supremum of the $e(\eta)^{-1}$, over all probabilities η on K. The theorem therefore establishes an upper bound of $16\Gamma(K)$ on the probability of hitting K. By Remark 2, the bound could be improved to $\Gamma(K)$ itself. We will see later that $\Gamma(K)/2$ is a lower bound. These are the kinds of estimates used in [BPP].

Remark 4. If the Green function g is *symmetric* (that is, $g(x,x') = g(x',x)$) and γ is constant on K, then this hitting probability turns out to be exactly $\Gamma(K)$. This is one reason the classical theory is usually developed under such assumptions. In fact, one usually chooses the initial probability μ so that γ is a constant c on E (see [KSK]), or, if this is not possible, one works with a sequence of μ's, whose γ's converge to c. The constant c is then absorbed into the left hand side by not requiring μ to be a probability measure. The energy expressions that result are therefore with respect to the Green function g, rather than the kernel g^μ.

This parallels what is done in continuous time. For example, the Newtonian capacity of a set K in Euclidean space is interpreted as the probability that a Brownian motion will hit K, "started from ∞". That is, we let B be a ball of radius $R > 1$ centred at the origin. We start Brownian motion off uniformly on ∂B and divide the hitting probability of K by that of the unit ball. If $K \subset B$ this gives back the capacity of K. Note that the rescaling guarantees that the unit ball has unit capacity. For unbounded sets K, we need to take limits as $R \to \infty$.

Proof of Theorem: Let

$$\mathcal{F}_n = \sigma\{ X_m \,;\, m \leq n \}.$$

LEMMA 1. *Let v be a real valued function on E such that $v(x) \geq 0$ for each x. Set $\kappa(n) = v(X_n)/\gamma(X_n)$. Then*

$$v(x) = E^\mu\Big[\sum_{n\geq 0} 1_{\{x\}}(X_n)\kappa(n)\Big], \qquad \textit{and}$$

$$\sum_{x,x'} g^\mu(x,x')v(x)v(x') \leq E^\mu\Big[\Big(\sum_{n\geq 0}\kappa(n)\Big)^2\Big].$$

Proof. The first statement is almost the definition of κ. The second follows by the Markov property:

$$\begin{aligned}
E^\mu\Big[\Big(\sum_{n\geq 0}\kappa(n)\Big)^2\Big] &\geq E^\mu\Big[\sum_{n\geq m\geq 0}\kappa(m)\kappa(n)\Big]\\
&= E^\mu\Big[\sum_{m\geq 0}\kappa(m)E^\mu\Big[\sum_{n\geq m}\frac{v(X_n)}{\gamma(X_n)} \,|\, \mathcal{F}_m\Big]\Big]\\
&= E^\mu\Big[\sum_{m\geq 0}\kappa(m)E^{X_m}\Big[\sum_{n\geq 0}\frac{v(X_n)}{\gamma(X_n)}\Big]\Big]\\
&= \sum_{x'}\frac{v(x')}{\gamma(x')}E^\mu\Big[\sum_{m\geq 0}\kappa(m)g(X_m,x')\Big]\\
&= \sum_{x,x',x''}\frac{v(x)}{\gamma(x)}\frac{v(x')}{\gamma(x')}g(x'',x)g(x,x')\mu(x'')\\
&= \sum_{x,x'} g^\mu(x,x')v(x)v(x').
\end{aligned}$$

□

Remark 5. The reverse inequality holds, after doubling the left side. To prove this, note that the right side is $\le 2E^\mu[\sum_{n\ge m\ge 0}\kappa(m)\kappa(n)]$. In continuous time, and with κ a continuous additive functional, this reverse inequality becomes an equality, again with the left side doubled.

The following is the energy inequality of Burkholder, Davis and Gundy [BDG], which we prove for completeness. See also VI (95.3) of [DM]. The only property used in the proof is that the σ-fields $\mathcal{F}_n$ increase with n.

LEMMA 2. *Let $\kappa^0(n)\ge 0$ be a sequence of random variables, and set $\kappa^1(n) = E^\mu[\kappa^0(n)\,|\,\mathcal{F}_n]$. Then*

$$E^\mu\left[\left(\sum_{n\ge 0}\kappa^1(n)\right)^2\right] \le 4E^\mu\left[\left(\sum_{n\ge 0}\kappa^0(n)\right)^2\right].$$

Proof. The $\mathcal{F}_n$ increase, so $\kappa^1(m)\in\mathcal{F}_n$ for $n\ge m$. Thus

$$\begin{aligned}
E^\mu\left[\left(\sum_n\kappa^1(n)\right)^2\right] &\le E^\mu\left[2\sum_{n\ge m}\kappa^1(m)\kappa^1(n)\right]\\
&= 2\sum_{n\ge m}E^\mu\left[\kappa^1(m)E^\mu\left[\kappa^0(n)\,|\,\mathcal{F}_n\right]\right] = 2\sum_{n\ge m}E^\mu\left[\kappa^1(m)\kappa^0(n)\right]\\
&\le 2E^\mu\left[\sum_{n,m}\kappa^1(m)\kappa^0(n)\right] = 2E^\mu\left[\left(\sum_m\kappa^1(m)\right)\left(\sum_n\kappa^0(n)\right)\right]\\
&\le 2E^\mu\left[\left(\sum_m\kappa^1(m)\right)^2\right]^{1/2}E^\mu\left[\left(\sum_n\kappa^0(n)\right)^2\right]^{1/2}.
\end{aligned}$$

Squaring both sides gives the result. If necessary, we may add a truncation argument, to cope with the possibility that the left side is infinite. □

Now set $\kappa^0(n) = 1_{\{T=n\}}$ and $\kappa(n) = u(X_n)/\gamma(X_n)$. Then $\sum\kappa^0(n) = 0$ or 1, so

$$\begin{aligned}
E^\mu\left[\left(\sum\kappa^0(n)\right)^2\right] &= E^\mu\left[\sum\kappa^0(n)\right] = P^\mu(T<\infty)\\
&= P^\mu(X_n\in K \text{ for some } n\ge 0) = \sum_x u(x).
\end{aligned}$$

Thus by Lemma 1,

$$\begin{aligned}
P^\mu(X_n\in K \text{ for some } n\ge 0) &= \frac{\left(\sum u(x)\right)^2}{E^\mu\left[\left(\sum\kappa^0(n)\right)^2\right]}\\
&\le \frac{\left(\sum u(x)\right)^2}{E^\mu\left[\left(\sum\kappa^0(n)\right)^2\right]}\cdot\frac{E^\mu\left[\left(\sum\kappa(n)\right)^2\right]}{\sum g^\mu(x,x')u(x)u(x')}.
\end{aligned}$$

To show the theorem, we need therefore only show that

$$\frac{E^\mu\big[\big(\sum\kappa(n)\big)^2\big]}{E^\mu\big[\big(\sum\kappa^0(n)\big)^2\big]}\le 16.$$

In order to do so, we reverse time from ζ. Set

$$\begin{aligned}\hat{X}_n &= \begin{cases} X_{\zeta-n}, & 0\le n\le\zeta \\ \Delta, & \text{otherwise,}\end{cases} \qquad \hat{\mathcal{F}}_n=\sigma\{\hat{X}_m\,;\,m\le n\},\\ \kappa^1(n) &= E^\mu\big[\kappa^0(n)\,|\,\mathcal{F}_n\big], \qquad \hat\kappa^1(n)=\kappa^1(\zeta-n),\\ \hat\kappa^2(n) &= E^\mu\big[\hat\kappa^1(n)\,|\,\hat{\mathcal{F}}_n\big], \qquad \kappa^2(n)=\hat\kappa^2(\zeta-n).\end{aligned}$$

Note that $\kappa^0(n)=0$ if $n<0$ or $n>\zeta$, so the same will be true for $\kappa^i(n)$ and $\hat\kappa^i(n)$. By the energy inequality,

$$\begin{aligned}E^\mu\big[\big(\sum\kappa^2(n)\big)^2\big] &= E^\mu\big[\big(\sum\hat\kappa^2(n)\big)^2\big]\le 4E^\mu\big[\big(\sum\hat\kappa^1(n)\big)^2\big]\\ &= 4E^\mu\big[\big(\sum\kappa^1(n)\big)^2\big]\le 16E^\mu\big[\big(\sum\kappa^0(n)\big)^2\big].\end{aligned}$$

The following result therefore finishes the proof.

LEMMA 3. $\kappa^2=\kappa$

Proof. Define the chain killed after time n to be

$$X_m^n=\begin{cases} X_m, & m\le n\\ \Delta, & m>n.\end{cases}$$

Let Ω be the space of possible paths of the chain. Then, by definition, there is a measurable function $f(n,\omega)$ so that

$$\kappa^1(n)=f(n,X_\cdot^n).$$

Define a new function by $\hat f(n,\omega)=f(n,\omega_{n-\cdot})$, with the convention that $\omega_k=\Delta$ if $k<0$. Then

$$\hat\kappa^1(n)=\hat f(\zeta-n,\hat X_{n+\cdot}),$$

where we let $f(k,\omega)=0$ if $k<0$. Let $\phi(x)=E^x[\hat f(\zeta,\hat X)]$ and $\phi(\Delta)=0$.

An observation going back to [H] is that reversing a Markov chain from its lifetime produces another Markov chain. (An elementary computation shows that $\hat X$ will be Markov under P^μ (in general, on $\{\gamma>0\}$), with transition function $\hat p(x,x')=p(x',x)\gamma(x')/\gamma(x)$ and initial measure $\hat\mu(x)=\gamma(x)\big[1-\sum p(x,x')\big]$.) Noting that $\zeta-n$ is the lifetime of $\hat X_{n+\cdot}$, we may therefore use the Markov property of $\hat X$ at time n to get that

$$\hat\kappa^2(n)=E^\mu\big[\hat f(\zeta-n,\hat X_{n+\cdot})\,|\,\hat{\mathcal{F}}_n\big]=\phi(\hat X_n).$$

Therefore $\kappa^2(n) = \phi(X_n)$ as well. Finally, since $X_n \in \mathcal{F}_n$ and $\hat{X}_n \in \hat{\mathcal{F}}_n$,

$$\begin{aligned} u(x) &= E^\mu\big[\sum_n 1_{\{x\}}(X_n)\kappa^0(n)\big] = \sum_n E^\mu\big[1_{\{x\}}(X_n)\kappa^0(n)\big] \\ &= \sum_n E^\mu\big[1_{\{x\}}(X_n)\kappa^1(n)\big] = E^\mu\big[\sum_n 1_{\{x\}}(X_n)\kappa^1(n)\big] \\ &= E^\mu\big[\sum_n 1_{\{x\}}(\hat{X}_n)\hat{\kappa}^1(n)\big] = E^\mu\big[\sum_n 1_{\{x\}}(\hat{X}_n)\hat{\kappa}^2(n)\big] \\ &= E^\mu\big[\sum_n 1_{\{x\}}(X_n)\kappa^2(n)\big] = \phi(x)\gamma(x), \end{aligned}$$

so $\kappa(n) = u(X_n)/\gamma(X_n) = \phi(X_n) = \kappa^2(n)$. □

Remark 6. It is not hard to produce an inequality in the other direction. If $v \geq 0$ and $v = 0$ off K, let $\kappa(n) = v(X_n)/\gamma(X_n)$. Then by Remark 5,

$$\begin{aligned} \sum_x v(x) &= E^\mu\big[\sum_{n\geq 0} \kappa(n)\big] = E^\mu\big[1_{\{X_n \in K \text{ for some } n \geq 0\}} \sum_{n\geq 0}\kappa(n)\big] \\ &\leq P^\mu(X_n \in K \text{ for some } n \geq 0)^{1/2} E^\mu\Big[\Big(\sum_{n\geq 0}\kappa(n)\Big)^2\Big]^{1/2} \\ &\leq P^\mu(X_n \in K \text{ for some } n \geq 0)^{1/2}\big(2\sum_{x,x'} g^\mu(x,x')v(x)v(x')\big)^{1/2}. \end{aligned}$$

Therefore,

$$P^\mu(X_n \in K \text{ for some } n \geq 0) \geq \frac{1}{2}\frac{\left(\sum_x v(x)\right)^2}{\sum_{x,x'} g^\mu(x,x')v(x)v(x')}.$$

Taking the supremum over such v's produces the capacitary lower bound $\Gamma(K)/2$ on the probability of hitting K (see Remark 3).

3. Energy for two chains. Almost exactly the same arguments work with two time parameters. That is, let P be a probability measure under which X_m and Y_n are independent Markov chains, with discrete state spaces E and F, Green functions g and h (finite everywhere), filtrations $\mathcal{F}_n$ and $\mathcal{G}_n$, finite lifetimes ζ and ξ, and initial laws μ and ν. Set

$$\gamma(x',y') = \sum_{x,y} g(x,x')h(y,y')\mu(x)\nu(y)$$
$$\overline{g}(x,y\,;x',y') = g(x,x')h(y,y')/\gamma(x',y').$$

We assume that $\gamma > 0$ everywhere. Let K be a finite subset of $E \times F$, and let (S,T) be any random element of $\{\,(m,n)\,;\,(X_m,Y_n) \in K\,\}$, if this set is nonempty, and $S = T = \infty$ otherwise. Set

$$u(x,y) = P(X_S = x,\, Y_T = y,\, S \text{ and } T < \infty\}.$$

THEOREM 2. *Under the above conditions,*

$$P((X_m, Y_n) \in K \text{ for some } m, n \geq 0) \leq 256 \frac{\left(\sum_{x,y} u(x,y)\right)^2}{\sum_{x,x',y,y'} \overline{g}(x,y;x',y')u(x,y)u(x'y')}.$$

Proof. Let

$$\kappa^0(m,n) = 1_{\{S=m, T=n\}}, \quad \kappa(m,n) = u(X_m, Y_n)/\gamma(X_m, Y_n).$$

As before,

$$E\Big[\Big(\sum_{m,n} \kappa^0(m,n)\Big)^2\Big] = P((X_m, Y_n) \in K \text{ for some } m, n \geq 0) = \sum_{x,y} u(x,y),$$

and

$$\begin{aligned} E\Big[\Big(\sum_{m,n} \kappa(m,n)\Big)^2\Big] &\geq E\Big[\sum_{i \geq m, j \geq n} \frac{u(X_m, Y_n)}{\gamma(X_m, Y_n)} \cdot \frac{u(X_i, Y_j)}{\gamma(X_i, Y_j)}\Big] \\ &= \sum_{\substack{x, x' \\ y, y'}} \overline{g}(x, y; x', y') u(x,y) u(x', y'). \end{aligned}$$

The desired inequality will again follow, once we show that

$$\frac{E\big[(\sum \kappa(m,n))^2\big]}{E\big[(\sum \kappa^0(m,n))^2\big]} \leq 256.$$

To see this, let

$$\begin{aligned} \kappa^1(m,n) &= E\big[\kappa^0(m,n) \,|\, \mathcal{F}_m \vee \mathcal{G}_\infty\big], & \hat{\kappa}^1(m,n) &= \kappa^1(\zeta - m, n), \\ \hat{\kappa}^2(m,n) &= E\big[\hat{\kappa}^1(m,n) \,|\, \hat{\mathcal{F}}_m \vee \mathcal{G}_\infty\big], & \kappa^2(m,n) &= \hat{\kappa}^2(\zeta - m, n), \\ \kappa^3(m,n) &= E\big[\kappa^2(m,n) \,|\, \mathcal{F}_\infty \vee \mathcal{G}_n\big], & \hat{\kappa}^3(m,n) &= \kappa^3(m, \xi - n), \\ \hat{\kappa}^4(m,n) &= E\big[\hat{\kappa}^3(m,n) \,|\, \mathcal{F}_\infty \vee \hat{\mathcal{G}}_n\big], & \kappa^4(m,n) &= \hat{\kappa}^4(m, \xi - n). \end{aligned}$$

We let $\kappa_1^i(m) = \sum_n \kappa^i(m,n)$ and $\kappa_2^i(n) = \sum_m \kappa^i(m,n)$, with a similar convention for the $\hat{\kappa}$'s. Applying the energy inequality to $\hat{\kappa}_2^3$ and κ_2^2 yields

$$E\Big[\Big(\sum_{m,n} \kappa^4(m,n)\Big)^2\Big] \leq 16 E\Big[\Big(\sum_{m,n} \kappa^2(m,n)\Big)^2\Big],$$

as before. Applying it to $\hat{\kappa}_1^1$ and κ_1^0 makes this

$$\leq 256 E\Big[\Big(\sum_{m,n} \kappa^0(m,n)\Big)^2\Big].$$

Finally, since X and Y are independent, X_m will be Markov with respect to the filtration $\mathcal{F}_m \vee \mathcal{G}_\infty$. Therefore as before, there is a function $\phi(x,\omega)$ such that $\kappa^2(m,n) = \phi(X_m, Y_\cdot)$. Applying the same argument to this expression produces a function $\psi(x,y)$ such that $\kappa^4(m,n) = \psi(X_m, Y_n)$. Then

$$\begin{aligned}\gamma(x,y)\psi(x,y) &= E\Big[\sum_{m,n} 1_{\{(x,y)\}}(X_m, Y_n)\kappa^4(X_m, Y_n)\Big] \\ &= E\Big[\sum_{m,n} 1_{\{(x,y)\}}(X_m, Y_n)\kappa^0(X_m, Y_n)\Big] = u(x,y)\end{aligned}$$

shows that $\kappa^4 = \kappa$, finishing the proof. □

Theorem 2 produces a capacitary upper bound for hitting probabilities. As in the case of a single time parameter, there is a corresponding lower bound. The following gathers both together.

COROLLARY 1. *Let P be a probability measure under which X_m and Y_n are independent Markov chains, with discrete state spaces E and F, finite Green functions g and h, and initial laws μ and ν. Assume that both X and Y have finite lifetimes almost surely. Set*

$$\gamma(x',y') = \sum_{x'',y''} g(x'',x')h(y'',y')\mu(x'')\nu(y'')$$

and assume that $\gamma > 0$ everywhere. Put

$$\overline{g}(x,y\,;x',y') = \frac{g(x,x')h(y,y')}{\gamma(x',y')}.$$

Let K be a finite subset of $E \times F$ and set

$$\Gamma(K) = \sup\Big\{ \Big(\sum_{\substack{x,x'\\ y,y'}} \overline{g}(x,y\,;x',y')v(x,y)v(x',y')\Big)^{-1};\ v \geq 0,\ v = 0 \text{ off } K,\ \textstyle\sum_{x,y} v(x,y) = 1 \Big\}$$

Then

$$\frac{\Gamma(K)}{4} \leq P((X_m, Y_n) \in K \text{ for some } m,n \geq 0) \leq 256\Gamma(K).$$

4. Intersections of random walks. Let X_m and Y_n be independent copies of simple random walk in $\mathbb{Z}^d$, $d \geq 3$, with $X_0 = Y_0 = 0$. For $K \subset \mathbb{Z}^d$ (no longer finite), the zero-one law implies that

$$P(X_m = Y_n \in K \text{ for infinitely many } m \text{ and } n) = 0 \text{ or } 1.$$

We are interested in differentiating between these two alternatives. This is, or course, only an issue when $d = 3$ or $d = 4$, since it is in these dimensions that simple random walk is transient yet two independent random walk paths intersect infinitely often.

Let g be the Green function of X and let

$$\gamma(x', y') = g(0, x')g(0, y')$$

$$G(x, x') = |x - x'|^{-2(d-2)} \wedge 1, \qquad \tilde{G}(x, x') = g(x, x')^2/\gamma(x', x'),$$

$$e(\eta) = \sum_{x, x' \in \mathbb{Z}^d} G(x, x')\eta(x)\eta(x'), \quad \tilde{e}(\eta) = \sum_{x, x' \in \mathbb{Z}^d} \tilde{G}(x, x')\eta(x)\eta(x'),$$

$$\mathcal{C}(K) = \sup\{1/e(\eta);\ \eta \geq 0,\ \eta = 0 \text{ off } K,\ \sum \eta(x) = 1\}.$$

THEOREM 3. *Let $K \subset \mathbb{Z}^d$, where $d > 2$, and let $\lambda > 1$. Let $K_j = K \cap \{x\,;\, \lambda^j \leq |x| < \lambda^{j+1}\}$. Then $P(X_m = Y_n \in K \text{ i.o.}) = 0$ or 1 according to whether*

$$\sum_j \lambda^{-2(d-2)j}\mathcal{C}(K_j) < \infty \text{ or } = \infty.$$

Remark 7. It can be shown that $\mathcal{C}$ is subadditive, so that $\mathcal{C}(K_j) \leq c_d \lambda^{dj}$. If $d \geq 5$, the above sum is therefore always finite.

Proof. We need our chains to have finite lifetimes, so kill X after the last time m that $|X_m| \leq \lambda^{j+1}$. Then do the same for Y. This doesn't alter the Green function $g(x, x')$ for $|x|, |x'| \leq \lambda^{j+1}$. Let $\tilde{K}_j = \{\,(x, x)\,;\, x \in K_j\,\}$. Then

$$P(X_m = Y_n \in K_j \text{ for some } m,\ n \geq 0) \\ = P\big((X_m, Y_n) \in \tilde{K}_j \text{ for some } m,\ n \geq 0\big).$$

By Corollary 1,this is bounded between constant multiples of the capacity $\Gamma(\tilde{K}_j)$, with respect to the kernel $\overline{g}(x, y\,; x', y') = g(x, x')g(y, y')/\gamma(x', y')$. On $\tilde{K}$ we have $x = y$ and $x' = y'$, so this capacity is the same as that of K_j with respect to $\tilde{G}$. The Green function $g(x, x')$ is bounded between constant multiples of $|x - x'|^{-(d-2)}$, for $x, x' \in \mathbb{Z}^d$ (see [L]), so that on $K_j \times K_j$, $\tilde{G}(x, x')$ is bounded between constant multiples of $\lambda^{2(d-2)j}G(x, x')$. Thus $\Gamma(\tilde{K}_j)$ is itself bounded between constant multiples of $\lambda^{-2(d-2)j}\mathcal{C}(K_j)$, independently of j.

Now suppose that $\sum \lambda^{-2(d-2)j}\mathcal{C}(K_j) < \infty$. Then $\sum P(X_m = Y_n \in K_j \text{ for some } m, n \geq 0) < \infty$ too. By Borel-Cantelli we conclude that $P(X_m = Y_n \in K \text{ i.o.}) = 0$.

The brief argument just given used the upper bound in Corollary 1 and hence the projection arguments that are the main thrust of the paper. The converse uses the lower bound and is only slightly longer. One approach would be to use Borel-Cantelli again, supplementing it by some form of asymptotic independence (see for example Theorem 3.3.2 of [PS]). Instead we'll argue directly.

Suppose that $\sum \lambda^{-2(d-2)j}\mathcal{C}(K_j) = \infty$. We may assume that the sum over the even indices j is infinite (if not, argue with the odd indices instead). To show that $P(X_m = Y_n \in K \text{ i.o.}) = 1$ it will suffice (by the zero-one law) to show that

$$P\big(X_m = Y_n \in \bigcup_{j\geq k} K_{2j} \text{ for some } m, n \geq 0\big)$$

remains bounded from 0 as $k \to \infty$.

Given k, we may find an l so that

$$\delta = \sum_{j=k}^{l} \lambda^{-4(d-2)j}\mathcal{C}(K_{2j}) \geq 1.$$

For those j with $K_{2j} \neq \emptyset$ we let η_j be a probability on K_j so that $\mathcal{C}(K_{2j}) \leq 2/e(\eta_j)$. Let

$$\eta = \delta^{-1}\sum_{j=k}^{l} \lambda^{-4(d-2)j}\mathcal{C}(K_{2j})\eta_j.$$

Then η is a probability on $\bigcup_{j\geq k} K_{2j}$. We have already estimated $\tilde{G}$ on $K_{2j} \times K_{2j}$, from which it follows that $\tilde{e}(\eta_j) \leq c\lambda^{4(d-2)j}e(\eta_j)$. On $K_{2i} \times K_{2j}$ with $i \neq j$, $\tilde{G}$ is bounded by a constant (here we use that the odd indices have been omitted). Thus

$$\begin{aligned}
\tilde{e}(\eta) &= \sum_{k\leq i,j\leq l}{}' \sum_{\substack{x \in K_i \\ x' \in K_j}} \tilde{G}(x,x')\eta(x)\eta(x') \\
&\leq c \sum_{\substack{k \leq i,j \leq l \\ i \neq j}} \eta(K_{2i})\eta(K_{2j}) + \sum_{j=k}^{l} \left(\frac{\lambda^{-4(d-2)j}\mathcal{C}(K_{2j})}{\delta}\right)^2 \tilde{e}(\eta_j) \\
&\leq c\Big[\eta(K)^2 + \sum_{j=k}^{l} \left(\frac{\lambda^{-4(d-2)j}\mathcal{C}(K_{2j})}{\delta}\right)^2 \lambda^{4(d-2)j}e(\eta_j)\Big] \\
&\leq c\Big[1 + \delta^{-2}\sum_{j=k}^{l} \lambda^{-4(d-2)j}\mathcal{C}(K_{2j})\Big] = c(1+\delta^{-1}) \leq 2c.
\end{aligned}$$

The desired lower bound now follows from Corollary 1. □

5. Reversing from a fixed time. The argument given can, of course, be applied to more general objects. In trying to produce a very general version, several simplifications can be realized by using Kuznetsov measures (as in [FS]), at the cost of working in a more sophisticated setting. For example, this avoids the kind of limiting procedure mentioned at the end of Remark 4. Reversing a continuous time Markov process from its lifetime can also create severe technical problems, even if that lifetime turns out to be finite. Kuznetsov measures simplify such time reversal, without destroying time homogeneity (a related construction was used for this purpose as early as [H]). We are about to consider an even simpler version of the projection argument, the price for which is exactly the loss of this homogeneity in time.

Go back to the context of a single Markov chain X_n, now with infinite lifetime. If we reverse time from a fixed time N, then the arguments of section 2 go through exactly as before. We let $\hat{X}_n = X_{N-n}$, for $0 \le n \le N$, and $\hat{\mathcal{F}}_n = \sigma\{X_n, \dots, X_N\}$. Starting with $\kappa^0(n)$, we let

$$\kappa^1(n) = E[\,\kappa^0(n) \,|\, \mathcal{F}_n\,] \text{ and } \kappa^2(n) = E[\,\kappa^1(n) \,|\, \hat{\mathcal{F}}_n\,].$$

In the general theory of processes, κ^1 would be called the dual optional projection of κ^0, and κ^2 the dual co-optional projection of κ^1.

The Markov property amounts to the commutation of the conditional expectations $E[\,\cdot\,|\,\mathcal{F}_n\,]$ and $E[\,\cdot\,|\,\hat{\mathcal{F}}_n\,]$, so that $\kappa^2(n) = E[\kappa^0(n)\,|\,X_n]$. Following [D1] we could call κ^2 the *dual central projection* of κ^0. As before, $E[(\sum \kappa^2(n))^2] \le 16E[(\sum \kappa^0(n))^2]$. The only drawback is that now κ^2 may depend on n (that is, $\kappa(n) = \phi(n, X_n)$).

In general, if $\mathcal{H} = (\mathcal{H}_n)$ is any sequence of σ-fields and $\kappa(n)$ is any sequence of positive random variables, we write

$$\kappa^{\mathcal{H}}(n) = E[\,\kappa(n) \,|\, \mathcal{H}_n\,].$$

We say that $\mathcal{H}$ is *projectively bounded* if there is a constant c such that

$$E\Big[\Big(\sum \kappa^{\mathcal{H}}(n)\Big)^2\Big] \le cE\Big[\Big(\sum \kappa(n)\Big)^2\Big]$$

for every $\kappa \ge 0$. We let $||\mathcal{H}||^2$ be the least such constant c. If the $\mathcal{H}_n$ increase with n then $||\mathcal{H}||^2 \le 4$. We have just seen an example where $||\mathcal{H}||^2 \le 16$. Similarly, if we use $\mathbb{Z}^2$ as an index set, and $\mathcal{H}_{m,n} = \sigma\{X_m, Y_n\}$ for independent Markov chains X and Y, then $||\mathcal{H}||^2 \le 256$. On the other hand, the following (a minor modification of an example shown to me by Robin Pemantle) shows that such inequalities will not hold without some restriction on $\mathcal{H}$.

EXAMPLE 1. There is an $\mathcal{H}$ which is not projectively bounded.

Proof. Let M be geometric, with $P(M = m) = \theta^m(1-\theta)$ for $m \ge 0$. Conditional on M, let N be uniform on $\{2^M, 2^M + 1, \dots, 2^{M+1} - 1\}$. Let $\kappa(n) = 1_{\{N\}}(n)$, so that $E[\sum \kappa(n)] = 1 = E[(\sum \kappa(n))^2] = 1$.

Write $m(n)$ for the integer m for which $2^m \le n < 2^{m+1}$, and take $\mathcal{H}_n$ to be the σ-field generated by the event $\{N = n \text{ or } N = 2^{m(n)}\}$. Then

$$\kappa^{\mathcal{H}}(n) = \begin{cases} 1, & \text{if } N = n \text{ is a power of 2} \\ 1/2, & \text{if } N = n \text{ or } N = 2^{m(n)}, \text{ but } n \text{ isn't a power of 2} \\ 0, & \text{otherwise.} \end{cases} .$$

If $N = 2^M$ then there are 2^M values of n for which $\kappa^{\mathcal{H}}(n)$ equals 1 or 1/2. Thus $\sum \kappa^{\mathcal{H}}(n) \ge 2^{M-1} 1_{\{N=2^M\}}$, so

$$E\big[\big(\sum \kappa^{\mathcal{H}}(n)\big)^2\big] \ge \sum_m \frac{\theta^m(1-\theta)}{2^m}\left(2^{m-1}\right)^2 .$$

As long as $\theta > 1/2$, this sum diverges. □

It would be interesting to find more general conditions on $\mathcal{H}$ than those given above, under which $\mathcal{H}$ will be projectively bounded. For example, the following relates to Dynkin's theory of capacity for superprocesses [D2]

Problem. Let $\mathcal{H}_\tau = \sigma\{X_\tau\}$, where $(X_\tau)_{\tau \in T}$ is a Markov chain indexed by a binary tree T. Under what conditions will $\mathcal{H}$ be projectively bounded?

REFERENCES

[A] J. Azéma, *Une remarque sur les temps de retour. Trois applications*, Séminaire de Probabilités VI, Lecture Notes in Mathematics **258**, Springer-Verlag, Berlin-Heidelberg-New York, 1972, pp. 35–50.

[BPP] I. Benjamini, R. Pemantle and Y. Peres, *Martin capacity for Markov chains and random walks in varying dimension*, Preprint.

[BDG] D. L. Burkholder, B. J. Davis, and R. F. Gundy, *Integral inequalities for convex functions of operators on martingales*, Proc. Sixth Berkeley Symp. Math. Statist. Prob. **2** (1972), 223-240.

[DM] C. Dellacherie and P.-A. Meyer, *Probabilités et Potentiel*, Ch. V–VII, Hermann, Paris, 1980.

[D1] E. B. Dynkin, *Additive functionals of several time-reversible Markov processes*, J. Funct. Anal. **42** (1981), 64–101.

[D2] E. B. Dynkin, *Superdiffusions and parabolic nonlinear differential equations*, Ann. Probab. **20** (1992), 942–962.

[FS] P. J. Fitzsimmons and T. S. Salisbury, *Capacity and energy for multiparameter Markov processes*, Ann. Inst. Henri Poincaré **25** (1989), 325–350.

[G] R. K. Getoor, *Excessive Measures*, Birkhäuser, Boston, 1990.

[H] G. A. Hunt, *Markoff chains and Martin boundaries*, Ill. J. Math. **4** (1960), 313–340.

[KSK] J. G. Kemeny, J. L. Snell, and A. W. Knapp, *Denumerable Markov Chains*, Springer-Verlag, Berlin-Heidelberg-New York, 1972.

[L] G. F. Lawler, *Intersections of Random Walks*, Birkhäuser, Boston, 1991.

[P] Y. Peres, *Intersection-equivalence of Brownian paths and certain branching processes*, Preprint.

[PS] S. C. Port and C. J. Stone, *Brownian Motion and Classical Potential Theory*, Academic Press, New York, 1978.